T0177130

Rigid Body Kinematics

Master the conceptual, theoretical, and practical aspects of kinematics with this exhaustive text, which provides a rigorous analysis and description of general motion in mechanical systems, with numerous examples, from spinning tops to wheeled ground vehicles. Over 400 figures illustrate the main ideas and provide a geometrical interpretation and a deeper understanding of concepts, and quiz questions and problems throughout the text provide additional hands-on practice. Ideal for students taking courses on rigid body kinematics, and an invaluable reference for researchers.

Joaquim A. Batlle is Emeritus Professor of the Universitat Politècnica de Catalunya (UPC). He is a member of the Royal Academy of Sciences and Arts of Barcelona.

Ana Barjau Condomines is Associate Professor at the Department of Mechanical Engineering at the Universitat Politècnica de Catalunya (UPC).

Rigid Body Kinematics

JOAQUIM A. BATLLE
Universitat Politècnica de Catalunya

ANA BARJAU CONDOMINES
Universitat Politècnica de Catalunya

CAMBRIDGE
UNIVERSITY PRESS

CAMBRIDGE
UNIVERSITY PRESS

University Printing House, Cambridge CB2 8BS, United Kingdom

One Liberty Plaza, 20th Floor, New York, NY 10006, USA

477 Williamstown Road, Port Melbourne, VIC 3207, Australia

314–321, 3rd Floor, Plot 3, Splendor Forum, Jasola District Centre, New Delhi – 110025, India

79 Anson Road, #06–04/06, Singapore 079906

Cambridge University Press is part of the University of Cambridge.

It furthers the University's mission by disseminating knowledge in the pursuit of education, learning, and research at the highest international levels of excellence.

www.cambridge.org
Information on this title: www.cambridge.org/9781108479073
DOI: 10.1017/9781108782319

© Joaquim A. Batlle and Ana Barjau Condomines 2020

This publication is in copyright. Subject to statutory exception and to the provisions of relevant collective licensing agreements, no reproduction of any part may take place without the written permission of Cambridge University Press.

First published 2020

Printed in the United Kingdom by TJ International, Padstow Cornwall

A catalogue record for this publication is available from the British Library.

Library of Congress Cataloging-in-Publication Data
Names: Agulló i Batlle, Joaquim, author. | Barjau Condomines, Ana, author.
Title: Rigid body kinematics / Joaquim A. Batlle, Universitat Politècnica de Catalunya, Barcelona, Ana Barjau Condomines, Universitat Politècnica de Catalunya, Barcelona.
Description: First edition. | Cambridge ; New York : Cambridge University Press, 2020. | Includes index.
Identifiers: LCCN 2020009230 (print) | LCCN 2020009231 (ebook) | ISBN 9781108479073 (hardback) | ISBN 9781108782319 (epub)
Subjects: LCSH: Dynamics, Rigid. | Kinematics.
Classification: LCC QA861 .A38 2020 (print) | LCC QA861 (ebook) | DDC 531/.3–dc23
LC record available at https://lccn.loc.gov/2020009230
LC ebook record available at https://lccn.loc.gov/2020009231

ISBN 978-1-108-47907-3 Hardback

Cambridge University Press has no responsibility for the persistence or accuracy of URLs for external or third-party internet websites referred to in this publication and does not guarantee that any content on such websites is, or will remain, accurate or appropriate.

Contents

Preface

Origin and Scope

This textbook, together with *Rigid Body Dynamics* (Cambridge University Press, forthcoming), is the result of a whole professional life devoted to Newtonian Mechanics, both from an educational and a research point of view. Through more than 40 years, our way of teaching this subject to second-year undergraduate students (UG) in Industrial Engineering at the Universitat Politècnica de Catalunya (UPC) has evolved progressively. To find the right words and examples to make every concept clear, to discover the misconceptions that lead to the most frequent errors, and to kindle the students' curiosity for this discipline have been our major concerns throughout all these years.

We have found our personal point of view to present this classical subject, and have generated hundreds of exercises, questionnaires, and interesting case studies. Sharing this material with others involved in the higher education field seems nearly an obligation!

Our regular course on Newtonian Mechanics at UPC spans over 15 weeks and contains three main parts: Kinematics, Dynamics, and Energy. Students enrolling in this course already have a background in fundamental physics (mainly focused on 2D cases) and mathematics, and they will not be studying any other course on general mechanics (though they may follow some particular applications as "Machine and mechanism theory" and "Vibrations"). This context conditions the syllabus of our course: it has to be complete and provide the main tools to face any problem that may be encountered in Mechanical Engineering.

A percentage of our regular students are really enthusiastic about this discipline. For them we propose an additional and optional 15-week course where some concepts presented in the regular course are revisited and enlarged, and some well-known methods to perform dynamical analysis of mechanical systems are introduced. The course includes a final section devoted to an interesting problem: percussive dynamics.

Our goal when writing the book was not just to cover the syllabus of both courses but also to provide answers to questions raised by the best students (often out of scope of the course). The result has been a long text that includes more material than is likely to be covered in formal lectures to second-year undergraduates (including both the regular and the optional courses), and thus may serve as a reference also for graduate students, professional engineers working in the industry, and even researchers in robotics and biomechanics.

Though the main target are the undergraduate students, enlarging the potential audience was the main reason for splitting it into two volumes (*Rigid Body Kinematics* and *Rigid Body Dynamics*). We are aware that some people working in the fields of robotics and biomechanics and game developers are mainly interested in the kinematical description of mechanical systems. Thus, devoting a first volume dealing exclusively with kinematics seemed appropriate.

The contents of the UG standard course on Mechanics is not the same in the United States and in Europe. In the United States, this text may be better suited for intermediate (and even master) courses rather than for UG courses. Conversely, classical US textbooks may correspond to first-year UG students in Europe but may fall below the level we seek in second-year students.

Main Features

This textbook presents the general problem of rigid body kinematics in a very rigorous and precise way, and with a particularly suitable approach for engineering students.

Rigor is not at odds with the very visual and geometric approach we pursue throughout the book. We have included hundreds of illustrations which emphasize the geometrical properties of rigid body kinematics and help understand the concepts visually.

The mathematical notation throughout the book is highly explicit not only to avoid any ambiguity but also to allow an automatic and straightforward interpretation.

Every chapter starts with a short introduction that provides an overview of its content. Then, the different concepts and methods are introduced. The proofs are clearly isolated from the main text so that the reader may skip them. Every single concept is illustrated through figures (expressly designed for this work) and many fully worked-out examples (nothing is left to the reader as a further exercise!). Some of them are strictly pedagogical exercises, but many others correspond to simplified models of usual mechanical systems.

Advanced topics are presented in appendices, and they may be overlooked without impairing the understanding of the following chapters.

The book includes an extensive collection of multiple-choice quiz questions (with answers) and exercises (with final results). The quiz questions do not require long calculations, but a clear understanding of the fundamental concepts. Each of them is illustrated through a figure showing the mechanical system under study plus the precisions an engineer would add to complete the description (as dimensions and motion). Thus, it is not strictly necessary to read the text as the figures are self-explanatory.

The exercises are pencil-and-paper problems to be solved symbolically: symbolic results allow to detect possible errors (undetectable in numerical solutions) and give the whole picture of the system behavior (showing which parameters may be responsible for a bifurcation or transition from one regime to another). The student may implement the results in a computer, but that is not the main purpose of the exercises. Computer simulations may be time-consuming, and the required skills (programming, numerical calculation, etc.) do not fall within the objectives of our textbook.

The results for the exercises are given at the end of the chapters, but we do not provide the detailed resolution. Thus, the students are obliged to work on their approach and find by themselves the mistakes leading to wrong results: you have to do kinematics to learn kinematics!

A few puzzles complete the application material given at the end of the chapters. They confront the student with formally simple systems and situations found in everyday life. Justifying their design and/or understanding their kinematic behavior may be far from intuitive and calls for a rigorous application of the concepts introduced in the chapter. A complete explanation is given for each puzzle.

Content Description

The book presents a rigorous analysis of the kinematics of rigid body systems. Emphasis is made in 3D motion, though planar (2D) motion is reviewed as a particular case.

As the potential students have already acquired the basic notions in a fundamental physics course, units and error propagation are not included. However, some basic mathematical concepts are explained to make the book self-contained.

Chapter 1 sets the basic concept of reference frame and the mathematical representations of space and time. We review the usual mathematical vector operations, and provide both a geometrical and an analytical method to carry them out. The main concern in that chapter is the orientation of rigid bodies (and reference frames and vector bases) in 3D. We introduce the Euler angles, which will be extensively used throughout the book, and the angular rotation vector. Two appendices present alternative methods to describe the rotation and orientation of rigid bodies.

Chapter 2 reviews the kinematics of particles with a special focus on the composition of movements, which establishes the relationship between the kinematics of the same particle in two different reference frames with a general 3D relative motion. Absolute, relative, and transportation movements are introduced when considering the velocity composition. The acceleration composition adds a third term with no particularly useful physical interpretation: the Coriolis acceleration. The inertial guidance is presented in an appendix as an interesting application problem.

Chapter 3 is the core of the book. It is devoted to the general 3D motion of rigid bodies. The definition of rigid body (as a set of points mutually fixed) implies the existence of a relationship between the velocity and the acceleration of any pair of points in the rigid body. The geometry of the velocity distribution is thoroughly analyzed, and the concepts of Instantaneous Screw Axis (ISA), fixed axode, and moving axode are introduced as interesting tools to understand the general motion and the kinematical equivalence of mechanical systems (regardless their particular shape).

Planar motion is treated as a particular case, and the ISA and the axodes are substituted by the Instantaneous Centre of Rotation (ICR) and the centrodes.

The appendices present a very complete and unique coverage of the kinematics and maneuvering possibilities of wheeled vehicles (both with conventional and omnidirectional wheels), modeled as systems of rigid bodies, with non-sliding wheels rolling on a flat ground.

As the usual mechanical systems are multibody systems formed by rigid bodies with mutual kinematic restrictions, Chapter 4 is an introduction to the kinematics of such systems. The concepts of generalized coordinates and velocities, independent coordinates, degrees of freedom, holonomy, and constraint equations are presented. The main focus, however, is on the rigorous and systematic analysis of the kinematic constraints between pairs of rigid bodies. We provide a powerful tool to discover redundancy and all the associated drawbacks based on the analysis of kinematic torsors and the geometrical designs not included in any standard textbook.

In short, this textbook sets a powerful framework that will allow the student to implement a complete kinematic analysis of mechanical systems, which is the starting point to undertake their dynamic description.

Acknowledgments

This book is the result of many years of interactions with other professors in our teaching team. The regular discussions at the end of the academic year to assess what had been a success and what could be improved have been precious to us. The interaction with the students who have followed our course has also been of invaluable help. They have not only forced us to be precise and find the right examples but also contributed with interesting questions that have compelled us to address problems not specifically treated in the literature so far. Our deepest gratitude to all of them.

1 Space and Time
Orientation and Euler Angles

Kinematics deals with the *geometry of motion of material bodies along time* – change of position in space and over time – without regard to the physical phenomena on which it depends. Such description requires mathematical models for space and time (which is the physical framework of mechanical phenomena) and mathematical models for bodies.

There is a range of models of increasing complexity for material objects: mass point (or particle), rigid body, continuous media (including deformable bodies and fluids). This text deals exclusively with the first two (Chapters 2 and 3), whose mathematical complexity is much lower than that required by continuous media.

The mathematical operation *time derivative*, to evaluate rates of change (of position, orientation, etc.), is a fundamental tool in mechanics. The time derivative of vectors is more complex than that of scalars, since it has to take into account both the change of value and of orientation. While the former is identical for all observers (or all *reference frames*), the latter is not. For example, a radius marked on a rotating platform (and taken as a vector) has the same value and orientation for observers fixed to the platform but variable orientation for observers fixed to the ground.

Orientation is a key concept, and its mathematical description is not trivial when 3D motion is addressed. In this text, only the Euler angles are used as they are the most appropriate procedure for a first study. However, Appendix 1B presents a brief review of some alternative rotation parameters due to its importance in engineering branches such as robotics and spacecraft dynamics.

Operations on vectors can be done through their graphic representation (an arrow with value indication[1]) or their components (projections on a vector basis). In mechanics, *vector bases of variable orientation relative to the reference frame* (*moving bases*) are usual because they facilitate the vectors' projection and simplify the expressions of their components and their physical interpretation. For instance, when studying the kinematics of a vehicle, the basis with constant orientation relative to the chassis is particularly interesting.

The time derivative of vectors expressed through their components in a moving basis leads to the concept of *angular velocity* of the vector basis, which describes its rate of change of orientation relative to the reference frame. The angular velocity is a key concept in rigid body kinematics (Chapter 3).

[1] *Value* and *module* (or *magnitude*) of a vector are not synonyms: while the former may be positive or negative, the latter is strictly positive.

1.1 The Absolute Time of Newtonian Mechanics

In mechanics, the concept of time is linked to that of *ordered succession of instants*. An *instant t* is what two simultaneous events have in common. Let us suppose that an observer detects two events (for example, the collision of particles P_1 and P_2 – event A – and the collision of particles P_3 and P_4 – event B). The observer can establish the following without ambiguity:

- If they are simultaneous, $t_A = t_B$
- If A precedes B, $t_A < t_B$
- If B precedes A, $t_A > t_B$

From the ordering of events, we can establish an ordered sequence of instants that we call *time*.

The **principle of absolute simultaneity** states that this sequence is the same for all observers whose relative speed is much lower than that of light (which is the case considered in Newtonian mechanics): Newtonian time is an **absolute time**. A same clock ticks the time instants for all observers – for example, by showing the succession of coincidences of a needle with marks on a dial. Since it is an ordered and dense succession of points (between two instants it is always possible to insert another), the mathematical model for time is the one-dimensional space of real numbers R^1.

The definition of time must be completed with its measurement (that is, the assessment of the time interval between two instants). This is not just a kinematic issue but a dynamic one (**principle of inertia**, in section 1.3, chapter 1 of *Rigid Body Dynamics* [Cambridge University Press, forthcoming]).

In relativistic mechanics, which holds when the relative speed between observers is close to that of light, the principle of absolute simultaneity is no longer valid: two events A and B can be simultaneous for an observer but be "A before B" for a second observer and "B before A" for a third observer. For this reason, each observer needs its own clock.

1.2 Space and Reference Frame

In Newtonian mechanics, the physical space is modeled as an affine Euclidean three-dimensional point space E^3. To define the location of points of E^3, we need to choose a point **O** (called **origin**) and three noncoplanar axes (Fig. 1.1). Vectors $\overline{\mathbf{OP}}$ (where **P** is any point in E^3) are position vectors, and they belong to the vector space associated with the affine point space.

An origin **O** and a set of three noncoplanar axes define a **mathematical frame of reference** in E^3. A different origin or a different set of axes yields a different description of those position vectors and hence defines a different frame of reference from the mathematical point of view.[2]

[2] The mathematical concept of reference frame (MRF) and that of coordinate system are not equivalent: for the same MRF, different coordinate systems (Cartesian, polar, cylindrical, etc.) can be used.

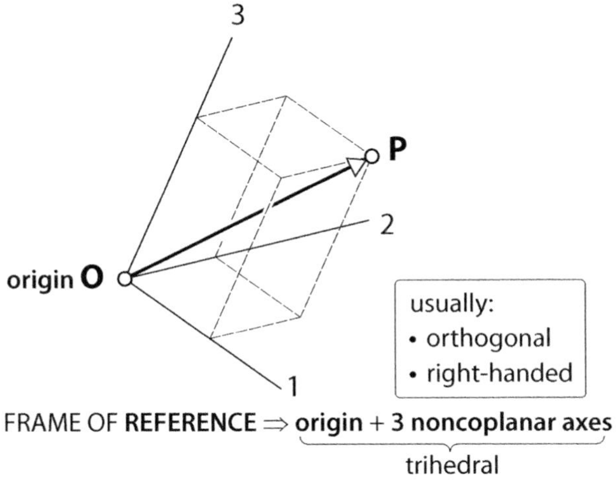

Fig. 1.1

In mechanics, the definition of frame of reference is different: it is a set of "rest points" (points whose mutual distances do not change – they are **mutually fixed**) in E^3. Note that this definition contains implicitly the idea of **state of motion** and thus the concept of time (not taken into account in the mathematical reference frame). The constitutive elements of physical reference frames are space and time.

This physical notion is intrinsically linked to that of the **observer**. The observer (which is not a point) may be located anywhere, and what he/she measures is the movement of points relative to the reference frame.

Movement is always associated with a reference frame: movements are **relative** to reference frames. If a point **P** passes through different points of a reference frame R, it is a moving point relative to R. The set of those points constitutes the **trajectory of P in R**. In principle, a same point describes different trajectories in different reference frames.

A reference frame R can be graphically represented either through a set of points mutually fixed or through a **trihedral** (single rest point **O** in R and three noncoplanar axes, Fig. 1.2). Sometimes, the observer is added to the representation of the reference frame (Fig. 1.3). Note that no clock is included in those compact representations. This is so because we will be dealing with Newtonian mechanics, and in that context the *principle of absolute simultaneity* holds. In other words, a same clock is shared by all observers, and so it can be suppressed from the representation without generating any confusion.

Reference frame (R) and **rigid body** are close concepts: both are sets of mutually fixed points, though those in the rigid body are just a subset in E^3, and they are **particles** (or mass points) (Fig. 1.4). This is why we usually describe reference frames through rigid body names: rotating platform frame, chassis frame, etc.

A consequence of that equivalence is that the concept of orientation (and that of angular velocity) applies both to reference frames and rigid bodies.

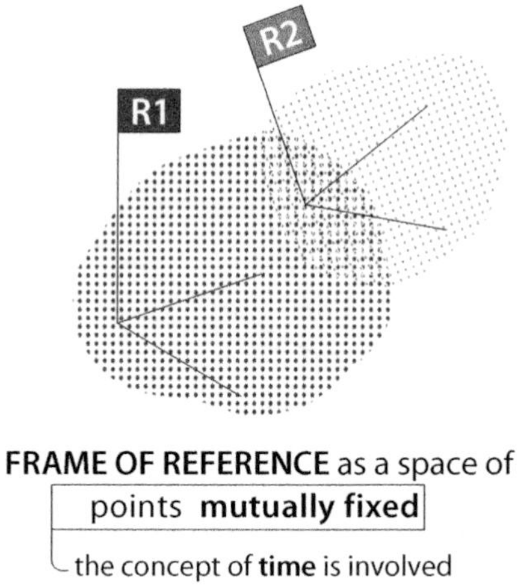
R2
R1
FRAME OF REFERENCE as a space of
points mutually fixed
the concept of time is involved

Fig. 1.2

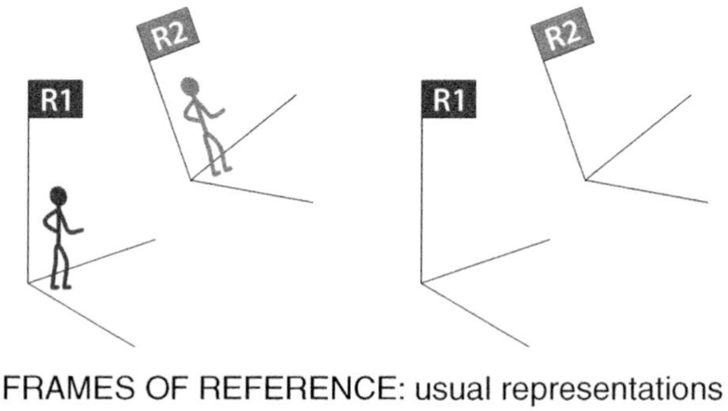
R2
R1
R2
R1
FRAMES OF REFERENCE: usual representations

Fig. 1.3

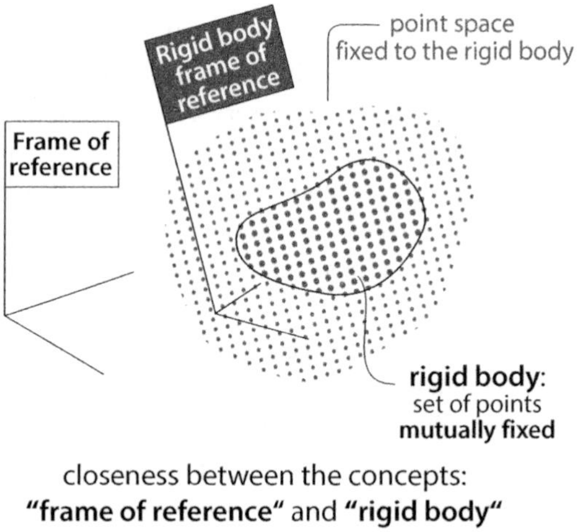
Rigid body frame of reference
point space
fixed to the rigid body
Frame of reference
rigid body:
set of points
mutually fixed
closeness between the concepts:
"frame of reference" and "rigid body"

Fig. 1.4

1.3 Representation of Vectors and Operations Involving a Single Time Instant

In a vector space of three dimensions (as E^3), vectors can be represented in a simple and intuitive graphical way as an **arrow** (which defines a **positive generic direction**) and a **value** (defined by a real number). When the value is positive, the vector direction corresponds to that of the arrow; when it is negative, it is the opposite.

The description (arrow, value) is more efficient than that of (**orientation, sense, magnitude**): as the value incorporates magnitude and sense, there is no need of two analytical formulations (one for positive sense and another one for negative).

Graphic representations played an important role before the advent of computers: **graphic statics** and **graphic kinematics** constituted two important branches of mechanics applied to engineering. Easy access to computational power has led to a partial disregard of graphic techniques in favor of numerical treatments.

The graphic representation is widely used in this book as an auxiliary element. In very simple cases, it allows operations (sum, vector product, derivation, etc.) without resorting to analytical treatments based on the representation of the vector through its components in a vector basis.

Vector operations carried out in the same time instant – algebraic operations such as addition, scalar product, and vector product – have a simple geometric description and can be performed from their graphical representation (Fig. 1.5).

The time derivative of a vector, which is an operation over time – it involves the vector in two time instants separated by a time interval that tends to zero – can also be treated directly through its graphic representation (Section 1.4).

All previous operations can be performed through the description of the vectors by their components in a vector basis. The basis orientation can be either constant (**fixed basis**) or variable in the reference frame (**moving basis**). **Whether a basis is fixed or moving is irrelevant in operations** that involve vectors at a same time instant.

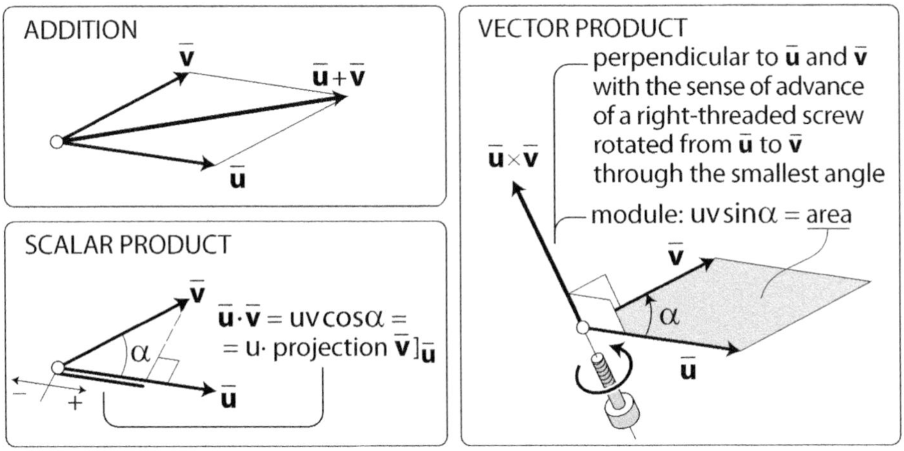

Fig. 1.5

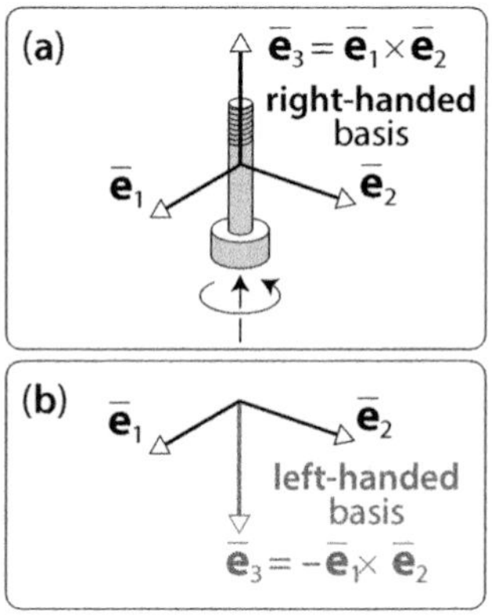

Fig. 1.6

In a vector basis B with versors $(\bar{\mathbf{e}}_1, \bar{\mathbf{e}}_2, \bar{\mathbf{e}}_3)$, a vector $\bar{\mathbf{u}}$ is represented by the column vector of its components $(\mathrm{u}_1, \mathrm{u}_2, \mathrm{u}_3)$:

$$\bar{\mathbf{u}} = \sum_{i=1}^{3} \mathrm{u}_i \bar{\mathbf{e}}_i \rightarrow \{\bar{\mathbf{u}}\}_B = \begin{Bmatrix} \mathrm{u}_1 \\ \mathrm{u}_2 \\ \mathrm{u}_2 \end{Bmatrix}. \tag{1.1}$$

In this text, all the bases are:

- Orthonormal: $\bar{\mathbf{e}}_i \cdot \bar{\mathbf{e}}_j = \begin{cases} 1, & i = j \\ 0, & i \neq j \end{cases}$
- Right-handed: $\bar{\mathbf{e}}_3 = \bar{\mathbf{e}}_1 \times \bar{\mathbf{e}}_2$ (Fig. 1.6a); if $\bar{\mathbf{e}}_3$ had the opposite direction, it would be an inverse (or left-handed) basis (Fig. 1.6b)

From that analytical representation, the addition and product operations are solved simply as:

$$\{\bar{\mathbf{u}}+\bar{\mathbf{v}}\}_B = \begin{Bmatrix} \mathrm{u}_1 \\ \mathrm{u}_2 \\ \mathrm{u}_3 \end{Bmatrix} + \begin{Bmatrix} \mathrm{v}_1 \\ \mathrm{v}_2 \\ \mathrm{v}_3 \end{Bmatrix} = \begin{Bmatrix} \mathrm{u}_1+\mathrm{v}_1 \\ \mathrm{u}_2+\mathrm{v}_2 \\ \mathrm{u}_3+\mathrm{v}_3 \end{Bmatrix}, \{\bar{\mathbf{u}}\cdot\bar{\mathbf{v}}\}_B = \begin{Bmatrix} \mathrm{u}_1 \\ \mathrm{u}_2 \\ \mathrm{u}_3 \end{Bmatrix} \cdot \begin{Bmatrix} \mathrm{v}_1 \\ \mathrm{v}_2 \\ \mathrm{v}_3 \end{Bmatrix} = \mathrm{u}_1\mathrm{v}_1 + \mathrm{u}_2\mathrm{v}_2 + \mathrm{u}_3\mathrm{v}_3. \tag{1.2}$$

The cross product can be calculated from a determinant:

$$\bar{\mathbf{u}} \times \bar{\mathbf{v}} = \mathrm{DET}\begin{vmatrix} \bar{\mathbf{e}}_1 & \bar{\mathbf{e}}_2 & \bar{\mathbf{e}}_3 \\ \mathrm{u}_1 & \mathrm{u}_2 & \mathrm{u}_3 \\ \mathrm{v}_1 & \mathrm{v}_2 & \mathrm{v}_3 \end{vmatrix} \Rightarrow \{\bar{\mathbf{u}} \times \bar{\mathbf{v}}\}_B = \begin{Bmatrix} \mathrm{u}_1 \\ \mathrm{u}_2 \\ \mathrm{u}_3 \end{Bmatrix} \times \begin{Bmatrix} \mathrm{v}_1 \\ \mathrm{v}_2 \\ \mathrm{v}_3 \end{Bmatrix} = \begin{Bmatrix} \mathrm{u}_2\mathrm{v}_3 - \mathrm{u}_3\mathrm{v}_2 \\ \mathrm{u}_3\mathrm{v}_1 - \mathrm{u}_1\mathrm{v}_3 \\ \mathrm{u}_1\mathrm{v}_2 - \mathrm{u}_2\mathrm{v}_1 \end{Bmatrix}. \tag{1.3}$$

The $\bar{\mathbf{u}} \times \bar{\mathbf{v}}$ product can be understood as a linear application $(\bar{\mathbf{u}}\times)$ transforming vector $\bar{\mathbf{v}}$ into vector $\bar{\mathbf{u}} \times \bar{\mathbf{v}}$. Therefore, it can be represented by a matrix (in this case, antisymmetric):[3]

[3] This matrix representation has a practical interest in certain analytical processes (Section 1.5).

$$\{\bar{\mathbf{u}} \times \bar{\mathbf{v}}\}_B = [\bar{\mathbf{u}}\times]_B \{\bar{\mathbf{v}}\}_B = \begin{bmatrix} 0 & -u_3 & u_2 \\ u_3 & 0 & u_1 \\ -u_2 & u_1 & 0 \end{bmatrix} \begin{Bmatrix} v_1 \\ v_2 \\ v_3 \end{Bmatrix}. \quad (1.4)$$

The expressions in Eqs. (1.2, 1.3) are valid for any basis B (fixed or moving). This is not the case in operations involving vectors at different time instants (such as time derivatives and integrations). As will be seen in Section 1.6, the analytical time derivative (from the vectors' projection on moving bases) introduces a term that depends on the rate of change of orientation of the basis in the reference frame. Time integration through moving bases is presented in Appendix 1C.

The criterion for choosing a vector basis is of a practical nature: it is convenient that the obtaining of the vectors components is simple, and that the result is simple enough (if possible) to facilitate its physical interpretation.

1.4 Geometric Time Derivative of Vectors

The time derivative of a variable (•) evaluates its change per time unit:

$$\frac{d(\bullet)}{dt} = \lim_{\Delta t \to 0} \frac{\Delta(\bullet)}{\Delta t}. \quad (1.5)$$

If the variable is a scalar function ρ (for instance, the distance ρ between two points), the result is identical in all reference frames, as all observers detect the same $\Delta\rho$ for a same Δt (because of the principle of absolute simultaneity):

$$\frac{d\rho}{dt} = \lim_{\Delta t \to 0} \frac{\Delta\rho}{\Delta t}, \text{ same unique value for all observers.} \quad (1.6)$$

However, the change of a vector $\Delta\bar{\mathbf{u}}$ is not the same for all observers in principle. The change of its value is indeed the same for everyone (since it is a scalar), but there may be discrepancies when it comes to evaluating the change of its orientation. For example, if we consider the earth frame RA and the platform frame RB (rotating about an axis fixed in RA, Fig. 1.7), a vector $\bar{\mathbf{u}}$ constant for observer B (with zero time derivative for observer B) is a rotary vector for observer A as the $\bar{\mathbf{u}}$ orientation is not constant. The origin of this discrepancy is the relative rotation between the two reference frames.

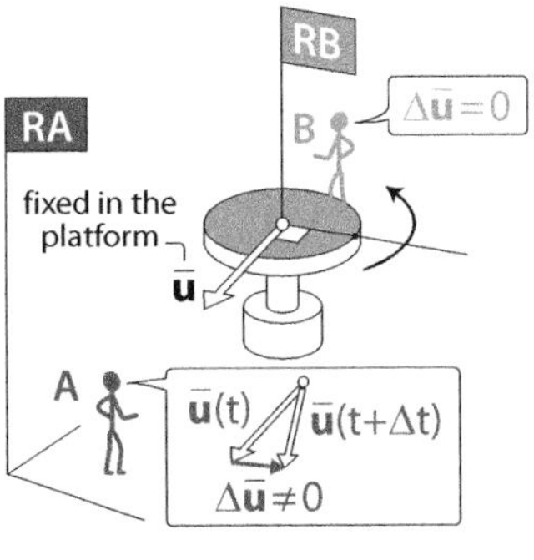

Fig. 1.7

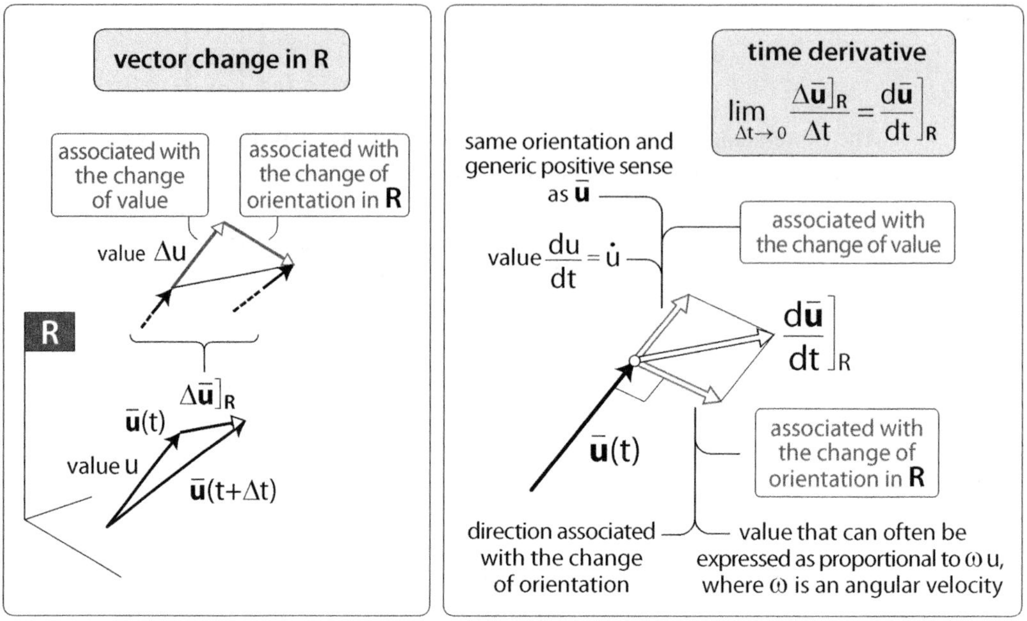

Fig. 1.8

The time derivative of a vector, then, is reference-dependent. It is advisable to explicitly include this dependency in the notation:

$$\left.\frac{d\bar{\mathbf{u}}}{dt}\right]_R = \lim_{\Delta t \to 0} \frac{\Delta \bar{\mathbf{u}}]_R}{\Delta t} = \lim_{\Delta t \to 0} \left.\frac{\Delta \bar{\mathbf{u}}}{\Delta t}\right]_R, \tag{1.7}$$

where subscript R is the reference frame where it is calculated. A frequent notation for the time derivative is a simple dot on the variable:

$$\dot{\rho} \equiv \frac{d\rho}{dt} \quad ; \quad \dot{\bar{\mathbf{u}}}\Big]_R \equiv \left.\frac{d\bar{\mathbf{u}}}{dt}\right]_R. \tag{1.8}$$

When there is no doubt about the reference frame where the time derivative is being calculated, the subscript R will be disregarded, and we will just write $\dot{\bar{\mathbf{u}}}$ to simplify the notation.

The time derivative of a vector in a given reference frame R is nonzero when there is a change in value (equal in all R) or a change in orientation (which is R-dependent in principle, Fig. 1.8). When $\Delta t \to 0$, those changes can be associated with two different vectors (whose addition yields the total time derivative):

- The vector associated with the change in value: it is parallel to $\bar{\mathbf{u}}$ and with the same positive generic sense; its value is $\dot{u} = du/dt$
- The vector associated with the change of orientation: it is perpendicular to $\bar{\mathbf{u}}$, with direction associated with the change of orientation; its value can often be expressed as a product of value u (or of some projection of $\bar{\mathbf{u}}$) and an angular velocity ω

This procedure will be called **geometric time derivative** (to distinguish it from the analytic time derivative, based on the components of the vector in a vector basis). It is highly recommended in simple cases not just because it is straightforward but because it is a description close to the physical meaning of the time derivative, so it may help to avoid mistakes that slip easily in the analytical derivative. Chapter 2 presents some examples of this procedure.

1.5 Analytical Time Derivative of Vectors: Angular Velocity Vector

The geometric time derivative described in the previous section is a powerful tool, but it may be difficult to use when vectors evolve in a 3D space. In those cases, it is advisable to shift to an analytical calculation (through a vector basis). However, this can be tricky, mainly for two reasons:

- The time derivative calls for the information of the vector in *two different time instants* (whose separation tends to zero); in the analytical calculation, this may go unnoticed (while it never does in the geometric one)
- If the vector is projected in a moving basis MB (relative to a reference frame R), the change of the value of its components does not describe the change of the vector in R as the latter may also be influenced by the change of orientation of MB relative to R

In a vector basis FB with constant orientation (fixed basis) in R, the components of vector $\left.\dot{\bar{\mathbf{u}}}\right]_{\mathrm{R}}$ are just the time derivative of the components of $\bar{\mathbf{u}}$:

$$\{\bar{\mathbf{u}}\}_{\mathrm{FB}} = \begin{Bmatrix} u_1 \\ u_2 \\ u_3 \end{Bmatrix} \rightarrow \left\{\left.\dot{\bar{\mathbf{u}}}\right]_{\mathrm{R}}\right\}_{\mathrm{FB}} = \begin{Bmatrix} \dot{u}_1 \\ \dot{u}_2 \\ \dot{u}_3 \end{Bmatrix} \equiv \frac{d}{dt}\{\bar{\mathbf{u}}\}_{\mathrm{FB}}. \tag{1.9}$$

♣ *Proof*

In a fixed basis with versors $(\bar{\mathbf{e}}_1, \bar{\mathbf{e}}_2, \bar{\mathbf{e}}_3)$, vector $\bar{\mathbf{u}}$ is:

$$\bar{\mathbf{u}} = \sum_{i=1}^{3} u_i \bar{\mathbf{e}}_i. \tag{1.10}$$

Its time derivative in a reference frame R is:

$$\left.\dot{\bar{\mathbf{u}}}\right]_{\mathrm{R}} = \sum_{i=1}^{3} \dot{u}_i \bar{\mathbf{e}}_i + \sum_{i=1}^{3} u_i \left.\dot{\bar{\mathbf{e}}}_i\right]_{\mathrm{R}}. \tag{1.11}$$

As the components u_i are scalar functions, there is no need of explicit indication of the reference frame when calculating $\dot{u}_i$. This explicit indication is mandatory when performing the time derivative of versors $\bar{\mathbf{e}}_i$. However, as FB is a fixed basis, the time derivative of $\bar{\mathbf{e}}_i$ is zero, so:

$$\dot{\bar{\mathbf{u}}}\Big]_R = \sum_{i=1}^{3} \dot{u}_i \bar{\mathbf{e}}_i. \tag{1.12}$$

Note that this calculation requires the knowledge of the components along time $(u_i(t))$, or, equivalently, their expression has to be completely general (that is, valid for any time instant). ♣

Let's consider now the time derivative in R of a vector $\bar{\mathbf{u}}$ projected in a moving basis MB. In that case, a nonzero value of $\dot{u}_i$ does not imply that $\dot{\bar{\mathbf{u}}}\big]_R$ is nonzero. In general, that time derivative contains two terms:

$$\dot{\bar{\mathbf{u}}}\Big]_R = \sum_{i=1}^{3} \dot{u}_i \bar{\mathbf{e}}_i + \sum_{i=1}^{3} u_i \dot{\bar{\mathbf{e}}}_i\Big]_R, \quad \dot{\bar{\mathbf{e}}}_i\Big]_R \neq 0. \tag{1.13}$$

Equation (1.13) proves that a vector with constant $u_i (\dot{u}_i = 0)$ is not necessarily a vector constant in R:

$$\dot{u}_i = 0 \Rightarrow \dot{\bar{\mathbf{u}}}\Big]_R = \sum_{i=1}^{3} u_i \dot{\bar{\mathbf{e}}}_i\Big]_R \neq 0. \tag{1.14}$$

If the vector is constant in R:

$$\dot{\bar{\mathbf{u}}}\Big]_R = 0 \Rightarrow \sum_{i=1}^{3} \dot{u}_i \bar{\mathbf{e}}_i = -\sum_{i=1}^{3} u_i \dot{\bar{\mathbf{e}}}_i\Big]_R. \tag{1.15}$$

Equation (1.15) shows that now the only cause of variation of the vector components $(\dot{u}_i)$ is the change of orientation of the MB relative to R $\left(\dot{\bar{\mathbf{e}}}_i\big]_R\right)$.

When using moving bases, the time derivatives of the components are only a part of the vector time derivative.

The components of $\dot{\bar{\mathbf{u}}}\big]_R$ are obtained by adding the column vector of the time derivatives of the vector components and the vector product $\bar{\boldsymbol{\Omega}}_R^{MB} \times \bar{\mathbf{u}}$, where $\bar{\boldsymbol{\Omega}}_R^{MB}$ is the **angular velocity of MB relative to R**:

$$\left\{\dot{\bar{\mathbf{u}}}\Big]_R\right\}_{MB} = \frac{d}{dt}\{\bar{\mathbf{u}}\}_{MB} + \left\{\bar{\boldsymbol{\Omega}}_R^{MB} \times \bar{\mathbf{u}}\right\}_{MB}. \tag{1.16}$$

♣ *Proof*

The vector components of vector $\bar{\mathbf{u}}$ in a fixed basis FB and in a moving basis MB are related through the transformation matrix [S]:

$$\{\bar{\mathbf{u}}\}_{FB} = [S]\{\bar{\mathbf{u}}\}_{MB}. \tag{1.17}$$

The columns in [S] are the components of the MB basis projected in the FB basis. This matrix is variable in time. As the bases are orthonormal, it follows that:

$$[S]^{-1} = [S]^T. \tag{1.18}$$

The same matrix relates the components of the time derivatives of vector $\bar{\mathbf{u}}$ projected in the FB and in the MB:

$$\left\{\left[\dot{\bar{\mathbf{u}}}\right]_{\mathrm{R}}\right\}_{\mathrm{FB}} = [\mathrm{S}]\left\{\left[\dot{\bar{\mathbf{u}}}\right]_{\mathrm{R}}\right\}_{\mathrm{MB}}. \tag{1.19}$$

The time derivative of Eq. (1.17) yields:

$$\frac{\mathrm{d}}{\mathrm{dt}}\{\bar{\mathbf{u}}\}_{\mathrm{FB}} = [\mathrm{S}]\frac{\mathrm{d}}{\mathrm{dt}}\{\bar{\mathbf{u}}\}_{\mathrm{MB}} + \left[\dot{\mathrm{S}}\right]\{\bar{\mathbf{u}}\}_{\mathrm{MB}}, \tag{1.20}$$

where the elements of $\left[\dot{\mathrm{S}}\right]$ are the time derivative of the $[\mathrm{S}]$ elements. The left-hand side of Eq. (1.20) is the same as that of Eq. (1.19).

Premultiplication of Eq. (1.20) by $[\mathrm{S}]^{\mathrm{T}}$ yields:

$$\left\{\left[\dot{\bar{\mathbf{u}}}\right]_{\mathrm{R}}\right\}_{\mathrm{MB}} = \frac{\mathrm{d}}{\mathrm{dt}}\{\bar{\mathbf{u}}\}_{\mathrm{MB}} + [\mathrm{S}]^{\mathrm{T}}\left[\dot{\mathrm{S}}\right]\{\bar{\mathbf{u}}\}_{\mathrm{MB}}. \tag{1.21}$$

The matrix $[\mathrm{S}]^{\mathrm{T}}\left[\dot{\mathrm{S}}\right]$ is antisymmetric (that is, equal to its transpose but with opposite sign):

$$\begin{aligned} &[\mathrm{S}]^{\mathrm{T}}[\mathrm{S}] = [\mathrm{I}] \text{ (identity matrix)} \\ &\quad\Rightarrow \left[\dot{\mathrm{S}}\right]^{\mathrm{T}}[\mathrm{S}] + [\mathrm{S}]^{\mathrm{T}}\left[\dot{\mathrm{S}}\right] = 0 \Rightarrow \left[\dot{\mathrm{S}}\right]^{\mathrm{T}}[\mathrm{S}] = -[\mathrm{S}]^{\mathrm{T}}\left[\dot{\mathrm{S}}\right] = -\left(\left[\dot{\mathrm{S}}\right]^{\mathrm{T}}[\mathrm{S}]\right)^{\mathrm{T}}. \end{aligned} \tag{1.22}$$

As seen in Section 1.3, the mathematical operation $[\mathrm{S}]^{\mathrm{T}}\left[\dot{\mathrm{S}}\right]\{\bar{\mathbf{u}}\}_{\mathrm{MB}}$ (where $[\mathrm{S}]^{\mathrm{T}}\left[\dot{\mathrm{S}}\right]$ is an antisymmetric matrix) is equivalent to a cross product between a vector associated with $[\mathrm{S}]^{\mathrm{T}}\left[\dot{\mathrm{S}}\right]$ and vector $\bar{\mathbf{u}}$. In the present case, that vector is the **angular velocity vector of basis MB relative to the reference frame R** and will be written as $\bar{\mathbf{\Omega}}_{\mathrm{R}}^{\mathrm{MB}}$. Replacing $[\mathrm{S}]^{\mathrm{T}}\left[\dot{\mathrm{S}}\right]$ by $\left(\bar{\mathbf{\Omega}}_{\mathrm{R}}^{\mathrm{MB}}\times\right)$ in Eq. (1.21) yields Eq. (1.16). ♣

The $\bar{\mathbf{\Omega}}_{\mathrm{R}}^{\mathrm{MB}}$ components can be identified from $[\mathrm{S}]^{\mathrm{T}}\left[\dot{\mathrm{S}}\right]$:

$$[\mathrm{S}]^{\mathrm{T}}\left[\dot{\mathrm{S}}\right] = \begin{bmatrix} 0 & -\Omega_3 & \Omega_2 \\ \Omega_3 & 0 & -\Omega_1 \\ -\Omega_2 & \Omega_1 & 0 \end{bmatrix} \Rightarrow \left\{\bar{\mathbf{\Omega}}_{\mathrm{R}}^{\mathrm{MB}}\right\}_{\mathrm{MB}} = \begin{Bmatrix} \Omega_1 \\ \Omega_2 \\ \Omega_3 \end{Bmatrix}. \tag{1.23}$$

Consequently, it is always possible to obtain the vector $\bar{\mathbf{\Omega}}_{\mathrm{R}}^{\mathrm{MB}}$ as a function of the variables used to define the basis orientation. Those variables give the transformation matrix [S] at each time instant.

In this text, we use **Euler angles** for that purpose. From these angles, the basis angular velocity $\bar{\mathbf{\Omega}}_{\mathrm{R}}^{\mathrm{MB}}$ can be found in a very straightforward and intuitive way without calculating $[\mathrm{S}]^{\mathrm{T}}\left[\dot{\mathrm{S}}\right]$ in every application (as will be seen in Section 1.8).

Though introduced for vector bases, the angular velocity vector can be extended immediately to reference frames and rigid bodies.

The physical meaning of the angular velocity vector – of a basis B, a reference frame R′, or a rigid body S – is that of rate of change of orientation. A remarkable issue is that **the rate of change of orientation is a vector but the orientation itself is not**: $\bar{\boldsymbol{\Omega}}_{\mathrm{R}}^{\mathrm{B,R',S}}$ is not the time derivative of any "orientation vector" of B, R′, or S in R. This is discussed in more detail in Appendix 1A.

The time derivative of the angular velocity $\bar{\boldsymbol{\Omega}}_{\mathrm{R}}^{\mathrm{B,R',S}}$ is the **angular acceleration vector** $\bar{\boldsymbol{\alpha}}_{\mathrm{R}}^{\mathrm{B,R',S}}$:

$$\bar{\boldsymbol{\alpha}}_{\mathrm{R}}^{\mathrm{B,R',S}} \equiv \left.\frac{\mathrm{d}\bar{\boldsymbol{\Omega}}_{\mathrm{R}}^{\mathrm{B,R',S}}}{\mathrm{dt}}\right]_{\mathrm{R}} = \left.\dot{\bar{\boldsymbol{\Omega}}}_{\mathrm{R}}^{\mathrm{B,R',S}}\right]_{\mathrm{R}}. \tag{1.24}$$

This vector will pay an important role in the following chapters.

1.6 Time Derivative of a Vector in Different Reference Frames

In kinematics, it is interesting to relate the movement of a same object – for example, a vehicle – in two different reference frames – for example, the earth and a second vehicle. Since the time derivative of vectors plays a central role and it is reference-frame dependent (Section 1.4), we must investigate the relationship between the time derivatives of a same vector $\bar{\mathbf{u}}$ in different reference frames R1 and R2. Those derivatives are related through:

$$\left.\frac{\mathrm{d}\bar{\mathbf{u}}}{\mathrm{dt}}\right]_{\mathrm{R1}} = \left.\frac{\mathrm{d}\bar{\mathbf{u}}}{\mathrm{dt}}\right]_{\mathrm{R2}} + \bar{\boldsymbol{\Omega}}_{\mathrm{R1}}^{\mathrm{R2}} \times \bar{\mathbf{u}}, \tag{1.25}$$

where $\bar{\boldsymbol{\Omega}}_{\mathrm{R1}}^{\mathrm{R2}}$ is the angular velocity of R2 in R1. Equation (1.25) shows that the time derivatives in two different reference frames differ only if they are mutually rotating.

♣ *Proof*

The time derivatives of $\bar{\mathbf{u}}$ in R1 and R2 can be calculated using a basis B2 fixed in R2 but moving in R1 (Fig. 1.9).

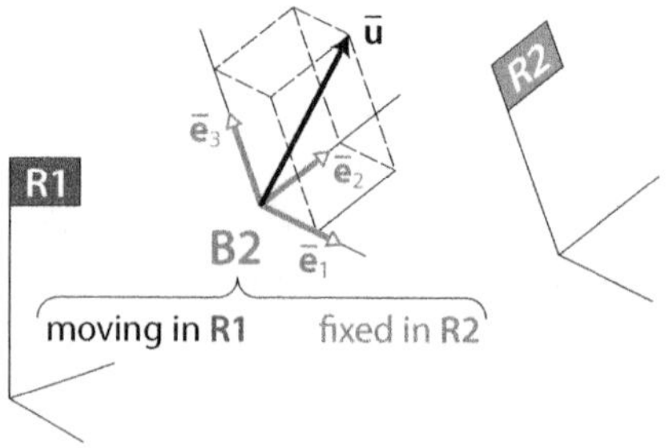

Fig. 1.9

According to Eq. (1.16):

$$\left.\begin{aligned}\left\{\left[\frac{d\bar{\mathbf{u}}}{dt}\right]_{R1}\right\}_{B2} &= \frac{d}{dt}\{\bar{\mathbf{u}}\}_{B2}+\left\{\bar{\boldsymbol{\Omega}}_{R1}^{B2}\times\bar{\mathbf{u}}\right\}_{B2}\\ \left\{\left[\frac{d\bar{\mathbf{u}}}{dt}\right]_{R2}\right\}_{B2} &= \frac{d}{dt}\{\bar{\mathbf{u}}\}_{B2}\end{aligned}\right]\Rightarrow\left\{\left[\frac{d\bar{\mathbf{u}}}{dt}\right]_{R1}\right\}_{B2}=\left\{\left[\frac{d\bar{\mathbf{u}}}{dt}\right]_{R2}\right\}_{B2}+\left\{\bar{\boldsymbol{\Omega}}_{R1}^{B2}\times\bar{\mathbf{u}}\right\}_{B2}. \tag{1.26}$$

Taking into account that $\bar{\boldsymbol{\Omega}}_{R1}^{B2}=\bar{\boldsymbol{\Omega}}_{R1}^{R2}$, Eq. (1.26) becomes:

$$\left\{\left[\frac{d\bar{\mathbf{u}}}{dt}\right]_{R1}\right\}_{B2}=\left\{\left[\frac{d\bar{\mathbf{u}}}{dt}\right]_{R2}\right\}_{B2}+\left\{\bar{\boldsymbol{\Omega}}_{R1}^{R2}\times\bar{\mathbf{u}}\right\}_{B2}, \tag{1.27}$$

which gives the vectorial relationship in Eq. (1.25) in the vector basis B2. ♣

1.7 Simple Rotation

There are different sets of variables to define the orientation of a moving or fixed basis (MB or FB) in a reference frame R. They all allow the calculation of the matrix [S] associated with the basis transformation (MB → FB). Among them, the rotation angles are the most used. They have the advantage of a very intuitive interpretation and a direct relationship with the associated angular velocity vector.

In mechanical systems, the relative orientation between contiguous rigid bodies is often introduced physically through joints that allow a mutual rotation about a common axis; it is a **simple rotation**. A simple rotation is defined through just one angle (Fig. 1.10).

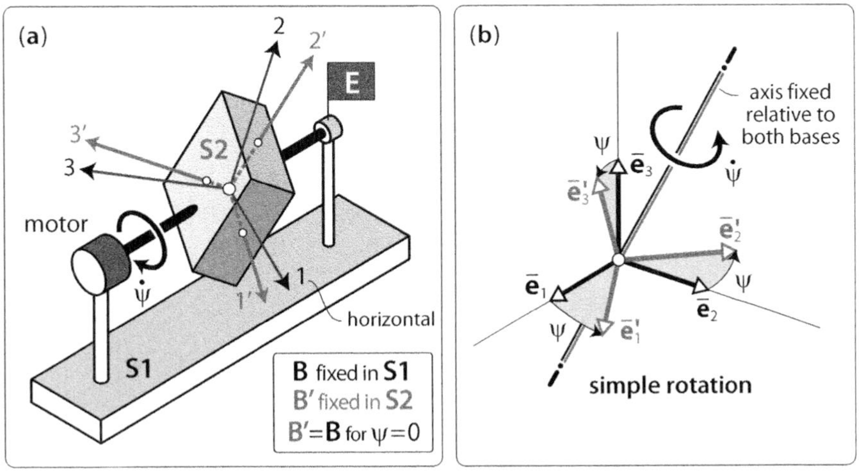

Fig. 1.10

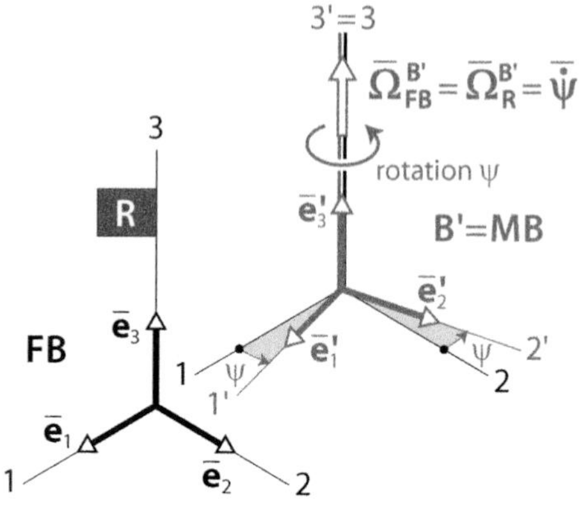

Fig. 1.11

Let's consider the MB basis, moving in R and with versors $(\bar{e}'_1, \bar{e}'_2, \bar{e}'_3)$, related to the FB fixed basis, with versors $(\bar{e}_1, \bar{e}_2, \bar{e}_3)$, through a simple rotation about the common axis 3 (Fig. 1.11). The angle ψ defines the orientation of versors $(\bar{e}'_1, \bar{e}'_2)$ relative to versors $(\bar{e}_1, \bar{e}_2)$. The angular velocity vector $\bar{\boldsymbol{\Omega}}_{R}^{MB}$ (or $\bar{\boldsymbol{\Omega}}_{FB}^{MB}$) is:

$$\left\{\bar{\boldsymbol{\Omega}}_{R}^{MB}\right\}_{FB} = \left\{\bar{\boldsymbol{\Omega}}_{FB}^{MB}\right\}_{FB} = \begin{Bmatrix} 0 \\ 0 \\ \dot{\psi} \end{Bmatrix} = \{\bar{\dot{\psi}}\}_{FB}. \tag{1.28}$$

Vector $\bar{\boldsymbol{\Omega}}_{R}^{MB}$ has the orientation of the rotation axis, the direction of advance of a screw (normal thread to the right) that rotates with increasing ψ, and with value $\dot{\psi}$.

In this particular case, as $\bar{\boldsymbol{\Omega}}_{R}^{MB}$ has the direction of axis 3 shared by FB and MB, its components are the same in both bases:

$$\left\{\bar{\boldsymbol{\Omega}}_{R}^{MB}\right\}_{FB} = \left\{\bar{\boldsymbol{\Omega}}_{R}^{MB}\right\}_{MB}. \tag{1.29}$$

♣ *Proof*

The transformation matrix [S] associated with the simple rotation about axis 3 of MB relative to FB (transformation (MB → FB)) is:

$$[S] = \begin{bmatrix} \cos\psi & -\sin\psi & 0 \\ \sin\psi & \cos\psi & 0 \\ 0 & 0 & 1 \end{bmatrix}. \tag{1.30}$$

Matrix $[S]^T[\dot{S}]$ associated with $\bar{\boldsymbol{\Omega}}_{R}^{MB}$ is:

$$[S]^T[\dot{S}] = \begin{bmatrix} \cos\psi & \sin\psi & 0 \\ -\sin\psi & \cos\psi & 0 \\ 0 & 0 & 1 \end{bmatrix} \begin{bmatrix} -\dot{\psi}\sin\psi & -\dot{\psi}\cos\psi & 0 \\ \dot{\psi}\cos\psi & -\dot{\psi}\sin\psi & 0 \\ 0 & 0 & 1 \end{bmatrix} = \begin{bmatrix} 0 & -\dot{\psi} & 0 \\ \dot{\psi} & 0 & 0 \\ 0 & 0 & 0 \end{bmatrix}. \tag{1.31}$$

The identification of the $\bar{\boldsymbol{\Omega}}_{R}^{MB}$ components according to Eq. (1.23) in Section 1.5 leads to Eq. (1.28). ♣

1.8 General Rotation: Euler Angles

The general orientation of a basis MB in a reference frame R (or a basis FB fixed in R) can be described by three independent values, since the nine elements of the transformation matrix [S] must fulfill the six conditions of orthonormalization (Section 1.3). Among the various options to describe the rotation of bases (or frames or rigid bodies), **Euler angles** are the most interesting for a first study because the final orientation depends univocally on the value of the angles and not on their relative time evolution. This is why they are very often implemented in orienting devices.

The Euler angles are a set of three simple rotations (ψ, θ, φ) introduced **in series**. This composition is shown in Fig. 1.12. Starting from basis FB, the first rotation orientates a basis B_ψ through a simple rotation of angle ψ. The second simple rotation (angle θ) orientates a new basis $B_{\psi\theta}$ with respect to B_ψ. Finally, the third simple rotation (angle φ) orientates the basis MB $\left(= B_{\psi\theta\varphi}\right)$, which has a general orientation, with respect to $B_{\psi\theta}$. Each axis of rotation is fixed to a pair of bases.

In order to achieve a totally general orientation of the MB relative to R (or to FB), each new axis has to be orthogonal to the previous one (Fig. 1.13). Otherwise, not all orientations would be attainable.

Euler axes (the axes of rotations ψ, θ, and φ) have different motions relative to R:

- The ψ axis (called "Euler first axis") is fixed in R
- The θ axis ("Euler second axis"), oriented through ψ, changes its orientation in a plane perpendicular to the ψ axis
- The φ axis ("Euler third axis"), oriented through ψ and θ, changes its orientation in a general nonrestricted way

Though the orientation of the Euler third axis in space is general, it can be identified very easily because it is fixed to the object being oriented. Like the first axis (fixed in R), once you have visualized it on a particular configuration – for instance, the $\psi = \theta = \varphi = 0$ orientation – it is easy to visualize it for any ψ and θ values. The direction of the Euler second axis, which is neither fixed to R nor to the object being oriented, can be found in principle through its perpendicularity to both the Euler first and third axes.

Despite the conditions of orthogonality $(1\text{st axis} \perp 2\text{nd axis})$ and $(2\text{nd axis} \perp 3\text{rd axis})$, the first and the third axes are not orthogonal in general. Their relative angle depends only on the θ value; if θ is varied from 0° to 360°, the directions of the first and third

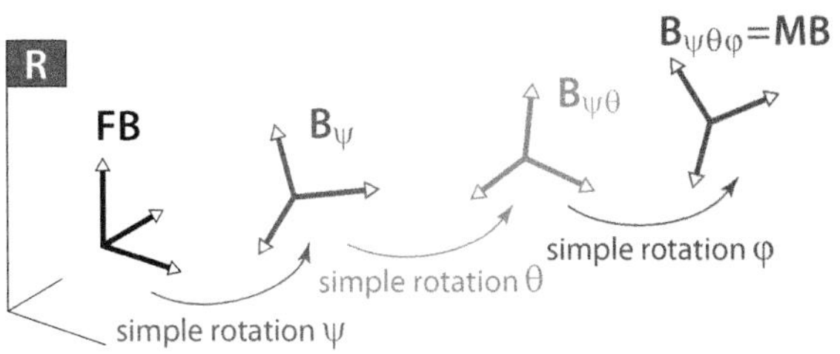

Fig. 1.12

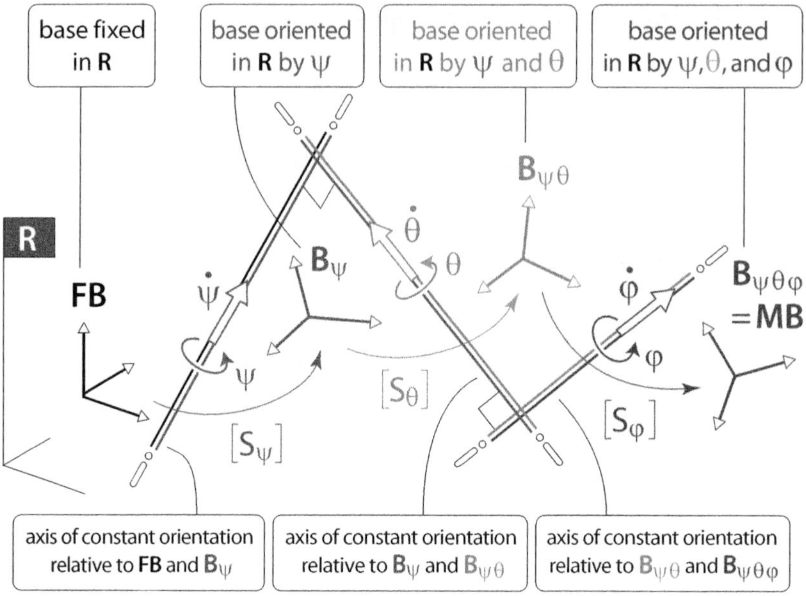

Fig. 1.13

axes coincide twice. For these two θ values, the ψ and φ angles are indeterminate; we may know the (ψ + φ) value but not the value of each of them separately. Moreover, the three Euler axes are coplanar, and the angular velocity $\bar{\mathbf{\Omega}}_{\mathrm{R}}^{\mathrm{MB}}$ is restricted to that plane.[4] These situations are called **singularities**, and all the associated drawbacks can be avoided by switching to a different collection of Euler angles when the first and the third axes are close to the singular configuration.

The angular velocity vectors that correspond to the (ψ, θ, φ) rotations are as follows:

- $\bar{\dot{\boldsymbol{\psi}}}$ is the vector of the relative rotation between B_{ψ} and FB; its direction is that of the axis of rotation of ψ, its positive sense is that of a screw turning toward increasing ψ, and its value is $\dot{\psi}$
- $\bar{\dot{\boldsymbol{\theta}}}$ is the vector of the relative rotation between $\mathrm{B}_{\psi\theta}$ and B_{ψ}, perpendicular to $\bar{\dot{\boldsymbol{\psi}}}$; its direction is that of the axis of rotation of θ, its positive sense is that of a screw turning toward increasing θ, and its value is $\dot{\theta}$
- $\bar{\dot{\boldsymbol{\varphi}}}$ is the vector of the relative rotation between $\mathrm{B}_{\psi\theta\varphi}(=\mathrm{MB})$ and $\mathrm{B}_{\psi\theta}$, perpendicular to $\bar{\dot{\boldsymbol{\theta}}}$; its direction is that of the axis of rotation of φ, its positive sense is that of a screw turning toward increasing φ, and its value is $\dot{\varphi}$

The angular velocity of $\mathrm{B}_{\psi\theta\varphi}(=\mathrm{MB})$, whose orientation relative to FB (or R) is totally general, is:

$$\bar{\mathbf{\Omega}}_{\mathrm{R}}^{\mathrm{B}\psi\theta\varphi} = \bar{\dot{\boldsymbol{\psi}}} + \bar{\dot{\boldsymbol{\theta}}} + \bar{\dot{\boldsymbol{\varphi}}}. \tag{1.32}$$

[4] If we wanted to find analytically the $(\dot{\psi}, \dot{\theta}, \dot{\varphi})$ values from the $\bar{\mathbf{\Omega}}_{\mathrm{R}}^{\mathrm{MB}}$ value, we would obtain an infinite number of values for $\dot{\psi}$ and $\dot{\varphi}$.

♣ *Proof*

There is a transformation matrix for each simple rotation:

$$\left.\begin{aligned}\{\bar{\mathbf{u}}\}_{FB} &= [S_\psi]\{\bar{\mathbf{u}}\}_{B_\psi}\\ \{\bar{\mathbf{u}}\}_{B_\psi} &= [S_\theta]\{\bar{\mathbf{u}}\}_{B_{\psi\theta}}\\ \{\bar{\mathbf{u}}\}_{B_{\psi\theta}} &= [S_\varphi]\{\bar{\mathbf{u}}\}_{B_{\psi\theta\varphi}}\end{aligned}\right\} \Rightarrow \{\bar{\mathbf{u}}\}_{FB} = [S_\psi][S_\theta][S_\varphi]\{\bar{\mathbf{u}}\}_{B_{\psi\theta\varphi}} \equiv [S]\{\bar{\mathbf{u}}\}_{B_{\psi\theta\varphi}} \quad (1.33)$$

The angular velocity of $B_{\psi\theta\varphi}$ relative to R is the vector associated with $[S]^T[\dot{S}]$:

$$\begin{aligned}[S]^T[\dot{S}] &= [S_\varphi]^T[S_\theta]^T[S_\psi]^T\left([\dot{S}_\psi][S_\theta][S_\varphi] + [S_\psi][\dot{S}_\theta][S_\varphi] + [S_\psi][S_\theta][\dot{S}_\varphi]\right)\\ &= \left([S_\theta][S_\varphi]\right)^T\left([S_\psi]^T[\dot{S}_\psi]\right)\left([S_\theta][S_\varphi]\right) + [S_\varphi]^T\left([S_\theta]^T[\dot{S}_\theta]\right)[S_\varphi] + [S_\varphi]^T[\dot{S}_\varphi].\end{aligned} \quad (1.34)$$

Taking into account that an arbitrary 3×3 matrix $[\Lambda]$ transforms according to:

$$\begin{aligned}[\Lambda]_{FB} &= [S_\psi]^T[\Lambda]_{B_\psi}[S_\psi],\\ [\Lambda]_{B_\psi} &= [S_\theta]^T[\Lambda]_{B_{\psi\theta}}[S_\theta],\\ [\Lambda]_{B_{\psi\theta}} &= [S_\varphi]^T[\Lambda]_{B_{\psi\theta\varphi}}[S_\varphi],\end{aligned} \quad (1.35)$$

and that:

$$\begin{aligned}[S_\psi]^T[\dot{S}_\psi] &= [\bar{\dot{\psi}}\times] \text{ expressed in basis } B_\psi,\\ [S_\theta]^T[\dot{S}_\theta] &= [\bar{\dot{\theta}}\times] \text{ expressed in basis } B_{\psi\theta},\\ [S_\varphi]^T[\dot{S}_\varphi] &= [\bar{\dot{\varphi}}\times] \text{ expressed in basis } B_{\psi\theta\varphi},\end{aligned} \quad (1.36)$$

Eq. (1.34) becomes:

$$\begin{aligned}[S]^T[\dot{S}] &= \left([S_\theta][S_\varphi]\right)^T[\bar{\dot{\psi}}\times]_{B_\psi}\left([S_\theta][S_\varphi]\right) + [S_\varphi]^T[\bar{\dot{\theta}}\times]_{B_{\psi\theta}}[S_\varphi] + [\bar{\dot{\varphi}}\times]_{B_{\psi\theta\varphi}}\\ &= [S_\varphi]^T[\bar{\dot{\psi}}\times]_{B_{\psi\theta}}[S_\varphi] + [\bar{\dot{\theta}}\times]_{B_{\psi\theta\varphi}} + [\bar{\dot{\varphi}}\times]_{B_{\psi\theta\varphi}} = \left([\bar{\dot{\psi}}\times] + [\bar{\dot{\theta}}\times] + [\bar{\dot{\varphi}}\times]\right)_{B_{\psi\theta\varphi}}.\end{aligned} \quad (1.37)$$

Equation (1.37) shows that $[S]^T[\dot{S}]$ expressed in basis $B_{\psi\theta\varphi}$ is equivalent to adding the three linear applications, thus proving Eq. (1.32). ♣

The most relevant aspect of the Euler angles is that **the MB final orientation is independent from the chronological order followed to introduce the angles values** (if done sequentially), or of their relative time history (if entered simultaneously), as it is easy to check analytically. Thus, although rotations (ψ, θ, φ) are usually called "first," "second," and "third" Euler rotations, these adjectives do not imply a time sequence. This "nonsequentiality" is one of the reasons why the rotations associated with most mechanisms correspond physically to Euler angles.

▶ **Example 1.1** The holder of focus F in Fig. 1.14, which allows a general orientation of the light axis relative to the ground (E), has two joints that are associated with the first and second Euler angles. Rotation ψ around the vertical axis corresponds to a first Euler angle (as it is around a fixed axis and affects the axis of the second rotation), and rotation θ (around the horizontal axis oriented by the angle ψ) corresponds to a second Euler angle.

The angular velocity of F relative to the ground is $\bar{\mathbf{\Omega}}_{\mathrm{E}}^{\mathrm{F}} = \bar{\dot{\boldsymbol{\psi}}} + \bar{\dot{\boldsymbol{\theta}}}$. Its components are expressed easily in the vector bases $\mathrm{B}_{\psi}(\equiv \mathrm{B}')$ and $\mathrm{B}_{\psi\theta}(\equiv \mathrm{B}'')$, oriented with the first Euler angle and the first two Euler angles, respectively:

$$\left\{\bar{\mathbf{\Omega}}_{\mathrm{E}}^{\mathrm{F}}\right\}_{\mathrm{B}'} = \begin{Bmatrix} \dot{\theta} \\ 0 \\ \dot{\psi} \end{Bmatrix} \; ; \; \left\{\bar{\mathbf{\Omega}}_{\mathrm{E}}^{\mathrm{F}}\right\}_{\mathrm{B}''} = \begin{Bmatrix} \dot{\theta} \\ \dot{\psi}\sin\theta \\ \dot{\psi}\cos\theta \end{Bmatrix}. \tag{1.38}$$

The calculation of the angular acceleration of F through the vector basis B′ yields:

$$\left\{\bar{\boldsymbol{\alpha}}_{\mathrm{E}}^{\mathrm{F}}\right\}_{\mathrm{B}'} = \frac{\mathrm{d}}{\mathrm{dt}}\left\{\bar{\mathbf{\Omega}}_{\mathrm{E}}^{\mathrm{F}}\right\}_{\mathrm{B}'} + \left\{\bar{\mathbf{\Omega}}_{\mathrm{E}}^{\mathrm{B}'} \times \bar{\mathbf{\Omega}}_{\mathrm{E}}^{\mathrm{F}}\right\}_{\mathrm{B}'} = \begin{Bmatrix} \ddot{\theta} \\ 0 \\ \ddot{\psi} \end{Bmatrix} + \begin{Bmatrix} 0 \\ 0 \\ \dot{\psi} \end{Bmatrix} \times \begin{Bmatrix} \dot{\theta} \\ 0 \\ \dot{\psi} \end{Bmatrix} = \begin{Bmatrix} \ddot{\theta} \\ \dot{\psi}\dot{\theta} \\ \ddot{\psi} \end{Bmatrix}. \tag{1.39}$$

If we use the basis B″, as $\bar{\mathbf{\Omega}}_{\mathrm{E}}^{\mathrm{F}} = \bar{\mathbf{\Omega}}_{\mathrm{E}}^{\mathrm{B}''}$, the vector product $\left(\bar{\mathbf{\Omega}}_{\mathrm{E}}^{\mathrm{B}''} \times \bar{\mathbf{\Omega}}_{\mathrm{E}}^{\mathrm{F}}\right)$ is zero:

$$\left\{\bar{\boldsymbol{\alpha}}_{\mathrm{E}}^{\mathrm{F}}\right\}_{\mathrm{B}''} = \frac{\mathrm{d}}{\mathrm{dt}}\left\{\bar{\mathbf{\Omega}}_{\mathrm{E}}^{\mathrm{F}}\right\}_{\mathrm{B}''} + \left\{\bar{\mathbf{\Omega}}_{\mathrm{E}}^{\mathrm{B}''} \times \bar{\mathbf{\Omega}}_{\mathrm{E}}^{\mathrm{F}}\right\}_{\mathrm{B}''} = \begin{Bmatrix} \ddot{\theta} \\ \ddot{\psi}\sin\theta + \dot{\psi}\dot{\theta}\cos\theta \\ \ddot{\psi}\cos\theta - \dot{\psi}\dot{\theta}\sin\theta \end{Bmatrix} + \begin{Bmatrix} 0 \\ 0 \\ 0 \end{Bmatrix}. \tag{1.40}$$

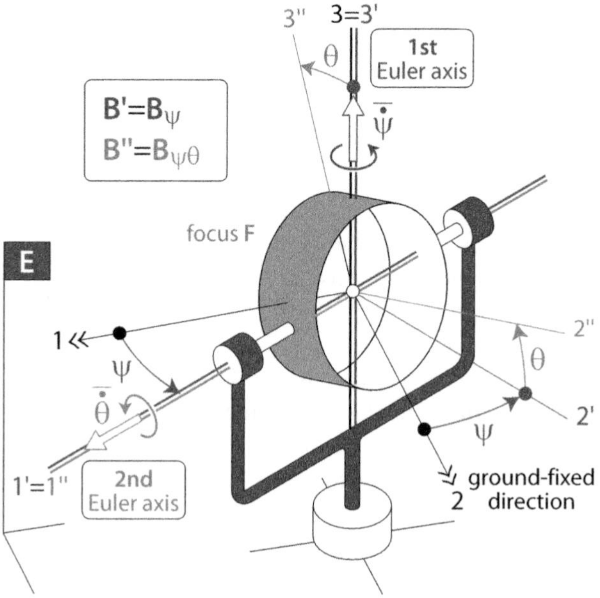

Fig. 1.14

The vector components in Eqs. (1.39, 1.40) do not coincide but correspond to the same vector (expressed in two different bases).

Note that the F angular acceleration is not zero when the $(\dot{\psi}, \dot{\theta})$ values are constant. The terms $(\dot{\psi}\dot{\theta})$ come from the change of orientation of the angular velocity $\overline{\dot{\boldsymbol{\theta}}}$ due to the rotation $\overline{\boldsymbol{\psi}}$. ◀

► **Example 1.2** The support in Fig. 1.15, which allows the gyro wheel to achieve any orientation relative to the ground (E), has three joints associated with three Euler angles. The rotation ψ around the fixed horizontal axis corresponds to the first Euler angle. The rotation θ around the axis common to the outer and the middle gimbal corresponds to the second, and the rotation φ around the axis of the wheel W corresponds to the third.

The first angle gives the outer gimbal orientation, whose angular velocity is $\overline{\boldsymbol{\psi}}$. The first and second angles give the middle gimbal orientation, whose angular velocity is $\overline{\boldsymbol{\psi}} + \overline{\dot{\boldsymbol{\theta}}}$. The gyro wheel orientation depends on all three angles, and its angular velocity is $\bar{\boldsymbol{\Omega}}_{\mathrm{E}}^{\mathrm{W}} = \overline{\boldsymbol{\psi}} + \overline{\dot{\boldsymbol{\theta}}} + \overline{\boldsymbol{\varphi}}$. The components of $\bar{\boldsymbol{\Omega}}_{\mathrm{E}}^{\mathrm{W}}$ are easily expressed in the vector bases $\mathrm{B}_{\psi}(\equiv \mathrm{B}')$ and $\mathrm{B}_{\psi\theta}(\equiv \mathrm{B}'')$, oriented by the first Euler angle and the first two Euler angles, respectively:

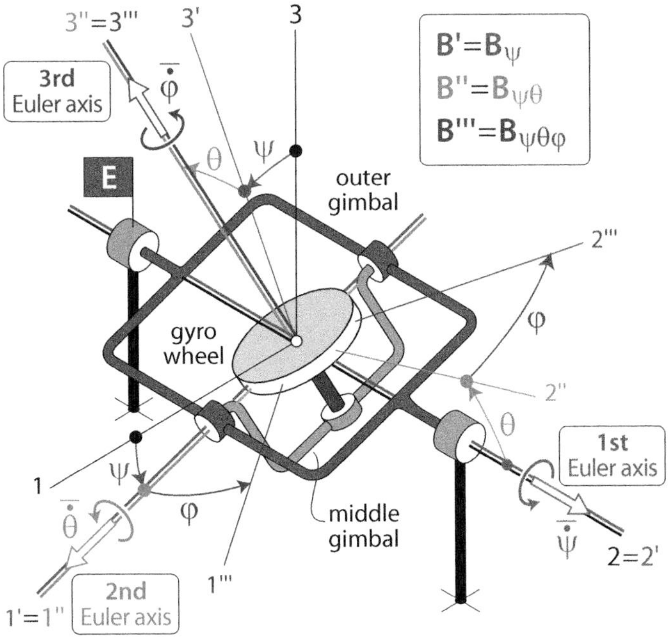

Fig. 1.15

$$\left\{\bar{\Omega}_{\mathrm{E}}^{\mathrm{W}}\right\}_{\mathrm{B}'} = \begin{Bmatrix} \dot{\theta} \\ \dot{\psi} - \dot{\varphi}\sin\theta \\ \dot{\varphi}\cos\theta \end{Bmatrix} \quad ; \quad \left\{\bar{\Omega}_{\mathrm{E}}^{\mathrm{W}}\right\}_{\mathrm{B}''} = \begin{Bmatrix} \dot{\theta} \\ \dot{\psi}\cos\theta \\ \dot{\varphi} - \dot{\psi}\sin\theta \end{Bmatrix}. \tag{1.41}$$

The fixed basis FB and the basis $\mathrm{B}_{\psi\theta\varphi}(\equiv \mathrm{B}''')$, whose orientation relative to the ground calls for the three angles, are less suitable to express the components of $\bar{\Omega}_{\mathrm{E}}^{\mathrm{W}}$ because they require double projections:

$$\left\{\bar{\dot{\varphi}}\right\}_{\mathrm{B}} = \begin{Bmatrix} \dot{\varphi}\cos\theta\sin\psi \\ -\dot{\varphi}\sin\theta \\ \dot{\varphi}\cos\theta\cos\psi \end{Bmatrix} \quad ; \quad \left\{\bar{\dot{\psi}}\right\}_{\mathrm{B}'''} = \begin{Bmatrix} \dot{\psi}\cos\theta\sin\varphi \\ \dot{\psi}\cos\theta\cos\varphi \\ -\dot{\psi}\sin\theta \end{Bmatrix}.$$

◀

1.9 The Two Families of Euler Angles

Although the Euler angles are suggested by the articulations in a mechanism or are conveniently defined when orientating a single rigid body (as will be seen in the following examples), they usually belong to the two families that are obtained when taking as axis of rotation an axis of each of the bases $(\mathrm{BF}, \mathrm{B}_{\psi}(\equiv \mathrm{B}'), \mathrm{B}_{\psi\theta}(\equiv \mathrm{B}''))$ generated when the three angles are introduced sequentially.

The procedure is shown in Fig. 1.16:

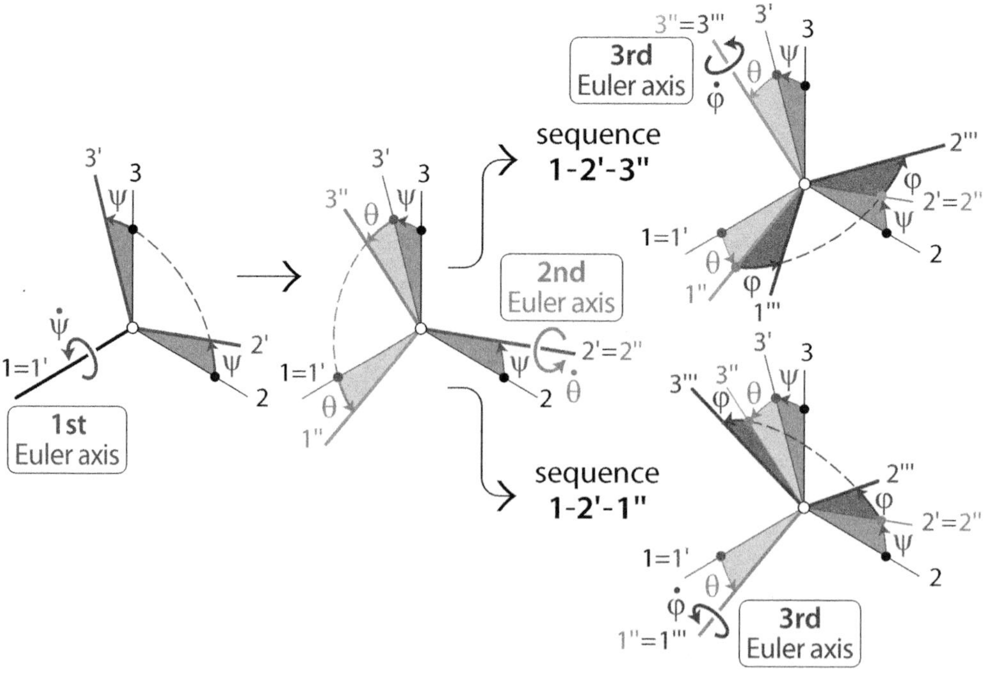

Fig. 1.16

- Start from the fixed basis FB with axes $(1, 2, 3)$, and take the first rotation about one of its axes (for example, 1); this generates a second basis $B_\psi (\equiv B')$, with axes $(1' \equiv 1, 2', 3')$
- Take the second rotation about one of the B_ψ axes (different from $1'$; for example, $2'$); this generates a third basis, with axes $\left(1'', 2'' \equiv 2', 3''\right)$
- The third rotation can be taken either about axis $1''$ or axis $3''$. In the former case, the axes sequence is $\left(1, 2', 3''\right)$. In the latter, the axes sequence is $\left(1, 2', 1''\right)$

So there are two possible schemes for the sequence of rotations:

- Sequence $\left(i, j', k''\right) \equiv (i - j - k)$ with $i \neq j \neq k$
- Sequence $\left(i, j', i''\right) \equiv (i - j - i)$ with $i \neq j$

In short, they will be called $(i - j - k)$ family ("three-body axis system") and $(i - j - i)$ family ("two-body axis system"), respectively.

The $(i - j - k)$ family is usually chosen to define the orientation of vehicles (chassis of a car, airplane, ship, etc.) and the three rotations have specific names. For the starting orientation of the vehicle $(\psi = \theta = \varphi = 0)$, the Euler axes are as follows:

- **Roll** axis = vehicle longitudinal axis (positive direction according to forward motion)
- **Pitch** axis = vehicle transverse axis (positive direction from right to left when looking forward)
- **Yaw** axis = axis perpendicular to the other two (positive direction from the vehicle floor to the ceiling)

When studying vehicles (not necessarily ground vehicles), the usual sequence is (yaw–pitch–roll). In that case, we establish the association ψ(yaw), θ(pitch), φ(roll) in order to be consistent with the nomenclature introduced in Section 1.8.

Figure 1.17 illustrates this sequence for a ship. For an arbitrary orientation, the yaw axis (first Euler rotation) remains vertical (and therefore does not coincide with the axis perpendicular to the deck). The roll axis (third Euler rotation) is the ship longitudinal axis. The pitch axis (second Euler rotation) is the horizontal axis perpendicular to the ship longitudinal axis (and does not coincide with the transverse direction of the ship).

As exercise, Fig. 1.18 shows the alternative sequence (roll–pitch–yaw) for an automobile (the coherent nomenclature would be ψ, θ, and φ, respectively).

The $(i - j - i)$ family is the usual choice to describe the rotation of a spinning top (Fig. 1.19). The inclination of the spinning top axis (relative to the vertical direction) is taken as the second Euler angle. Its change corresponds to a rotation around a horizontal direction – which is the Euler second axis – and therefore the Euler first axis is vertical. The spinning top axis is the Euler third axis. There are specific names associated with the angles: **precession** ψ, **nutation** θ, and **spin** φ.

These names are also used in the study of fast rotors, where the spin axis (Euler third axis) is the rotor axis. According to the starting orientation $(\psi = \theta = \varphi = 0)$, it is a $(i - j - k)$ family (Fig. 1.20).

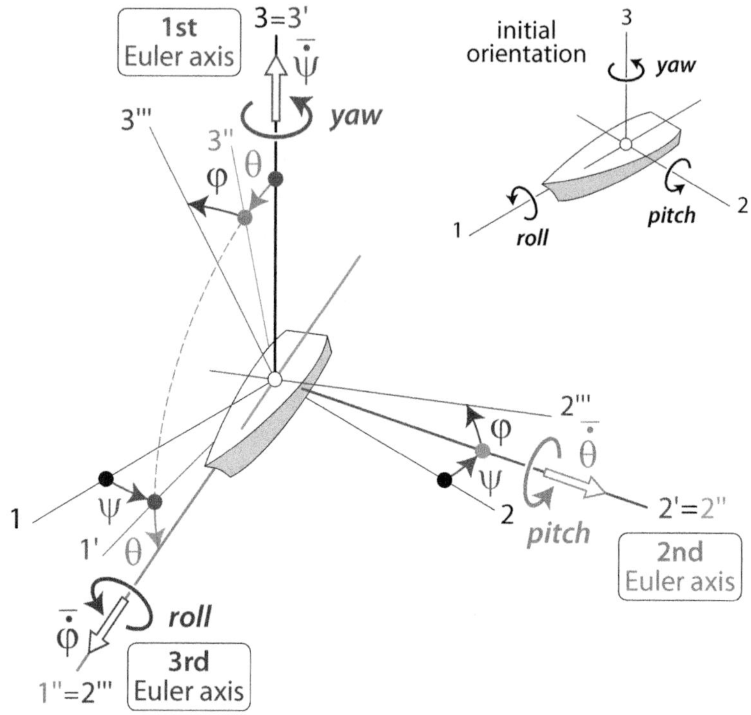

Fig. 1.17

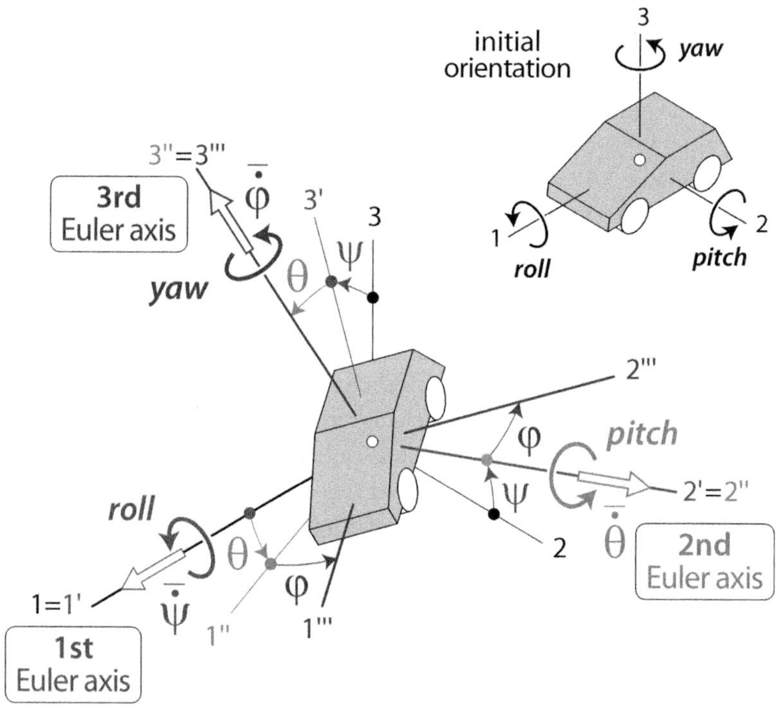

Fig. 1.18

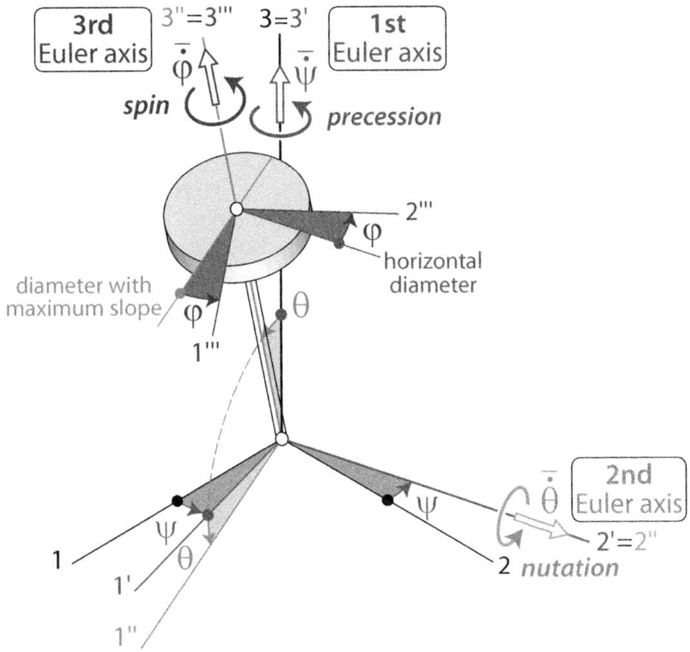

Fig. 1.19

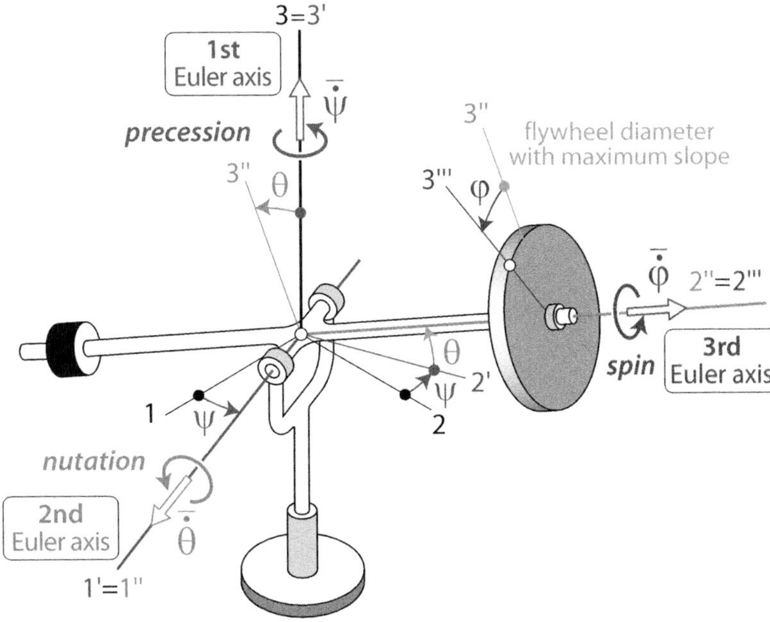

Fig. 1.20

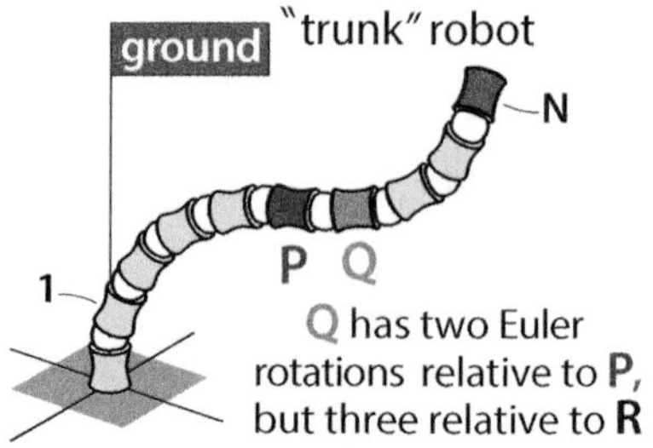
ground
"trunk" robot
N
1
P Q
Q has two Euler rotations relative to P, but three relative to R

Fig. 1.21

Multibody systems with more than three simple rotations chained in series are fairly common. For example, the robot manipulators of the "trunk" type (which have the same mobility as an elephant trunk, Fig. 1.21). The orientation of each element relative to the previous one can be controlled through two Euler rotations. However, the general orientation of any element with respect to the ground (except for the first moving element) requires three Euler angles. Increasing the number of elements does not imply an increase of possible orientations for the final element but a wider range of positions for the whole robot. This provides a greater accessibility in an environment with obstacles.

When we want to describe the orientation of a given body through rotation angles, there is often an angle particularly meaningful that can be identified with an Euler angle. If that angle corresponds to a rotation around a fixed axis, then it is the Euler first angle. If it corresponds to a rotation around an axis oriented by a single angle or one oriented by two angles, it will be, respectively, the Euler second or third angle. Whenever that angle corresponds to the second Euler angle, the first one is automatically defined.

► **Example 1.3** In the general movement of a wheel rotating on a horizontal plane, the inclination of the wheel plane with respect to the vertical plane changes over time. This inclination is relevant and can be taken as one of the Euler rotations (Fig. 1.22).

This inclination corresponds to a rotation about the direction of the horizontal diameter of the wheel (which has a variable orientation on the horizontal plane); it corresponds to the second Euler angle θ. The change of orientation of its axis (horizontal diameter) comes from the first rotation ψ, which has to be vertical.

The third rotation must correspond to a rotation φ of the wheel with respect to the trihedral formed by the wheel axis, the horizontal diameter, and the maximum slope diameter. As the inclination of the wheel plane is not modified by the third rotation, the third axis has to coincide with that of the wheel.

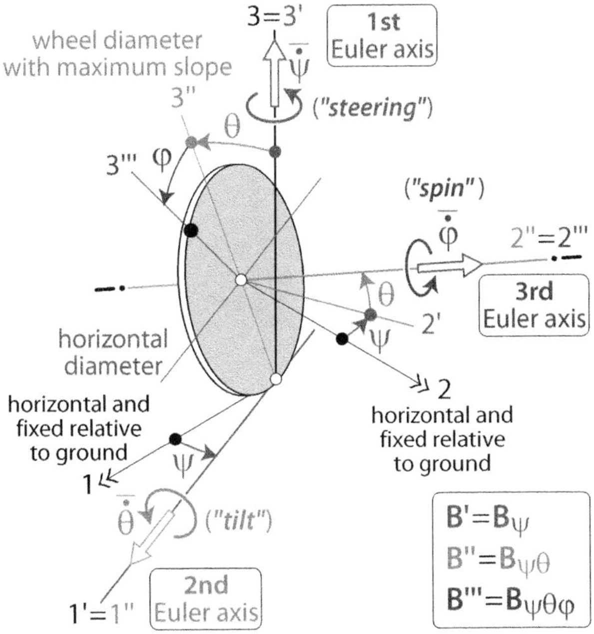

Fig. 1.22

Those angles correspond to a (i – j – k) sequence. There is no established nomenclature, but **steering**, **inclination** and **spin** could be appropriate names. ◀

▶ **Example 1.4** Projectiles are launched with a rapid rotation movement about their axis to prevent significant deviations from their nominal path. As that rotation is about a body-fixed axis, it corresponds to a third Euler angle φ.

Let us consider a portion of the trajectory of the projectile center **P**, short enough to accept that it is rectilinear (therefore, of constant direction). The angle of deviation of the projectile axis relative to the **P** trajectory (Fig. 1.23a) is the most relevant angle. The axis associated with this change of inclination is perpendicular to the plane containing the trajectory portion and the projectile axis. As this plane rotates about the trajectory portion, the inclination corresponds to a second Euler angle θ. The first Euler angle ψ defines the plane containing the trajectory portion where θ is defined.

The projectile still has φ the rotation about its axis relative to the trihedral formed by the projectile axis, the θ rotation axis, and the perpendicular to these two. The three angles considered correspond to a (i – j – i) family, and the names "precession–nutation–spin" are suitable.

A disadvantage of that choice is that ψ and φ are indeterminate when $\theta = 0$, as their respective rotation axes overlap. Its sum is determined but not the value of each of them

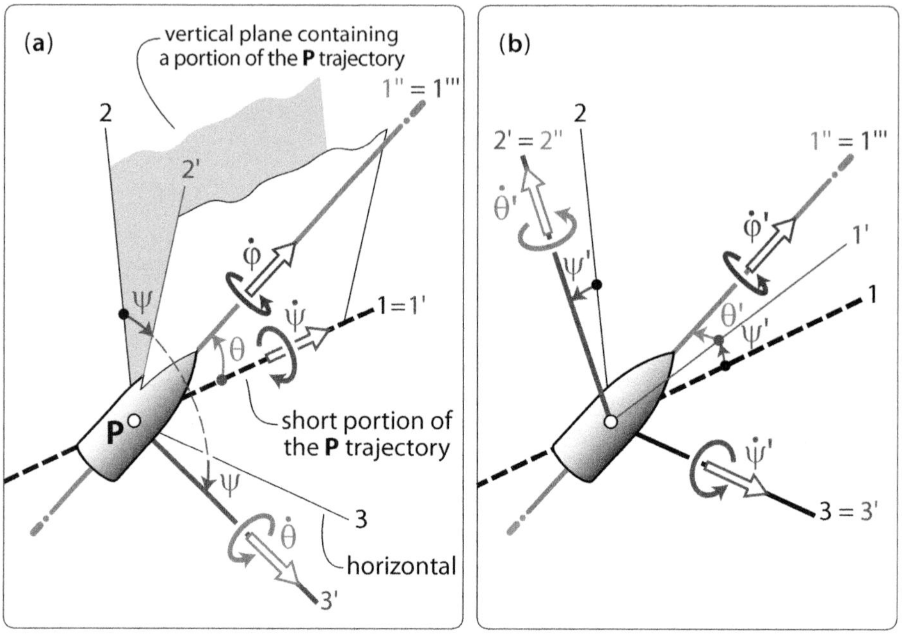

Fig. 1.23

separately because there is not a unique plane axial to the trajectory determining the angle ψ. If avoiding this singularity is more important than taking the inclination angle in the Euler family, we can shift to the $(i - j - k)$ family (Fig. 1.23b):

- Rotation ψ' around the horizontal axis perpendicular to the trajectory (axis 3)
- Rotation θ' around axis $2'$
- Rotation φ' around the projectile axis

In this case, the roll–pitch–yaw designations are not appropriate. ◄

Appendix 1A Composition of Rotations around Fixed Axes

A simple method (from a conceptual point of view) to orientate a vector basis MB in a reference frame R (or a basis FB fixed in R) through three angles consists in rotating MB sequentially around **three axes fixed relative to R**. A possible sequence would be (Fig. 1A.1):

- First rotation with angle β_1 around the fixed axis 1; it generates basis $B\beta_1$ from FB
- Second rotation with angle β_2 around the fixed axis 2; it generates basis $B\beta_1\beta_2$ from $B\beta_1$
- Third rotation with angle β_3 around the fixed axis 3; it generates basis $B\beta_1\beta_2\beta_3$ from $B\beta_1\beta_2$

The transformation matrix that determines the components of the vectors in the FB basis from the components in the MB basis $\left(\{\bar{\mathbf{u}}\}_{FB} = [S]\{\bar{\mathbf{u}}\}_{MB}\right)$ according to the previous sequence is:

$$[S] = [S\beta_3][S\beta_2][S\beta_1], \tag{1A.1}$$

where the $[S\beta_i]$ are the transformation matrices associated with each β_i rotation – generating the $B\beta_i$ basis from the FB basis – expressed in the FB:

$$\{\bar{\mathbf{u}}\}_{FB} = [S\beta_i]\{\bar{\mathbf{u}}\}_{B\beta_i}. \tag{1A.2}$$

Since the product of matrices is not commutative, the transformation matrix [S] in Eq. (1) (and therefore the final orientation) depends on the angles sequence.

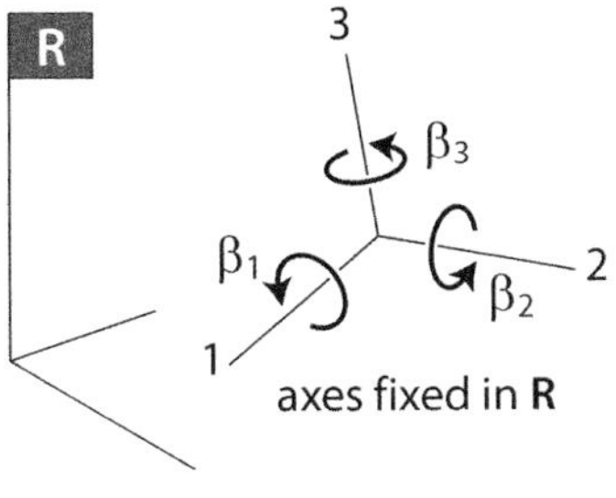

Fig. 1A.1

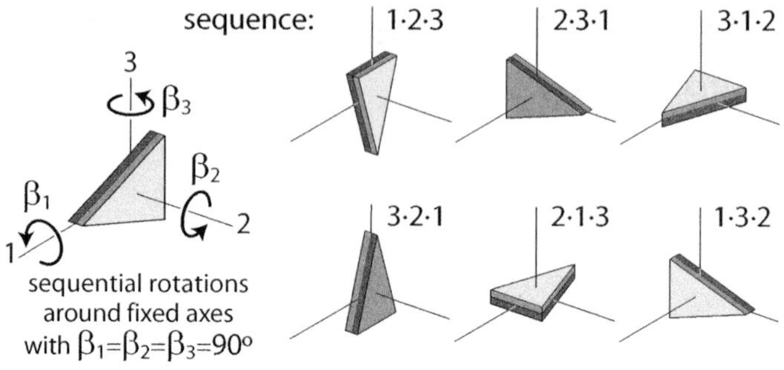

Fig. 1A.2

This is shown in Fig. 1A.2 through a triangular body undergoing three rotations around the three orthogonal axes 1, 2, 3. Each of the six possible sequences leads to a different final orientation.

It follows that for the three values $(\beta_1, \beta_2, \beta_3)$ to define the final orientation univocally, giving just their value is not enough: you must specify the sequence. This forced sequentiality is a drawback when using it in mechanisms to control bodies orientation over time.

Let us suppose that we introduce the fixed-axis rotations $(\beta_1, \beta_3, \beta_2)$ from the initial orientation $(\beta_1 = \beta_2 = \beta_3 = 0)$ according to a given time sequence. If we wanted to move to the new orientation $(\beta_1 + \Delta\beta_1, \beta_3, \beta_2)$, we would have to go back to the initial orientation $(\beta_1 = \beta_2 = \beta_3 = 0)$ and start again.

In graphic computing, fixed-axis sequential rotations are a common practice to inspect designed bodies from different visuals. But the sequence is neither predetermined and unique, nor is the number of rotations usually limited to three. As a consequence, the final orientation has no univocal relationship with the three total rotations that have been introduced (though this does not concern the user).

Nevertheless, the sequential composition of three fixed-axis rotations is a mathematical tool for defining univocally the final orientation of a vector basis MB relative to a reference frame R through the transformation matrix in Eq. (1A.1).

The equivalence between that sequential composition and the Euler angles can be found from the transformation matrix $[S]$ relating the components in the FB with the components in the MB. For a given sequence $(\beta_1, \beta_2, \beta_3)$, the matrix in Eq. (1A.1) can also be obtained as:

$$[S] = [S\beta_3][S\beta_2][S\beta_1] = [S_\psi][S_\theta][S_\varphi], \tag{1A.3}$$

where **the fixed-axis rotations have the same values as the Euler rotations** $(\beta_1 = \psi, \beta_2 = \theta, \beta_3 = \varphi)$, though **they have to be introduced in reverse order** (Fig. 1A.3). For $(\psi = \theta = \varphi = 0)$, the Euler first, second, and third axes coincide with the fixed axes of the first, second and third rotations, respectively.

One has to bear in mind that, unlike the Euler rotations – composition in series – where the matrices are expressed in different bases $\left([S_\psi]\right.$ in the FB, $[S_\theta]$ in the B_ψ and

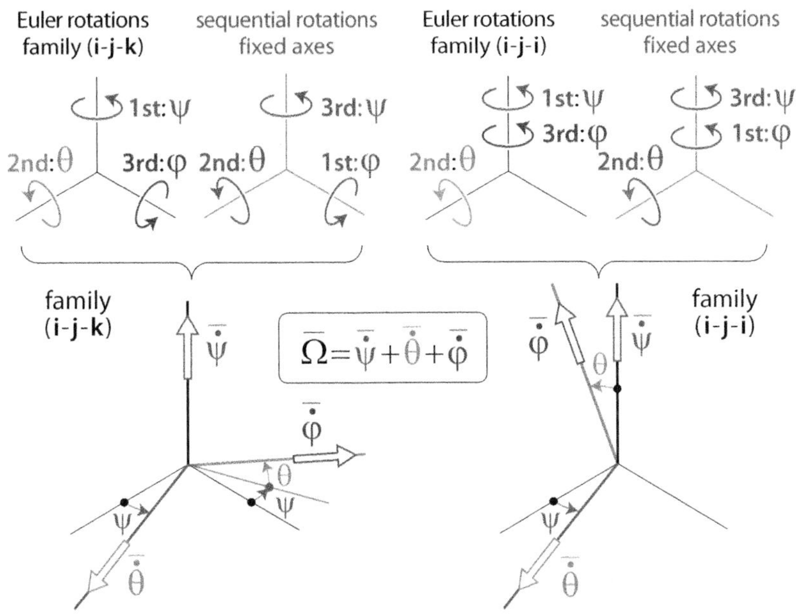

Fig. 1A.3

$[S_\varphi]$ in the $B_{\psi\theta}$), in the sequential fixed-axis rotations, all transformation matrices are expressed in the FB.

When introducing simultaneous rotations around the fixed axes (1, 2, 3), the final values $(\beta_1, \beta_2, \beta_3)$ do not define the final orientation univocally. For a given time history $(\beta_1(t), \beta_2(t), \beta_3(t))$, the transformation matrix [S] would be expressed as the product of an infinite number of matrices associated with rotation differentials $(d\beta_1, d\beta_2, d\beta_3 \rightarrow 0)$:

$$[S] = \underset{\text{last differentials}}{[Sd\beta_1][Sd\beta_2][Sd\beta_3]} \quad \ldots \quad \underset{\text{intermediate differentials}}{[Sd\beta_1][Sd\beta_2][Sd\beta_3]} \quad \ldots \quad \underset{\text{first differentials}}{[Sd\beta_1][Sd\beta_2][Sd\beta_3]}.^5 \tag{1A.4}$$

A different time history $(\beta_1(t), \beta_2(t), \beta_3(t))$ would yield different relative sizes of those $d\beta_i$, hence different $[Sd\beta_i]$ transformation matrices and different global matrix [S] for the same final values $(\beta_1, \beta_2, \beta_3)$.

For any time instant, the angular velocity $\bar{\boldsymbol{\Omega}}_R^{MB}$ is univocally associated with the matrix $[Sd\beta_1][Sd\beta_2][Sd\beta_3]$. As $d\beta_i \rightarrow 0$, $\bar{\boldsymbol{\Omega}}_R^{MB}$ is the superposition of $\bar{\dot{\boldsymbol{\beta}}}_1$ around axis 1, $\bar{\dot{\boldsymbol{\beta}}}_2$ around axis 2, and $\bar{\dot{\boldsymbol{\beta}}}_3$ around axis 3: $\bar{\boldsymbol{\Omega}}_R^{MB} = \bar{\dot{\boldsymbol{\beta}}}_1 + \bar{\dot{\boldsymbol{\beta}}}_2 + \bar{\dot{\boldsymbol{\beta}}}_3$ (Fig. 1A.4).

In many mechanical devices (as the ball of the old mechanical "mouse" of computers), the simultaneous rotations around fixed axes are limited to two axes. In that

[5] For three infinitesimal rotations, the sequence is irrelevant: $[Sd\beta_1][Sd\beta_2][Sd\beta_3] = [Sd\beta_3][Sd\beta_2][Sd\beta_1] = \ldots$

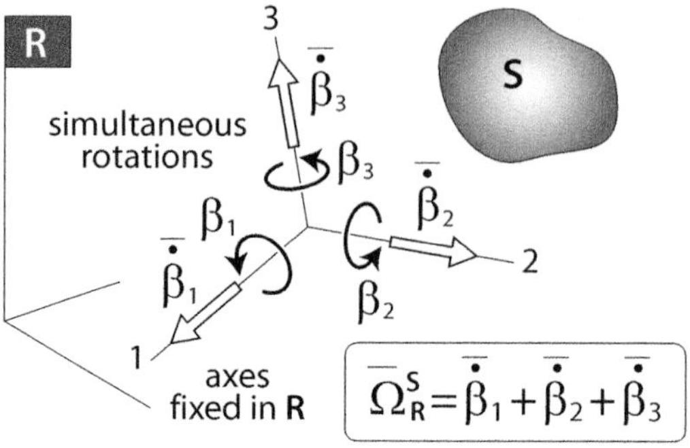

Fig. 1A.4

case, $\bar{\Omega}_R^{MB}$ is constrained to be in a plane, but the final orientation is general (though irrelevant for the purpose of the device).

In the mechanical "mouse," the angular velocity of the ball relative to the casing is measured as $\dot{\beta}_1$ and $\dot{\beta}_2$ around casing-fixed axes (which, in principle, do not have to change their orientation relative to the table). Example 1A.1 illustrates this procedure.

► **Example 1A.1** The sphere of radius R in Fig. 1A.5 rests on a support that allows a free rotation around its center **O**. At each time instant, the wheel P, which maintains a nonsliding contact with the sphere, rotates it around axis 1 with $\dot{\beta}_1 = \dot{\theta}_1 r/R$. The wheel Q, which also maintains a nonsliding contact with the sphere, rotates it around axis 2 with $\dot{\beta}_2 = \dot{\theta}_2 r/R$.

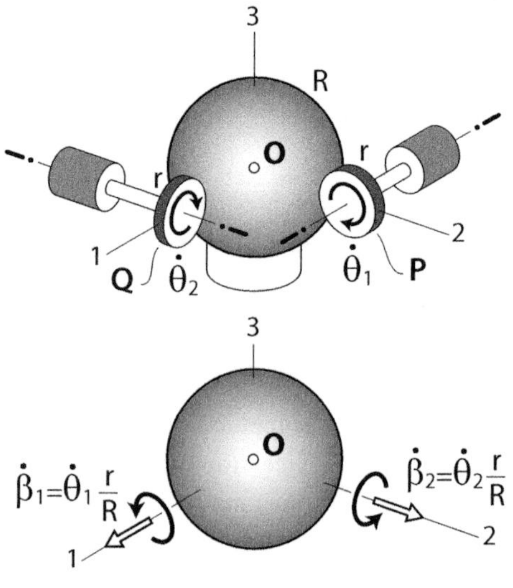

Fig. 1A.5

It is a composition of rotations around two fixed axes. If the two wheels rotate simultaneously, a pair of values β_1 and β_2 do not define univocally the sphere orientation; that orientation depends on the relative time evolution between β_1 and β_2. The sphere may achieve any orientation by maneuvering properly.

In a computer mouse, point **O** and axes (1, 2, 3) are fixed to the mouse casing (which has a translational motion relative to the table). The ball rotation causes the rotation of the wheels P and Q. The rotated angles $\Delta\theta_1$ and $\Delta\theta_2$ translate into proportional electrical voltage by means of a potentiometer. The components v_1 and v_2 of $\bar{v}_{TABLE}(\mathbf{O})$ relate to the wheels' angular velocities through $v_1 = (r/R)\dot{\theta}_2$ and $v_2 = -(r/R)\dot{\theta}_1$, and their integration yields $\Delta x_1 = (r/R)\Delta\theta_2$ and $\Delta x_2 = -(r/R)\Delta\theta_1$. ◀

Appendix 1B Alternatives to the Composition of Rotations

Let's consider the transformation matrix [S] associated with the (i − j − k) sequence of Euler angles:

$$[S] = \begin{bmatrix} 1 & 0 & 0 \\ 0 & \cos\psi & -\sin\psi \\ 0 & \sin\psi & \cos\psi \end{bmatrix} \begin{bmatrix} \cos\theta & 0 & \sin\theta \\ 0 & 1 & 0 \\ -\sin\theta & 0 & \cos\theta \end{bmatrix} \begin{bmatrix} \cos\varphi & -\sin\varphi & 0 \\ \sin\varphi & \cos\varphi & 0 \\ 0 & 0 & 1 \end{bmatrix} =$$

$$= \begin{bmatrix} \cos\theta\cos\varphi & -\cos\theta\sin\varphi & \sin\theta \\ \cos\psi\sin\varphi + \sin\psi\sin\theta\cos\varphi & \cos\psi\cos\varphi - \sin\psi\sin\theta\sin\varphi & -\sin\psi\cos\theta \\ \sin\psi\sin\varphi - \cos\psi\sin\theta\cos\varphi & \sin\psi\cos\varphi + \cos\psi\sin\theta\sin\varphi & \cos\psi\cos\theta \end{bmatrix} \tag{1B.1}$$

The S_{13} element depends exclusively on the second Euler angle θ, but there are no elements depending solely on the first or the third Euler angles. In other words, the Euler angles do not play the same role in the transformation matrix [S], and this can be seen as a drawback from the analytical point of view. The same thing would happen if we considered a (i − j − i) sequence.

There are other parameters whose role in [S] is more symmetrical, and as a consequence the associated analytical treatment is simpler. The disadvantage is that these parameters do not have the simple geometric interpretation of Euler angles.

The properties associated with the symmetry of the orientation variables are particularly interesting in the computer-aided kinematics of the rigid body. This has brought back orientation methods that had been developed in the past but had practically been disregarded. Astronautics and robotics have been the most prominent drivers of this recovery.

We review the three most representative alternatives to the composition of rotations:

- Euler parameters
- Rodrigues parameters
- Cayley–Klein parameters

Euler Parameters

Euler's theorem states that any change of orientation of a rigid body relative to a reference frame R can be described as a simple rotation around an axis fixed in

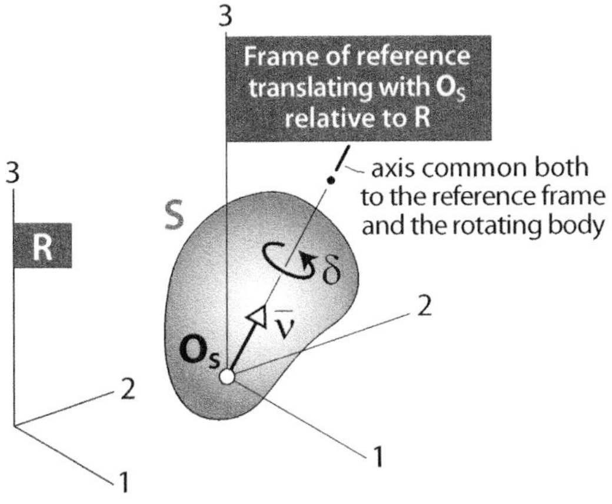

Fig. 1B.1

R and in the body reference frame. Hence a basis change can be described by a unit vector $\bar{\mathbf{v}}$ giving the direction of that axis and the angle δ rotated around it (Fig. 1B.1).

This procedure always starts from the initial orientation, and one has to go back to it to introduce any change of orientation (as is the case with rotations around fixed axes). For this reason, mechanical devices are unsuitable to implement this type of rotations.

From the components (v_1, v_2, v_3) of $\bar{\mathbf{v}}$ in the fixed base FB, the **Euler parameters** are defined as:

$$\begin{aligned} \varepsilon_1 &= v_1 \sin(\delta/2), \\ \varepsilon_2 &= v_2 \sin(\delta/2), \\ \varepsilon_3 &= v_3 \sin(\delta/2), \\ \varepsilon_4 &= \cos(\delta/2). \end{aligned} \tag{1B.2}$$

They fulfill the normalization property $\varepsilon_1^2 + \varepsilon_2^2 + \varepsilon_3^2 + \varepsilon_4^2 = 1$, so only three of them are independent.

The Euler parameters can be viewed as components of a Hamilton quaternion. Hamilton's quaternions are a generalization of complex numbers with three imaginary axes instead of just one. The quaternion $q = i\varepsilon_1 + j\varepsilon_2 + k\varepsilon_3 + \varepsilon_4$ contains three imaginary unities fulfilling:

$$\begin{aligned} & i^2 + j^2 + k^2 = 1, \\ & ij = -ji = k, \\ & jk = -kj = i, \\ & ki = -ik = j. \end{aligned} \tag{1B.3}$$

The most interesting analytical property of these parameters is the symmetrical role they play in the transformation matrix $[S]$ $\left(\{\bar{\mathbf{u}}\}_{FB} = [S]\{\bar{\mathbf{u}}\}_{MB}\right)$:

$$[S] = \begin{bmatrix} \varepsilon_1^2 - \varepsilon_2^2 - \varepsilon_3^2 + \varepsilon_4^2 & 2(\varepsilon_1\varepsilon_2 - \varepsilon_3\varepsilon_4) & 2(\varepsilon_1\varepsilon_3 + \varepsilon_2\varepsilon_4) \\ 2(\varepsilon_1\varepsilon_2 + \varepsilon_3\varepsilon_4) & -\varepsilon_1^2 + \varepsilon_2^2 - \varepsilon_3^2 + \varepsilon_4^2 & 2(\varepsilon_2\varepsilon_3 - \varepsilon_1\varepsilon_4) \\ 2(\varepsilon_1\varepsilon_3 - \varepsilon_2\varepsilon_4) & 2(\varepsilon_2\varepsilon_3 + \varepsilon_1\varepsilon_4) & -\varepsilon_1^2 - \varepsilon_2^2 + \varepsilon_3^2 + \varepsilon_4^2 \end{bmatrix}. \tag{1B.4}$$

The relationship between the Euler parameters and the Euler angles of the family $(i - j - k)$ in Eq. (1B.1) is:

$$\begin{aligned} \varepsilon_1 &= \sin\frac{\theta}{2}\sin\frac{\varphi - \psi}{2}, \\ \varepsilon_2 &= \sin\frac{\theta}{2}\cos\frac{\varphi - \psi}{2}, \\ \varepsilon_3 &= \cos\frac{\theta}{2}\sin\frac{\varphi + \psi}{2}, \\ \varepsilon_4 &= \cos\frac{\theta}{2}\cos\frac{\varphi + \psi}{2}. \end{aligned} \tag{1B.5}$$

Rodrigues Parameters

The Rodrigues parameters are expressed from those of Euler as:

$$\begin{aligned} \rho_1 &= \varepsilon_1/\varepsilon_4 = \nu_1 \tan(\delta/2), \\ \rho_2 &= \varepsilon_2/\varepsilon_4 = \nu_2 \tan(\delta/2), \\ \rho_3 &= \varepsilon_3/\varepsilon_4 = \nu_3 \tan(\delta/2). \end{aligned} \tag{1B.6}$$

Like Euler's, these parameters play a symmetric role in the transformation matrix:

$$[S] = \frac{1}{\rho_1^2 + \rho_2^2 + \rho_3^2 + 1}\begin{bmatrix} \rho_1^2 - \rho_2^2 - \rho_3^2 + 1 & 2(\rho_1\rho_2 - \rho_3) & 2(\rho_1\rho_3 + \rho_2) \\ 2(\rho_1\rho_2 + \rho_3) & -\rho_1^2 + \rho_2^2 - \rho_3^2 + 1 & 2(\rho_2\rho_3 - \rho_1) \\ 2(\rho_1\rho_3 - \rho_2) & 2(\rho_2\rho_3 + \rho_1) & -\rho_1^2 - \rho_2^2 + \rho_3^2 + 1 \end{bmatrix}. \tag{1B.7}$$

Compared to Euler's, the Rodrigues parameters have the advantage of being fewer in number (three instead of four) and the disadvantage of being unbounded (they can become infinite while Euler's never exceed unity).

Cayley–Klein Parameters

The Cayley–Klein parameters are expressed from those of Euler as:

$$\begin{aligned}\alpha &= \varepsilon_4 + i\varepsilon_3,\\ \beta &= \varepsilon_1 + i\varepsilon_2,\\ \gamma &= -\varepsilon_1 + i\varepsilon_2,\\ \delta &= \varepsilon_4 - i\varepsilon_3,\end{aligned} \tag{1B.8}$$

with $i = \sqrt{-1}$. They fulfill the relationship $(\alpha\delta - \beta\gamma = 1)$.

One advantage of the Cayley–Klein parameters over Euler's is that they retain part of the simplicity of the quaternion calculation using the ordinary imaginary unit $i = \sqrt{-1}$ instead of the three imaginary units (i, j, k) of Hamilton quaternions.

Regarding the symmetry, they are in an intermediate situation between Euler's (or Rodrigues') and the Euler angles. The transformation matrix is:

$$[S] = \begin{bmatrix} \frac{1}{2}(\alpha^2+\beta^2+\gamma^2+\delta^2) & \frac{i}{2}(\alpha^2-\beta^2+\gamma^2-\delta^2) & i(\alpha\beta+\gamma\delta) \\ \frac{i}{2}(-\alpha^2-\beta^2+\gamma^2+\delta^2) & \frac{1}{2}(\alpha^2-\beta^2-\gamma^2+\delta^2) & -\alpha\beta+\gamma\delta \\ i(\alpha\gamma+\beta\delta) & -\alpha\gamma+\beta\delta & \alpha\delta+\beta\gamma \end{bmatrix}. \tag{1B.9}$$

Appendix 1C Integration of Vectors through Moving Bases

Sometimes, the time derivative of a vector relative to a reference frame R is known through its components in a moving basis MB. This is the case, for example, when measuring the acceleration relative to the ground of a point of a vehicle by means of a triaxial accelerometer fixed on the chassis. The result is the components of the acceleration according to the vehicle-fixed basis (and therefore moving relative to the ground).

If we want to find the components of $\bar{\mathbf{u}}$ in a moving basis MB from those of its time derivative (also in MB), we cannot simply integrate the latter (for the same reason why the time derivative of a vector in a MB cannot be obtained by simply performing a time derivative of the vector components). Within the time interval dt considered both in the time derivative and in the integration, the basis MB undergoes a change of orientation that must be taken into account.

A possible procedure is to move to a fixed basis. Then, we only have to integrate the components:

$$\left\{\left[\frac{\mathrm{d}\bar{\mathbf{u}}}{\mathrm{dt}}\right]_{\mathrm{R}}\right\}_{\mathrm{FB}} = [\mathrm{S}]\left\{\left[\frac{\mathrm{d}\bar{\mathbf{u}}}{\mathrm{dt}}\right]_{\mathrm{R}}\right\}_{\mathrm{MB}},$$

$$\{\bar{\mathbf{u}}(\mathrm{t})\}_{\mathrm{FB}} = \{\bar{\mathbf{u}}(\mathrm{t_0})\}_{\mathrm{FB}} + \int_{\mathrm{t_0}}^{\mathrm{t}} \left\{\left[\frac{\mathrm{d}\bar{\mathbf{u}}}{\mathrm{dt}}\right]_{\mathrm{R}}\right\}_{\mathrm{FB}} \mathrm{dt} = \{\bar{\mathbf{u}}(\mathrm{t_0})\}_{\mathrm{FB}} + \int_{\mathrm{t_0}}^{\mathrm{t}} \left(\frac{\mathrm{d}}{\mathrm{dt}}\{\bar{\mathbf{u}}\}_{\mathrm{FB}}\right)\mathrm{dt}. \tag{1C.1}$$

Once $\bar{\mathbf{u}}(\mathrm{t})$ has been obtained in the FB, we can go back to the MB through multiplication by $[\mathrm{S}]^{\mathrm{T}}$:

$$\{\bar{\mathbf{u}}(\mathrm{t})\}_{\mathrm{MB}} = [\mathrm{S}]^{\mathrm{T}}\{\bar{\mathbf{u}}(\mathrm{t})\}_{\mathrm{FB}}. \tag{1C.2}$$

It is also possible to consider the integration of the vector directly in the moving basis MB. What must be integrated is the time derivative of the components in the MB. If we take into account Eq. (1.27) in Section 1.6:

$$\begin{aligned} \{\bar{\mathbf{u}}(\mathrm{t})\}_{\mathrm{MB}} &= \{\bar{\mathbf{u}}(\mathrm{t_0})\}_{\mathrm{MB}} + \int_{\mathrm{t_0}}^{\mathrm{t}} \frac{\mathrm{d}}{\mathrm{dt}}\{\bar{\mathbf{u}}(\mathrm{t})\}_{\mathrm{MB}}\mathrm{dt} = \\ &= \{\bar{\mathbf{u}}(\mathrm{t_0})\}_{\mathrm{MB}} + \int_{\mathrm{t_0}}^{\mathrm{t}} \left(\left\{\left[\frac{\mathrm{d}\bar{\mathbf{u}}(\mathrm{t})}{\mathrm{dt}}\right]_{\mathrm{R}}\right\}_{\mathrm{MB}} - \left\{\bar{\boldsymbol{\Omega}}_{\mathrm{R}}^{\mathrm{MB}} \times \bar{\mathbf{u}}(\mathrm{t})\right\}_{\mathrm{MB}}\right)\mathrm{dt}. \end{aligned} \tag{1C.3}$$

This expression presents the difficulty of including the integral of the vector to be determined. If one proceeds numerically (which is usually the case when starting from experimental information), this can be easily resolved. In the most elementary formulation of numerical integration, we would have:

$$\{\bar{\mathbf{u}}(t+\Delta t)\}_{MB} = \{\bar{\mathbf{u}}(t)\}_{MB} + \left(\left\{\left[\frac{d\bar{\mathbf{u}}(t)}{dt}\right]_R\right\}_{MB} - \left\{\bar{\boldsymbol{\Omega}}_R^{MB} \times \bar{\mathbf{u}}(t)\right\}_{MB}\right)\Delta t. \quad (1C.4)$$

This formulation can be improved by using more accurate estimates of the vectors at the right-hand side of this equation within the range $[t, t+\Delta t]$.

Quiz Questions

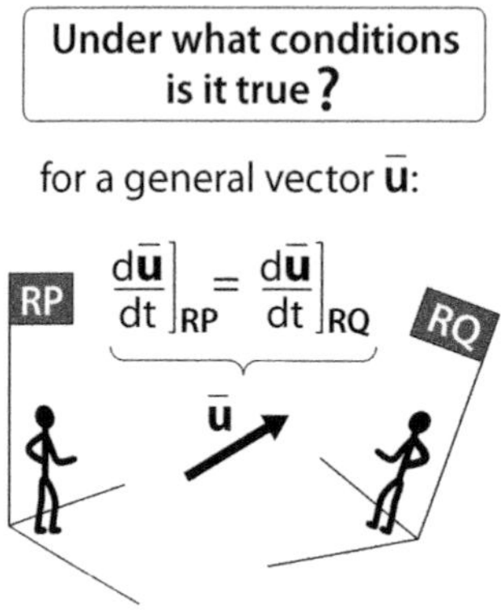

1.1 The time derivative of a same vector $\bar{\mathbf{u}}$ is calculated in two different reference frames (RP and RQ). Under what conditions will the result always be the same?

A Only when it is the same reference frame (RP = RQ)
B When the same vector basis is used in both reference frames
C When the reference frames have a uniform rectilinear relative motion
D When the reference frames have a translational relative motion, though not necessarily rectilinear and uniform
E When the vector bases used by the observers have a translational relative motion

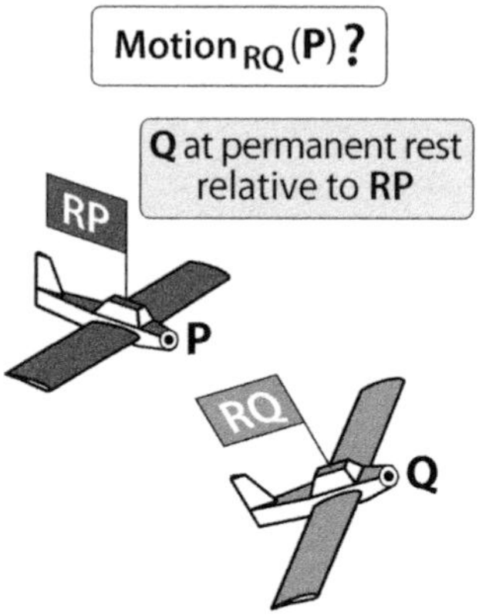

1.2 Point **Q** of aircraft RQ is seen permanently at rest relative to aircraft RP. How does point **P** of aircraft RP move relative to aircraft RQ?

A It is at rest
B It is a rectilinear motion
C It is a parabolic motion on a vertical plane
D It follows a general curve on a sphere
E It is either at rest or moving with a circular motion

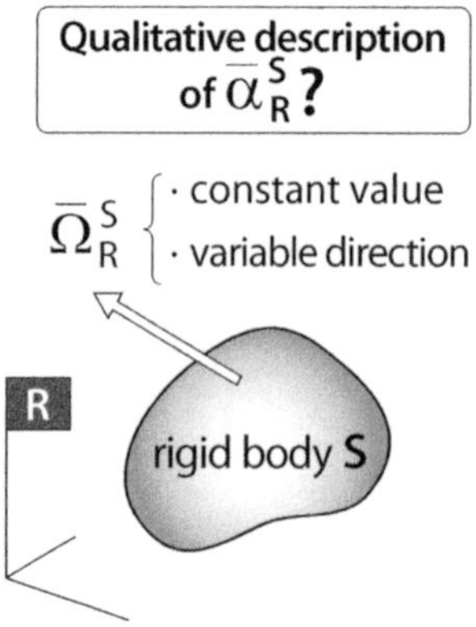

1.3 What is the qualitative description of the angular acceleration $\bar{\boldsymbol{\alpha}}_{\mathrm{R}}^{\mathrm{S}}$?

A $\bar{\boldsymbol{\alpha}}_{\mathrm{R}}^{\mathrm{S}} \neq \bar{\mathbf{0}}$ and always perpendicular to $\bar{\boldsymbol{\Omega}}_{\mathrm{R}}^{\mathrm{S}}$
B $\bar{\boldsymbol{\alpha}}_{\mathrm{R}}^{\mathrm{S}} \neq \bar{\mathbf{0}}$ and only parallel to $\bar{\boldsymbol{\Omega}}_{\mathrm{R}}^{\mathrm{S}}$ if the vector basis is fixed in R
C $\bar{\boldsymbol{\alpha}}_{\mathrm{R}}^{\mathrm{S}} \neq \bar{\mathbf{0}}$ and always parallel to $\bar{\boldsymbol{\Omega}}_{\mathrm{R}}^{\mathrm{S}}$
D $\bar{\boldsymbol{\alpha}}_{\mathrm{R}}^{\mathrm{S}} \neq \bar{\mathbf{0}}$ and with an unconstrained direction
E $\bar{\boldsymbol{\alpha}}_{\mathrm{R}}^{\mathrm{S}} = \bar{\mathbf{0}}$

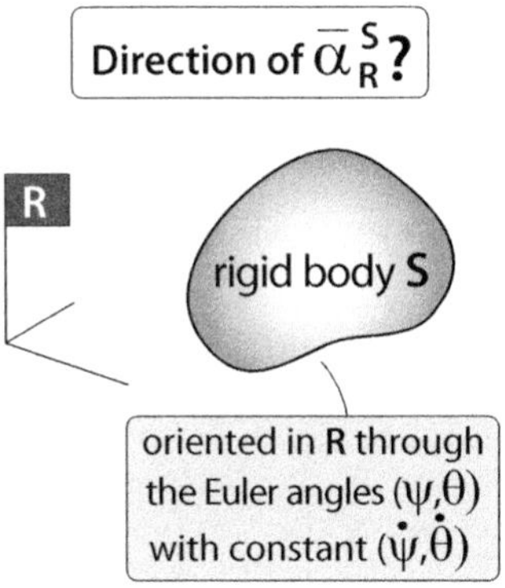
Direction of $\bar{\alpha}_R^S$?
R
rigid body S
oriented in R through the Euler angles (ψ,θ) with constant ($\dot\psi$,$\dot\theta$)

1.4 What is the direction of the angular acceleration $\bar{\alpha}_R^S$?

A It is perpendicular to the plane defined by $\bar{\dot{\psi}}$ and $\bar{\dot{\theta}}$
B It is contained in the plane defined by $\bar{\dot{\psi}}$ and $\bar{\dot{\theta}}$
C It is zero
D It can have any direction in a plane perpendicular to $\bar{\dot{\psi}}$
E It can have any direction in the plane perpendicular to $\bar{\dot{\theta}}$

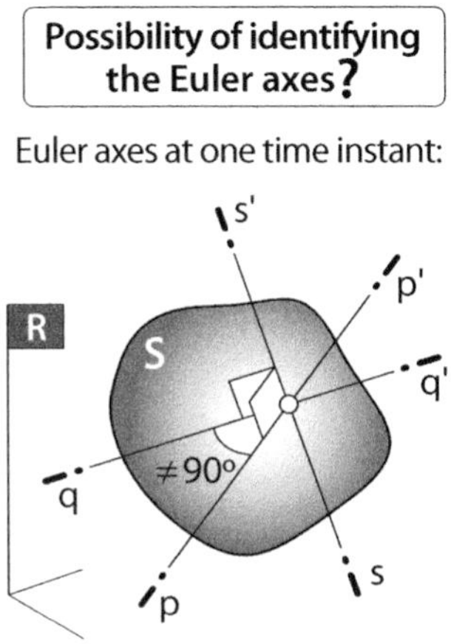
Possibility of identifying the Euler axes?
Euler axes at one time instant:
s'
p'
q'
R
S
≠90°
q
p
s

1.5 At a given time instant, the p–p′, q–q′, and s–s′ axes correspond to the three Euler axes of the rigid body S relative to R. Can you identify any of them?

A Not enough information
B They cannot correspond to Euler axes because they are not orthogonal
C The s–s′ direction corresponds to $\bar{\dot{\psi}}$
D The s–s′ direction corresponds to $\bar{\dot{\theta}}$
E The s–s′ direction corresponds to $\bar{\dot{\varphi}}$

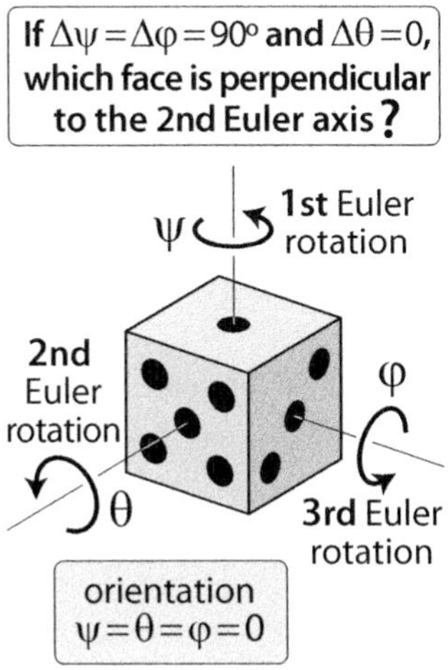
If Δψ=Δφ=90° and Δθ=0, which face is perpendicular to the 2nd Euler axis?
ψ
1st Euler rotation
2nd Euler rotation
θ
φ
3rd Euler rotation
orientation ψ=θ=φ=0

1.6 The three Euler angles for a die are shown for the initial ($\psi = \theta = \varphi = 0$) orientation. If we introduce the incremental rotations ($\Delta\psi = \Delta\varphi = 90^{\circ}, \Delta\theta = 0$), what will be the final direction of the second Euler axis? (**Remark:** the sum of points of opposite faces is 7.)

A Perpendicular to face 1
B Perpendicular to face 2
C Perpendicular to face 3
D Perpendicular to face 4
E Perpendicular to face 5

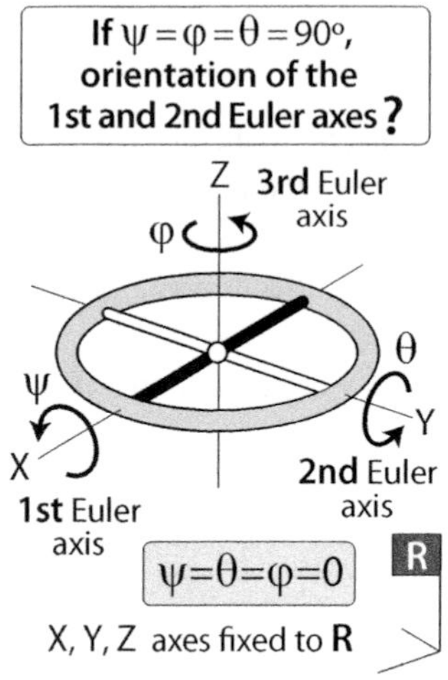

1.7 The three Euler angles for a ring are shown for the initial ($\psi = \theta = \varphi = 0$) orientation. If we introduce the incremental rotations ($\Delta\varphi = 90^{\circ}, \Delta\theta = 90^{\circ}, \Delta\psi = 90^{\circ}$), what will be the final directions of the first and second Euler axes?

	1st	2nd
A	X	X
B	Y	X
C	Z	Z
D	X	Z
E	Y	Y

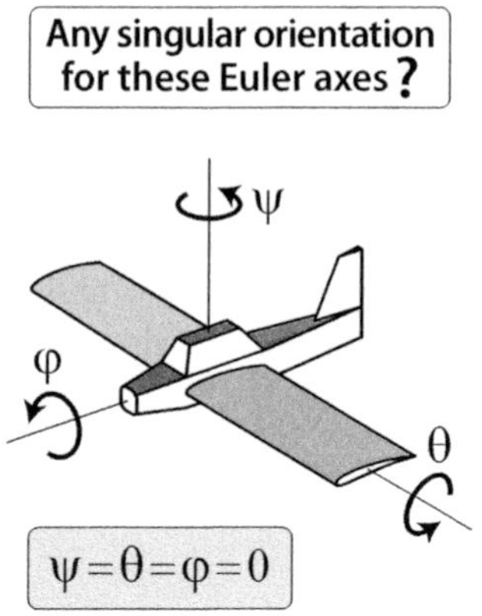

1.8 The three Euler angles for an airplane are shown for the ($\psi = \theta = \varphi = 0$) orientation. Are there any singular orientations?

A No
B Yes: $\varphi = \pm 90^{\circ}$
C Yes: $\theta = \pm 90^{\circ}$
D Yes: $\psi = \pm 90^{\circ}$
E Yes: $\varphi = \pm 180^{\circ}$

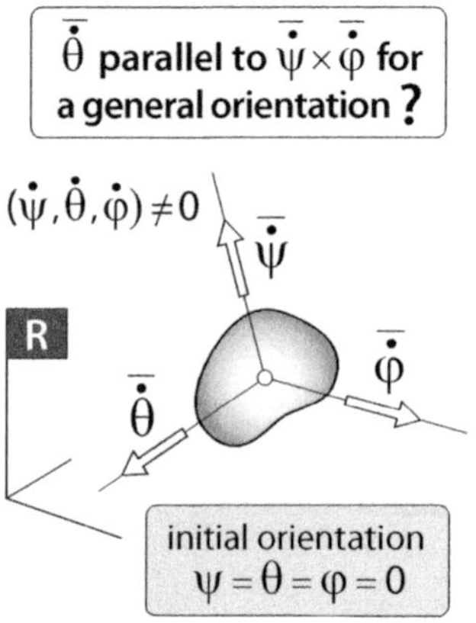

1.9 Someone says that if the angular velocities $(\dot{\psi}, \dot{\theta}, \dot{\varphi})$ associated with the Euler angles are nonzero, $\bar{\dot{\boldsymbol{\theta}}}$ is always parallel to $(\bar{\dot{\boldsymbol{\psi}}} \times \bar{\dot{\boldsymbol{\varphi}}})$. This assertion is:

A Always false
B True provided the $(\dot{\psi}, \dot{\theta}, \dot{\varphi})$ values are not constant
C False in general because Euler angles are independent
D Always true
E True provided the $(\dot{\psi}, \dot{\theta}, \dot{\varphi})$ values are constant

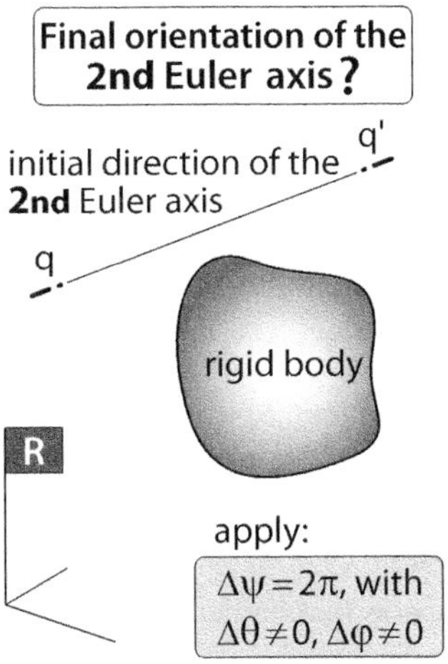

1.10 The orientation of a rigid body is described through three Euler angles. In a given initial orientation, the second axis is parallel to q–q′. If we introduce the incremental rotations ($\Delta\psi = 2\pi, \Delta\theta \neq 0, \Delta\varphi \neq 0$), what will be the final direction of the second Euler axis?

A It depends on the $\Delta\varphi$ value
B It depends on the $\Delta\theta$ value
C It is perpendicular to q–q′
D It is q–q′
E It is not defined when $\Delta\psi = 2\pi$

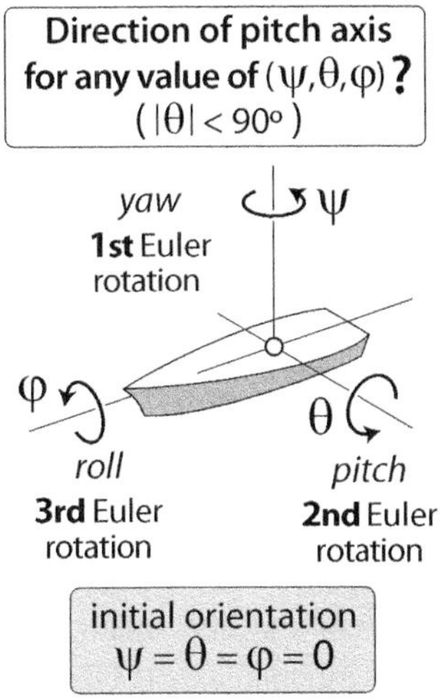

1.11 The three Euler angles for a ship are shown for the ($\psi = \theta = \varphi = 0$) orientation. What is the direction of the pitch axis for any (ψ, θ, φ) values with $|\theta| < 90^o$?

A The horizontal projection of the longitudinal axis of the ship
B The horizontal projection of the transverse axis of the ship
C The axis perpendicular to the symmetry plane of the ship
D The direction perpendicular to both the yaw and the roll axes
E The axis perpendicular to the symmetry plane of the ship for the ($\psi = \theta = \varphi = 0$) orientation

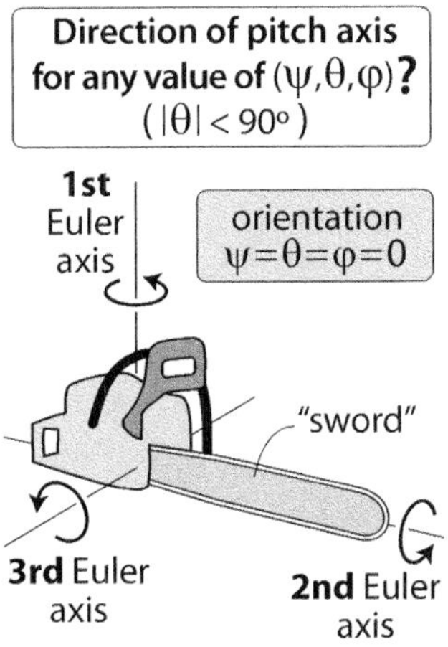

1.12 The three Euler angles for the sword of a chain saw are shown for the ($\psi = \theta = \varphi = 0$) orientation. What is the direction of the second Euler axis for any (ψ, θ, φ) values with $|\theta| < 90^o$?

A The longitudinal direction of the sword
B The horizontal projection of the longitudinal direction of the sword
C It is indeterminate
D The horizontal direction perpendicular to the longitudinal direction of the sword
E The horizontal direction in the plane of the sword

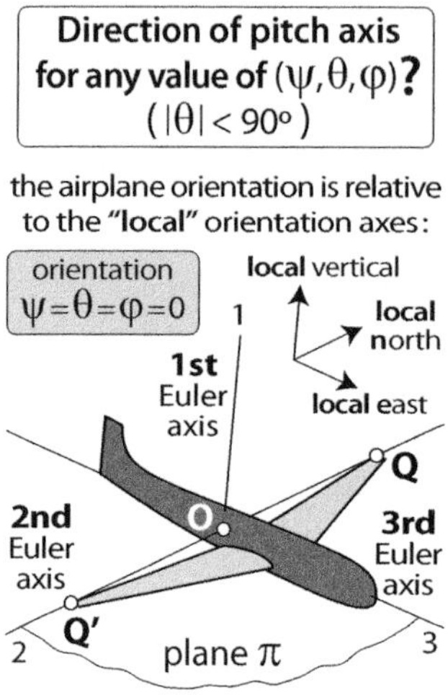

1.13 The orientation of an airplane, relative to the trihedral defined by the local vertical, north, and east directions in the initial orientation, is described through three Euler angles. The figure shows the Euler axes for the ($\psi = \theta = \varphi = 0$) orientation. What is the direction of the second Euler axis for any (ψ, θ, φ) values with $|\theta| < 90^{o}$?

A The local horizontal projection of $\overline{\mathbf{QQ}'}$

B It is not defined because the local trihedral is not fixed relative to the Earth

C That of $\overline{\mathbf{QQ}'}$

D The direction perpendicular to the Earth through **O**

E The intersection of plane π (fixed relative to the airplane) and the local horizontal plane

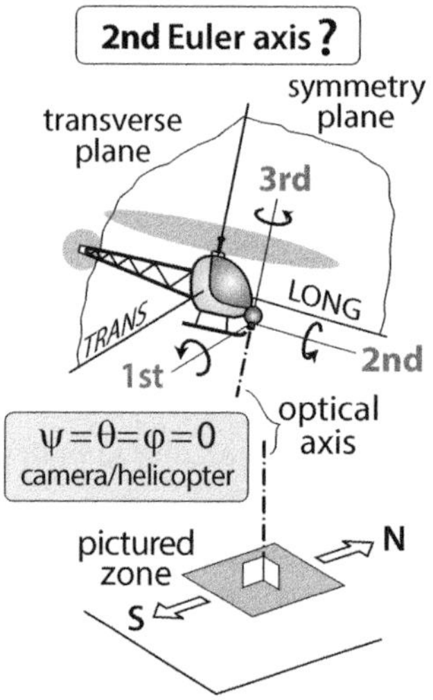

1.14 From a helicopter, we want to take pictures perpendicular to the ground in a way that the cardinal directions (north, south, east, west) are always the same in the pictures. The joint element between the camera and the helicopter allows three Euler relative rotations. The three Euler axes are shown for the ($\psi = \theta = \varphi = 0$) relative orientation. What is the general orientation of the second axis while taking a picture?

A It is indeterminate when the optical axis is vertical

B The direction of the horizontal projection of the helicopter longitudinal axis (LONG)

C The horizontal direction in the transverse plane of the helicopter

D The horizontal direction in the helicopter symmetry plane

E The longitudinal axis of the helicopter (LONG)

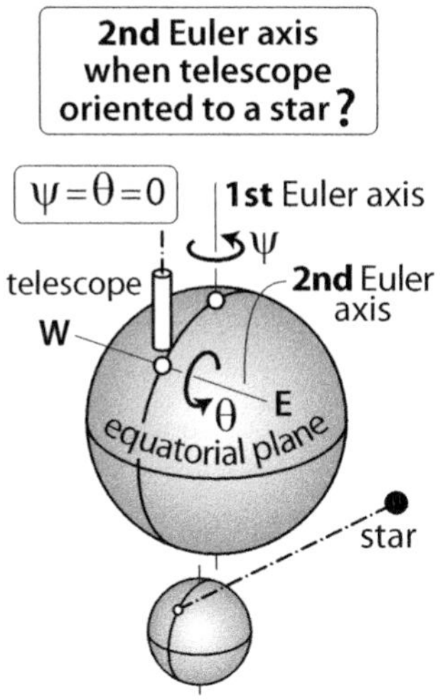

1.15 The orientation of the optical axis of a telescope relative to the Earth is described through two Euler angles. The Euler axes are shown for the ($\psi = \theta = 0$) orientation. When the telescope remains oriented toward a star, what is the direction of the second axis?

A The W–E direction
B The horizontal projection of the telescope axis
C The horizontal direction perpendicular to the horizontal projection of the telescope axis
D The projection of the telescope axis on the equatorial plane
E The direction on the equatorial plane perpendicular to the telescope axis

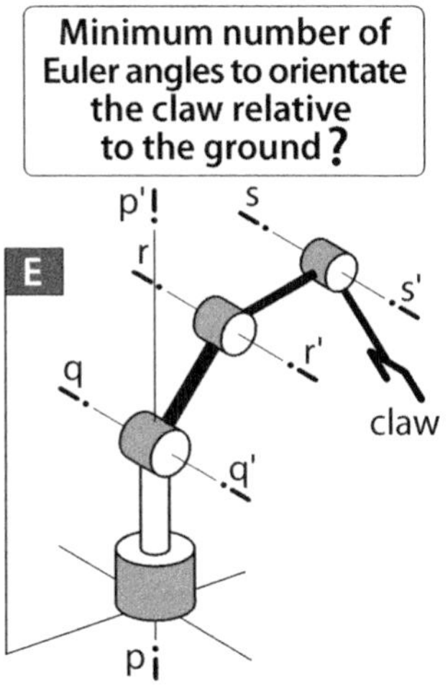

1.16 The elements of the robotic arm are connected through revolute joints with axes p–p′, q–q′, r–r′, and s–s′. What is the minimum number of Euler angles to describe the orientation of the claw relative to the ground (E)?

A 1
B 2
C 3
D 4
E 5

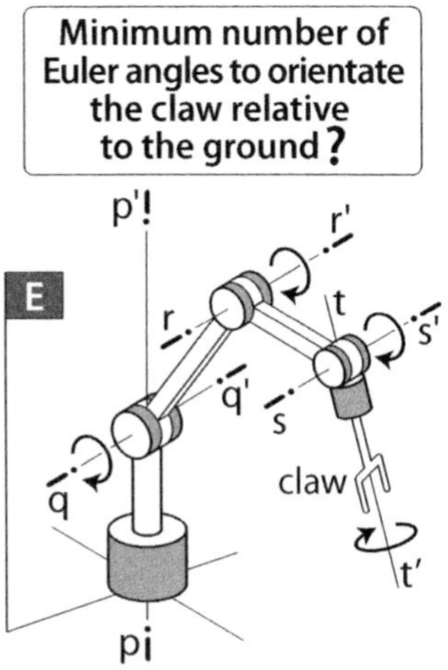

1.17 The elements of the robotic arm are connected through revolute joints with axes p–p′, q–q′, r–r′, and s–s′. What is the minimum number of Euler angles needed to describe the orientation of the claw relative to the ground (E)?

A 1
B 2
C 3
D 4
E 5

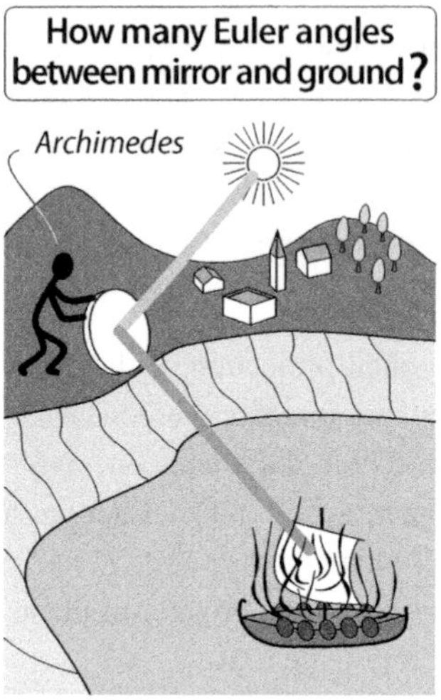
How many Euler angles between mirror and ground?
Archimedes

1.18 The legend says that Archimedes defeated the Turkish army that wanted to conquer Syracuse through a mirror reflecting the sun on the Turkish ships. If the articulation between mirror and ground corresponded to Euler angles, how many angles were necessary?

A 1
B 2
C 3
D Euler angles could not be used as they had not yet been invented
E Euler angles could not be used as the rotation axes had to be fixed to the ground

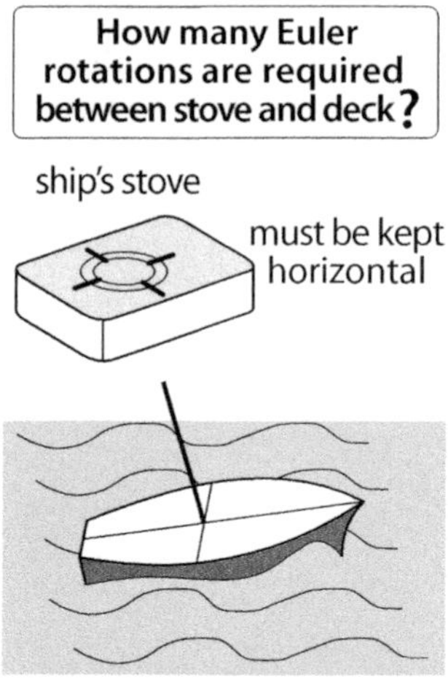
How many Euler rotations are required between stove and deck?
ship's stove
must be kept horizontal

1.19 In a sailboat sailing in heavy seas, the stove has to be articulated to the deck so that it can always be kept horizontal. How many Euler rotations are needed between stove and deck?

A 1
B 2
C 3
D 4: 3 to compensate for roll, pitch, and yaw, plus an extra angle to avoid the singularity encountered when the first and third axes coincide
E Euler rotations are not suitable; we need rotation axes fixed to the deck

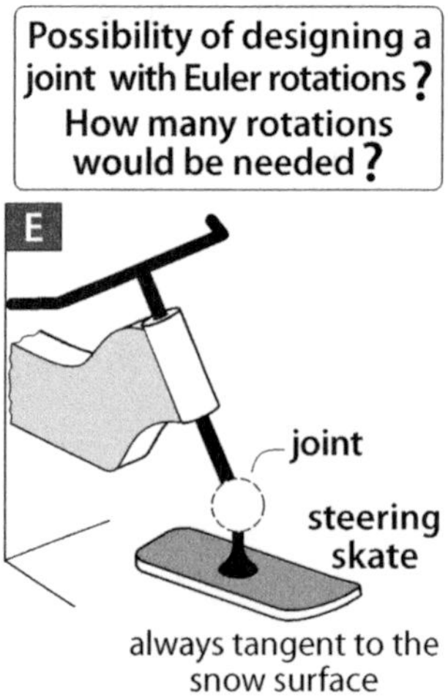
Possibility of designing a joint with Euler rotations?
How many rotations would be needed?
E
joint
steering skate
always tangent to the snow surface

1.20 A sled has a front steering skate that has to be kept tangent to the surface of the snow. Is it possible to design the sled skate joint by means of Euler rotations? If it is, how many rotations are required?

A It is possible, and it requires just one rotation because a horizontal joint is enough
B It is not possible; it calls for two sled-fixed axes
C It is possible, and it requires two rotations
D It is possible, and it requires three rotations because the sled is a rigid body
E It is not possible; it calls for two skate-fixed axes

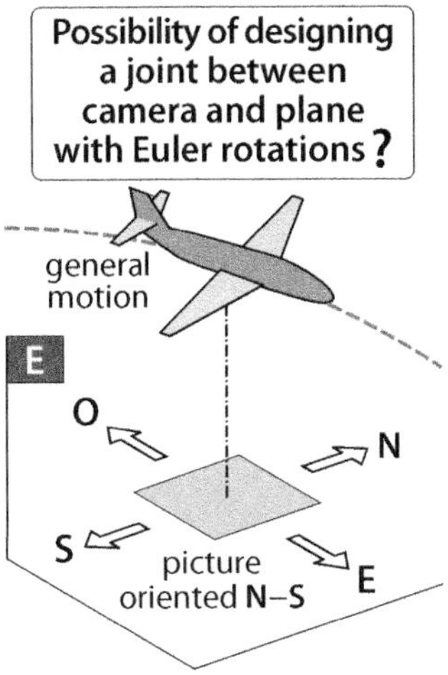
Possibility of designing a joint between camera and plane with Euler rotations?
general motion
E
O
N
S
E
picture oriented N–S

1.21 We want to take pictures of the ground from an airplane that follows a general trajectory. The center of the pictures must correspond to the lowest point in the vertical line through the center of the camera lens, and the cardinal directions (north, south, east, west) must be kept constant in the pictures. Can we meet all these conditions by articulating the camera to the plane with Euler rotations?

A No; two rotation axes have to be ground-fixed
B Yes, we need just one rotation
C Yes, we need just two rotations
D Yes, we need three rotations
E No; two rotation axes have to be plane-fixed

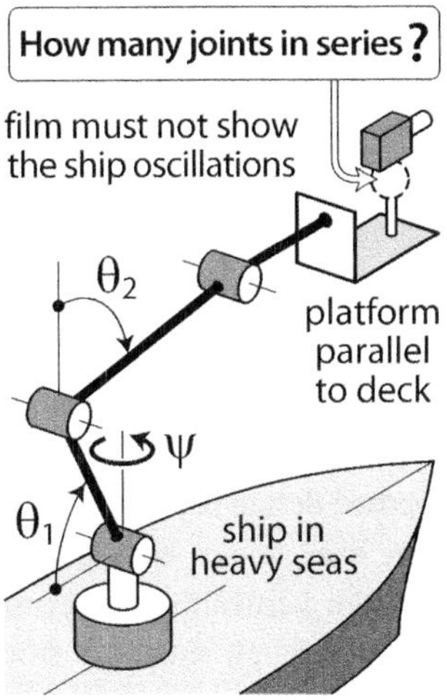
How many joints in series?
film must not show the ship oscillations
θ_2
platform parallel to deck
ψ
θ_1
ship in heavy seas

1.22 We want to shoot a film from a mobile platform articulated to the deck of a ship that moves in heavy seas. The film should not show the ship oscillations. The platform supporting the camera is parallel to the deck. How many joints in series are needed between platform and camera?

A 2
B 3
C 4
D 5
E 6

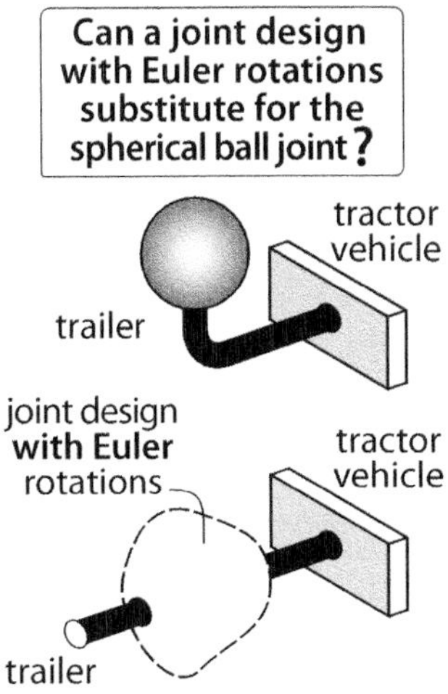
Can a joint design with Euler rotations substitute for the spherical ball joint?
tractor vehicle
trailer
joint design with Euler rotations
tractor vehicle
trailer

1.23 The usual link between tractors and trailers is based on a spherical ball joint. The inner sphere – the "ball" – is fixed to the tractor vehicle, while the outer element is fixed to the trailer. Could it be replaced by joints based on Euler rotations?

A No, because the trailer–tractor mobility would be insufficient
B Yes, we would need just one Euler rotation
C Yes, we would need just two Euler rotations
D Yes, we would need three Euler rotations
E Yes, we would need four Euler rotations

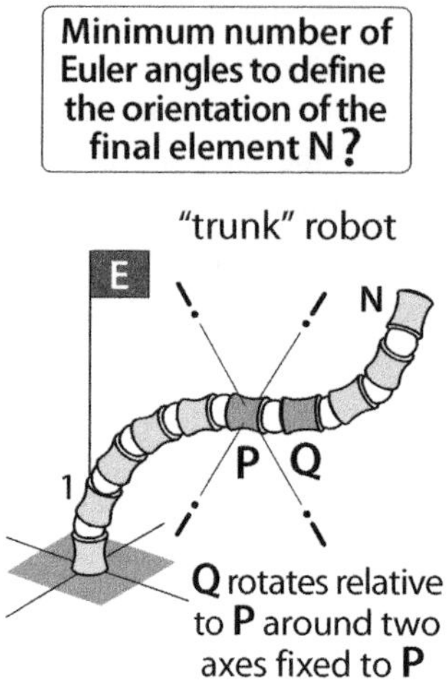

1.24 A robotic arm of the "trunk" type consists of a chain of mobile elements. Each element can rotate around two orthogonal axes fixed to the previous element. If there are N elements ($N \geq 2$), what is the minimum number of Euler angles needed to describe the general orientation of the final element relative to the ground (E)?

A 3
B 3N
C 2
D $3(N - 1)$
E $2(N - 1)$

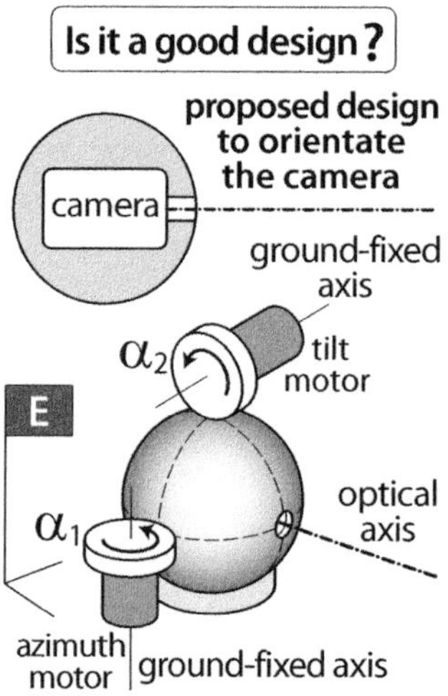

1.25 We want to orientate a film camera relative to the ground so that the verticality of the images is guaranteed. Can we achieve this with the camera inside a sphere oriented by two small wheels that maintain a nonsliding contact with it?

A Yes, the proposed design introduces two Euler rotations
B No, the proposed design introduces two Euler rotations but we need three
C No, the proposed design introduces two Euler but the corresponding axes are not suitable
D No, because the verticality of the images cannot be guaranteed through rotations around ground-fixed axes
E No, because the two wheels cannot rotate simultaneously without sliding

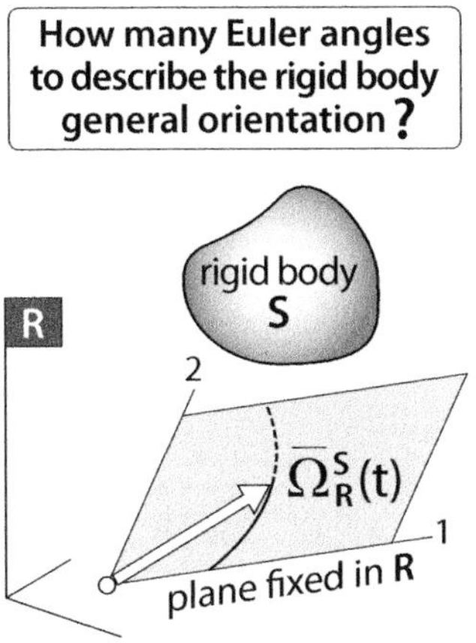

1.26 We modify the orientation of a rigid body S relative to a reference frame R in a way that $\bar{\mathbf{\Omega}}_R^S$ is variable but restricted to a plane fixed in R. The S final orientation relative to R is defined in principle by:

A Just an Euler angle
B Two independent Euler angles
C Three independent Euler angles
D Two nonsequential rotations around frame-fixed axes
E Three nonsequential rotations around frame-fixed axes

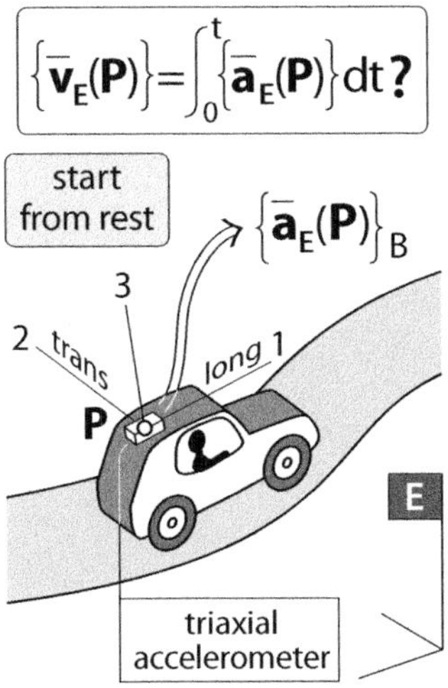

1.27 A triaxial accelerometer measures the acceleration relative to the ground (E) of a point **P** in a vehicle, $\bar{\mathbf{a}}_E(\mathbf{P})$. The result is given as three components on a vehicle-fixed basis B. Can we obtain the components of the velocity $\bar{\mathbf{v}}_E(\mathbf{P})$ in the same vector basis through the integration of the $\bar{\mathbf{a}}_E(\mathbf{P})$ components for a general movement of the vehicle?

A No
B Yes
C Only if the vehicle changes its orientation relative to the ground according to a rotation around a ground-fixed axis
D Only if the vehicle changes its orientation relative to the ground according to two Euler rotations
E Only if the acceleration evolves on a plane

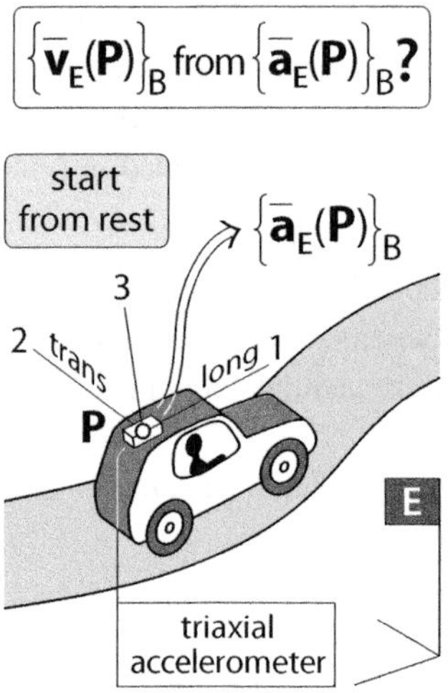

1.28 A triaxial accelerometer measures the acceleration relative to the ground (E) of a point **P** in a vehicle, $\bar{\mathbf{a}}_E(\mathbf{P})$. The result is given as three components on a vehicle-fixed basis B. If the vehicle starts from rest, how can we obtain the components of $\bar{\mathbf{v}}_E(\mathbf{P})$ in that same basis from those of $\bar{\mathbf{a}}_E(\mathbf{P})$ for a general movement of the vehicle?

A $\int_0^t \{\bar{\mathbf{a}}_E(\mathbf{P})\}_B dt$

B $\int_0^t \left(\{\bar{\mathbf{a}}_E(\mathbf{P})\}_B + \left\{\bar{\mathbf{\Omega}}_E^B \times \bar{\mathbf{a}}_E(\mathbf{P})\right\}_B\right) dt$

C $\int_0^t \left(\{\bar{\mathbf{a}}_E(\mathbf{P})\}_B - \left\{\bar{\mathbf{\Omega}}_E^B \times \bar{\mathbf{a}}_E(\mathbf{P})\right\}_B\right) dt$

D $\int_0^t \left(\{\bar{\mathbf{a}}_E(\mathbf{P})\}_B + \left\{\bar{\mathbf{\Omega}}_E^B \times \bar{\mathbf{v}}_E(\mathbf{P})\right\}_B\right) dt$

E $\int_0^t \left(\{\bar{\mathbf{a}}_E(\mathbf{P})\}_B - \left\{\bar{\mathbf{\Omega}}_E^B \times \bar{\mathbf{v}}_E(\mathbf{P})\right\}_B\right) dt$

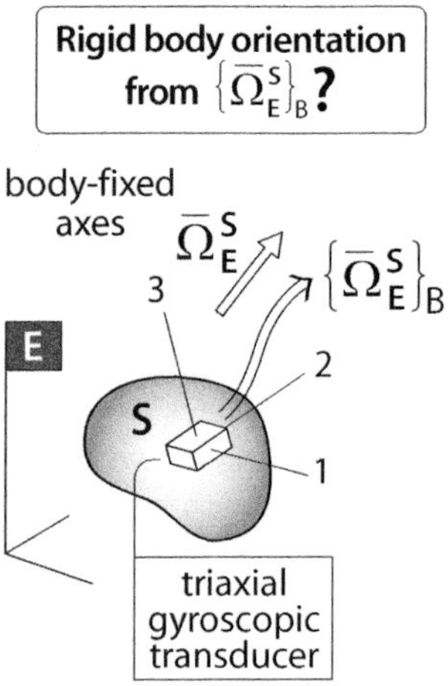

1.29 A triaxial gyroscopic transducer fixed to a rigid body S measures its angular velocity relative to the ground $\left(\bar{\mathbf{\Omega}}_E^S\right)$ over time. The result is given as three components $(\Omega_1, \Omega_2, \Omega_3)$ on an S-fixed basis B. Can we obtain the change of orientation of S over time from $\left\{\bar{\mathbf{\Omega}}_E^S\right\}_B$?

A No, because orientation is not a vector

B It is only possible if the axes in B are Euler axes

C Yes, we only have to integrate the components: $\int_0^t \left\{\bar{\mathbf{\Omega}}_E^S\right\}_B dt$

D First we have to translate the results in a ground-fixed basis FB, and then integrate the components: $\int_0^t \left\{\bar{\mathbf{\Omega}}_E^S\right\}_{FB} dt$

E First we have to define a family of Euler angles, obtain $\left(\dot{\psi}, \dot{\theta}, \dot{\varphi}\right)$ as linear forms of $(\Omega_1, \Omega_2, \Omega_3)$, and then proceed to the integrations

$$\Delta\psi = \int_0^t \dot{\psi} dt, \quad \Delta\theta = \int_0^t \dot{\theta} dt, \quad \Delta\varphi = \int_0^t \dot{\varphi} dt.$$

Puzzles

1.1 Design of a sweeping tool to clean the floor

The sweeping tool consists of a rectangular frame covered with felt and an articulated handle. The frame should have a sliding translation motion when pushing the handle.

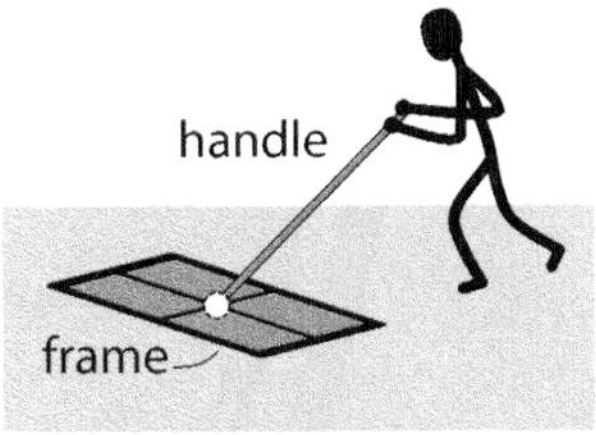

How many variable Euler angles should introduce the articulation between handle and frame?

1.2 Choosing a photography tripod

We propose three different designs for the camera–tripod joint: with two Euler joints (**P**), with three (**Q**), and with a ball-and-socket joint (**S**).

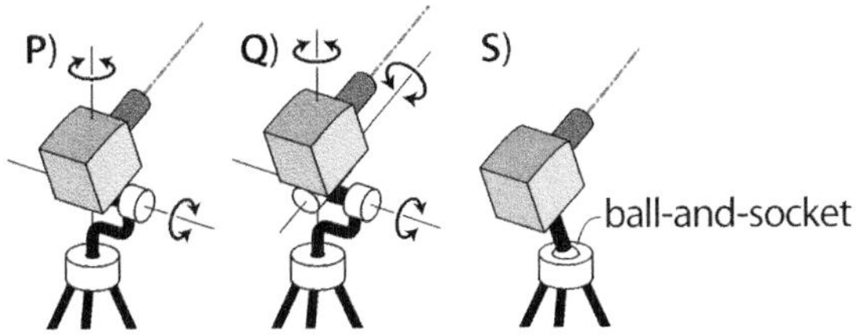

Which one would you choose to take simple pictures (with no special artistic orientation effects)? For what reasons?

1.3 Suitability of *flexo* lamps design

Consider the following *flexo* lamp design.

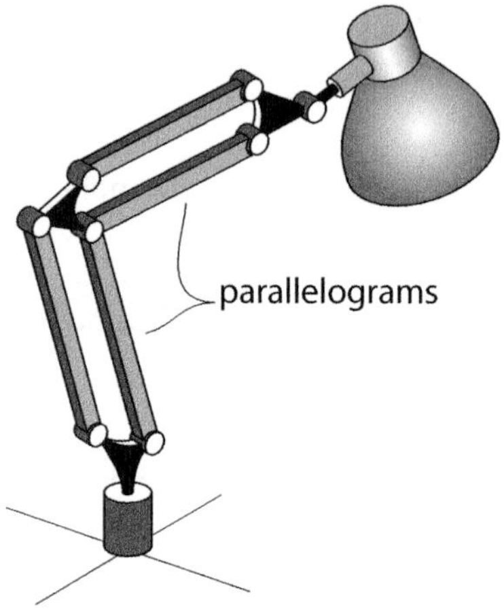

- **Can it illuminate any direction** (for a given position of the arm)?
- **Would it need any additional articulation for that purpose?**

Quiz Questions: Answers

	1	2	3	4	5	6	7	8	9	10
	D	D	A	A	D	A	D	C	D	D
+10	D	E	C	D	E	B	C	B	B	C
+20	D	B	D	A	D	C	A	E	E	

Puzzles: Solutions

1.1 Design of a sweeping tool to clean the floor

The joint must have just two Euler angles.
If there were a third Euler angle (rotation of the handle about its longitudinal direction), the orientation of the frame on the ground would no longer be controlled by the handle. The orientation of the frame on the ground is modified precisely by rotating the handle about its axis. The following figure shows a possible solution for this joint.

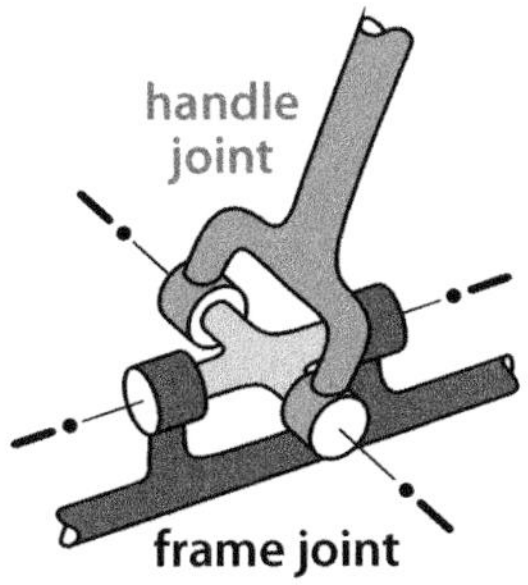

1.2 Choosing a photography tripod

We usually take pictures of people, landscapes, buildings, etc. where we want to **be level.**

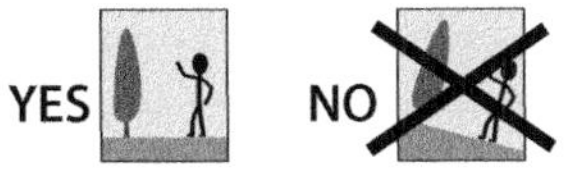

With the two-Euler-joint model (**P**):

- You can freely choose the direction of the camera's optical axis
- The pictures can be level. You just have to adjust the tripod column in vertical position

With a three-Euler-joint model (**Q**), you can rotate the camera around the optical axis. This is interesting for studio photographs (where you do not specifically want them to be level) or when you take pictures from the deck of a ship in rough seas or from a helicopter (in these cases, the rotation to guarantee that the pictures are level is automatic from vehicle orientation sensors).

For the usual photographs, you have to block the third Euler rotation; the tripod is then equivalent to the previous model (two Euler joints).

With the ball-and-socket model (**S**), the camera can have a totally general orientation – as with the three-Euler-joint tripod – but since the rotation around the optical axis cannot be blocked, you have to control constantly the horizontality and verticality of each photograph.

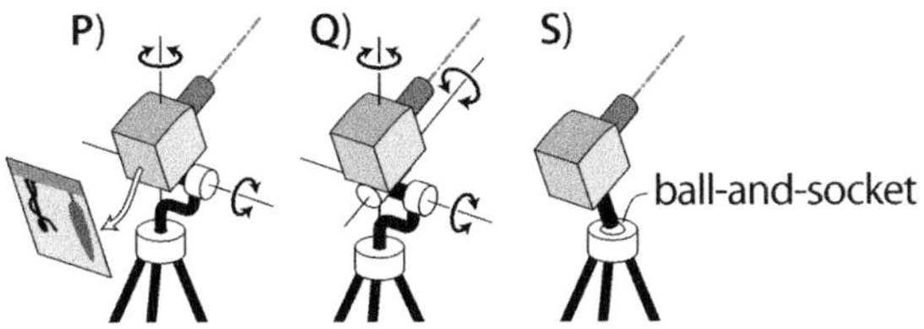

Consequently, the most suitable tripod for the usual photographs is the two-Euler-joint one (P). In addition, it has the advantage of lower cost, weight, and volume than the three-Euler-joint tripod (Q).

1.3 Suitability of *flexo* lamps design

The joints between the different lamp elements provide a chain of rotations.

- Rotation ψ around a ground-fixed axis. It drags all the other joints, and plays the role of the Euler first rotation for all of them
- Rotations $(\theta_1, \theta_2, \theta_3)$, around horizontal axes, are not linked in series because the bars form two parallelograms. θ_1 and θ_2 are the Euler second rotation of the first and second parallelogram, respectively, and θ_3 is that of the focus
- Rotation φ – whose axis rotates with ψ and θ_3 – is the Euler third rotation of the focus

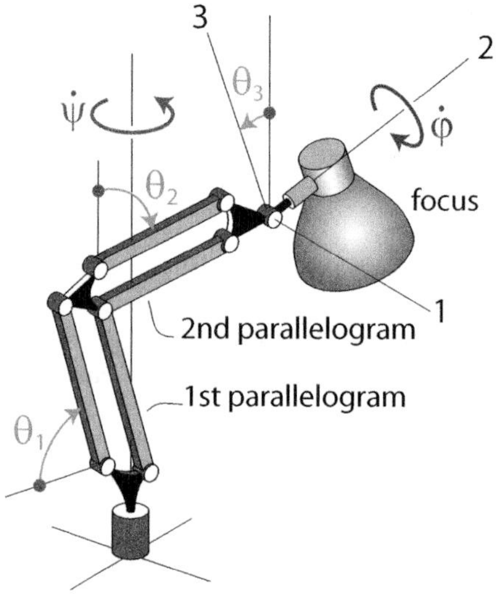

However, this is not the good interpretation of the articulations when it comes to analyzing the directionality of the light beam.

Rotations $(\psi, \theta_1, \theta_2)$ play the role of three positional coordinates to choose the location of point **O**. Once the chosen location is reached, the focus must be oriented relative to the arm to orientate the light beam conveniently. This is equivalent to orienting its axis, which requires two Euler rotations relative to the plane of the arm (reference frame R). In this interpretation, θ_3 and φ play the role of Euler first and second rotations, respectively. The focus axis can be orientated freely with the only constraint given by the maximum values of those two angles.

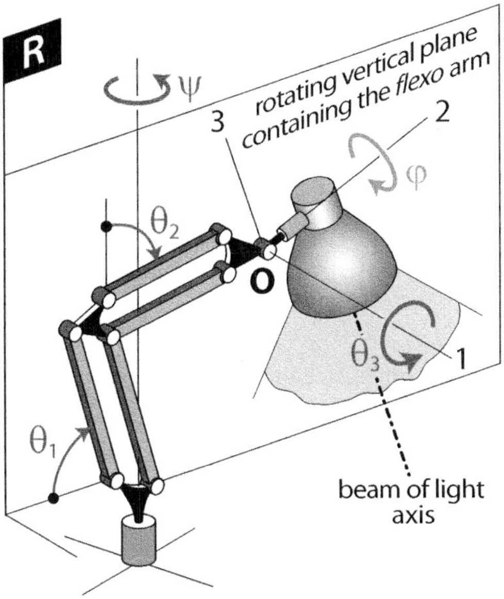

Coordinates (θ_1, θ_2) (**arm coordinates**) are widely used in positioning mechanisms – for example, robots – because they allow greater accessibility than those based on Cartesian, cylindrical, or spherical coordinates.

2 Point Kinematics

In this chapter, we apply the time derivative presented in Chapter 1 to study the kinematics of the point (**mass point** or **particle**), which is the simplest model for material bodies.

The location of a point in a reference frame R is given through a position vector (Section 2.1) – which is time dependent in general. The velocity and acceleration vectors are its first and second time derivatives (Section 2.2). These vectors evaluate the rate of change of the position and the velocity vectors relative to R, respectively. Higher-order derivatives are not considered because the formulation of Newtonian dynamics does not need them.

The succession of positions of a point in a reference frame defines a curve, so there has to be a relationship between the point kinematics and the geometry of curves. This relationship becomes evident when the kinematic vectors are projected on the **intrinsic** or **Frenet basis** (associated with the curve geometry). The components are then called **intrinsic components** (Section 2.3). The angular velocity of this vector basis is studied in Appendix 2A.

On many occasions, it is interesting to analyze the movement of mechanical systems in two different reference frames. For example, in vehicle kinematics, the ground and the chassis frames are interesting because some aspects are better known (or described) in the former and others in the latter. The study of the relationship between the kinematic variables of a same point in two different frames (Sections 2.4, 2.5, and 2.6) is called **composition of movements**. The concept of **transportation movement** is a key in that composition, as well as in the formulation of the kinematics of the rigid body (Chapter 3). Although the composition of velocities is a very intuitive superposition, this is not the case for accelerations because of a complementary term: **Coriolis acceleration**.

Appendix 2B presents a problem particularly relevant in aeronautical and aerospace engineering: **inertial guidance**.

2.1 The Mass Point: Position Vector

The **mass point** or **particle** is the simplest model of material body. In kinematics, it is described by a point **P** in space. The validity of this model has no relation with the body's size. Whenever the body's orientation is irrelevant to what is studied, the particle model is adequate.

For example, talking about the elliptical path of the earth around the sun is implicitly treating the earth as a particle. However, the study of the alternation night/day calls for a rigid body model (as a first approximation).

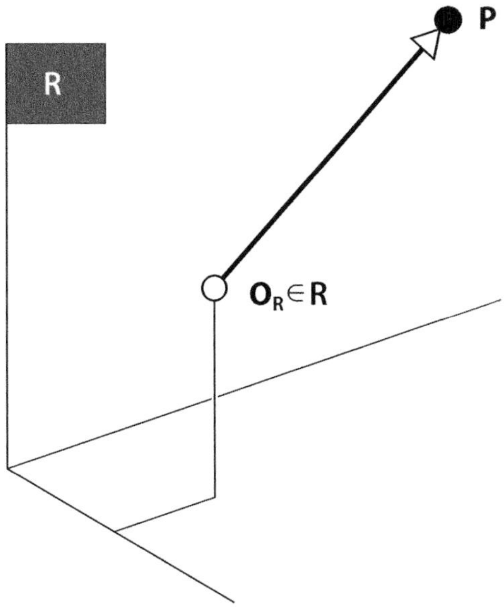

Fig. 2.1

The position of a particle **P** in a reference frame R is time-dependent when **P** moves in R. It is defined by the **position vector** $\overline{\mathbf{O}_R\mathbf{P}}$ (Fig. 2.1). Point $\mathbf{O}_R$ can be any point fixed in R. The criterion to choose it is that of maximum simplicity in the description of the position vector (value and orientation). When the $\overline{\mathbf{O}_R\mathbf{P}}$ description is not simple, it is better to express it as a sum of simple-description vectors. This does not complicate the mathematical operations where $\overline{\mathbf{O}_R\mathbf{P}}$ may intervene (products, sums, derivatives, etc.) given their distributive nature. It is advisable that this sum does not contain any vectors constant in R. The set of positions of **P** defines the **trajectory** of **P** relative to R.

In a time derivative, constant terms do not contribute, and this makes the origin $\mathbf{O}_R$ irrelevant. Thus, in the study of the kinematics relative to the ground (E) of a point **P** fixed to the periphery of a platform that rotates relative to E about its vertical axis (fixed to E, Fig. 2.2), it is irrelevant whether you take as origin the platform center $\mathbf{O}_E$, the intersection of the platform axis with the ground $\mathbf{O}'_E$, or the origin $\mathbf{O}''_E$ of the trihedral representing E:

- $\overline{\mathbf{O}_E\mathbf{P}}$, vector variable in E
- $\overline{\mathbf{O}'_E\mathbf{P}} = \overline{\mathbf{O}'_E\mathbf{O}_E} + \overline{\mathbf{O}_E\mathbf{P}}$, where $\overline{\mathbf{O}'_E\mathbf{O}_E}$ is a constant vector (zero time derivative)
- $\overline{\mathbf{O}''_E\mathbf{P}} = \overline{\mathbf{O}''_E\mathbf{O}'_E} + \overline{\mathbf{O}'_E\mathbf{O}_E} + \overline{\mathbf{O}_E\mathbf{P}}$, where $\overline{\mathbf{O}''_E\mathbf{O}'_E}$ and $\overline{\mathbf{O}'_E\mathbf{O}_E}$ are constant vectors (zero time derivative)

If we proceed to the geometric time derivative, constant terms are excluded from the very beginning. If we proceed analytically, a constant vector added to $\overline{\mathbf{O}_E\mathbf{P}}$ can have a high cost in the expression of the components and their subsequent use (since we usually use moving vector bases).

In the example in Fig. 2.2, the most suitable basis is B with axes $(1, 2, 3)$ (Fig. 2.3) because the vector projection is entirely on one of the axes, and the interpretation of the

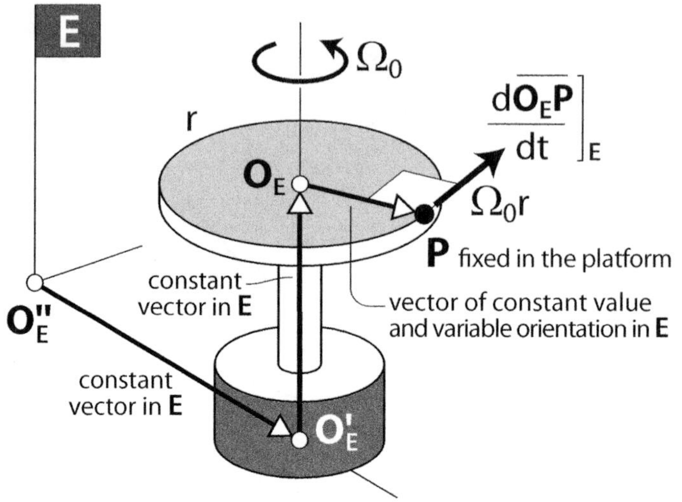

Fig. 2.2

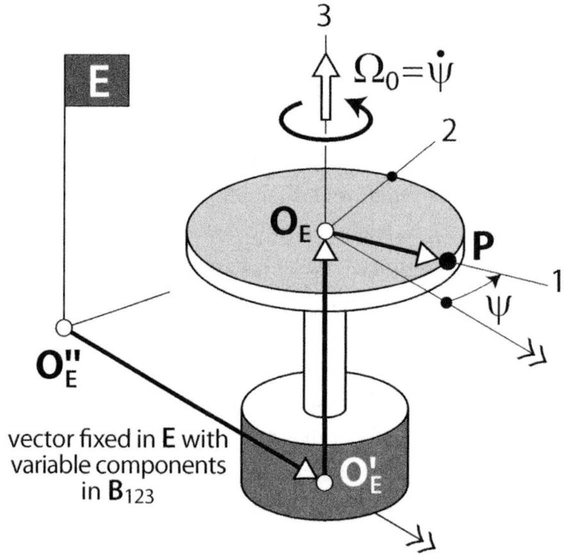

Fig. 2.3

result is straightforward. This facilitates the detection of possible errors – for instance, in the sign of the product vector. It is a platform-fixed basis and therefore moving relative to the ground. Its angular velocity is $\overline{\boldsymbol{\Omega}}_{\mathrm{E}}^{\mathrm{B}} = \overline{\boldsymbol{\Omega}}_{\mathbf{0}} = \overline{\dot{\boldsymbol{\psi}}}$ (Section 1.8):

$$\left\{\overline{\mathbf{O}_{\mathrm{E}}\mathbf{P}}\right\}_{\mathrm{B}} = \begin{Bmatrix} \mathrm{r} \\ 0 \\ 0 \end{Bmatrix}; \quad \left\{\overline{\boldsymbol{\Omega}}_{\mathrm{E}}^{\mathrm{B}}\right\}_{\mathrm{B}} = \begin{Bmatrix} 0 \\ 0 \\ \Omega_0 \end{Bmatrix};$$
$$\left\{\left.\frac{\mathrm{d}\overline{\mathbf{O}_{\mathrm{E}}\mathbf{P}}}{\mathrm{dt}}\right]_{\mathrm{E}}\right\}_{\mathrm{B}} = \begin{Bmatrix} 0 \\ 0 \\ 0 \end{Bmatrix} + \begin{Bmatrix} 0 \\ 0 \\ \Omega_0 \end{Bmatrix} \times \begin{Bmatrix} \mathrm{r} \\ 0 \\ 0 \end{Bmatrix} = \begin{Bmatrix} 0 \\ \Omega_0 \mathrm{r} \\ 0 \end{Bmatrix}. \tag{2.1}$$

The zero in the first component corresponds to the lack of radial movement. The **P** circular movement (associated with the platform rotation) is responsible for the term $\Omega_0 r$ (with a positive sign) in the second component.

If $\mathbf{O'_E}$ is taken as origin, as vector $\overline{\mathbf{O'_E O_E}}$ has constant components and is parallel to $\bar{\mathbf{\Omega}}_{\mathrm{E}}^{\mathrm{B}}$, the time derivative is unaffected:

$$\left\{\overline{\mathbf{O'_E O_E}}\right\}_B = \begin{Bmatrix} 0 \\ 0 \\ h \end{Bmatrix} \Rightarrow \left\{\left.\frac{d\overline{\mathbf{O'_E O_E}}}{dt}\right]_E\right\}_B = \begin{Bmatrix} 0 \\ 0 \\ 0 \end{Bmatrix} + \begin{Bmatrix} 0 \\ 0 \\ \Omega_0 \end{Bmatrix} \times \begin{Bmatrix} 0 \\ 0 \\ h \end{Bmatrix} = \begin{Bmatrix} 0 \\ 0 \\ 0 \end{Bmatrix}. \tag{2.2}$$

But if $\mathbf{O''_E}$ is taken as origin, vector $\overline{\mathbf{O''_E O'_E}}$ has variable components that have to be determined, time derived, and added to $\bar{\mathbf{\Omega}}_{\mathrm{E}}^{\mathrm{B}} \times \overline{\mathbf{O''_E O'_E}}$ to yield, in the end, a zero additive vector:

$$\left\{\overline{\mathbf{O''_E O'_E}}\right\}_B = \begin{Bmatrix} L\cos\psi \\ -L\sin\psi \\ h \end{Bmatrix},$$

$$\left\{\left.\frac{d\overline{\mathbf{O''_E O'_E}}}{dt}\right]_E\right\}_B = \begin{Bmatrix} -L\dot{\psi}\sin\psi \\ -L\dot{\psi}\cos\psi \\ 0 \end{Bmatrix} + \begin{Bmatrix} 0 \\ 0 \\ \dot{\psi} \end{Bmatrix} \times \begin{Bmatrix} L\cos\psi \\ -L\sin\psi \\ 0 \end{Bmatrix}$$

$$= \begin{Bmatrix} -L\dot{\psi}\sin\psi + L\dot{\psi}\sin\psi \\ -L\dot{\psi}\cos\psi + L\dot{\psi}\cos\psi \\ 0 \end{Bmatrix} = \begin{Bmatrix} 0 \\ 0 \\ 0 \end{Bmatrix}. \tag{2.3}$$

2.2 The Velocity Vector and the Acceleration Vector

The **velocity** and **acceleration** of **P** with respect to a reference frame R are the two successive time derivatives of the position vector of **P** in R:

$$\bar{\mathbf{v}}_R(\mathbf{P}) \equiv \left.\frac{d\overline{\mathbf{O}_R\mathbf{P}}}{dt}\right]_R, \quad \bar{\mathbf{a}}_R(\mathbf{P}) = \left.\frac{d\bar{\mathbf{v}}_R(\mathbf{P})}{dt}\right]_R. \tag{2.4}$$

Equation (2.4) shows the double dependency on R of those two vectors:

- They are the time derivative of two vectors ($\overline{\mathbf{O}_R\mathbf{P}}$ and $\bar{\mathbf{v}}_R(\mathbf{P})$) *relative to R*
- Their rate of change is *relative to R*

In order to simplify the notation, the subscript R will be dropped when the problem under study involves just one single reference frame.

▶ **Example 2.1** Point **P** moves radially on a platform (Fig. 2.4a). The platform rotates with variable angular velocity $\bar{\dot{\psi}}(t)$ relative to the ground (E) about its axis (which goes through its ground-fixed center **O**). We want to calculate $\bar{\mathbf{v}}_E(\mathbf{P})$ and $\bar{\mathbf{a}}_E(\mathbf{P})$.

The starting point is the **P** position vector. As $\mathbf{O} \in E$, $\overline{\mathbf{OP}}$ is the best choice.

This example is very simple, and the geometric time derivative yields the results in a very straightforward way. Both the value r(t) and the orientation ψ(t) of vector $\overline{\mathbf{OP}}$ are

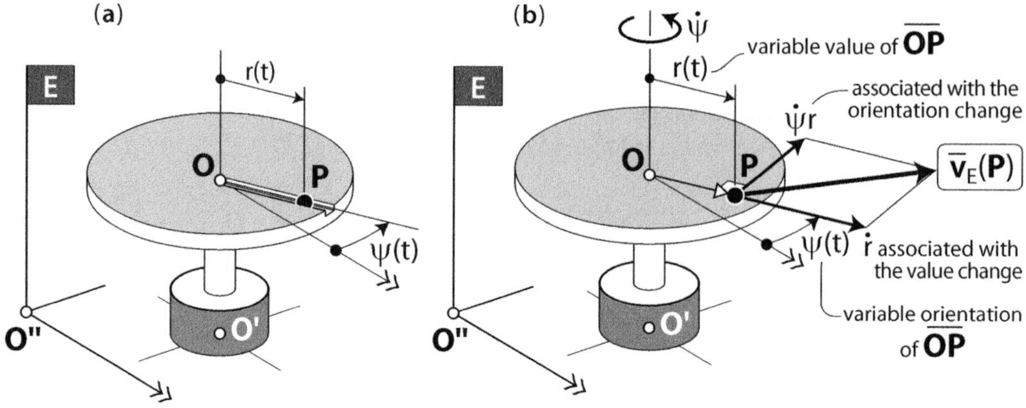

Fig. 2.4

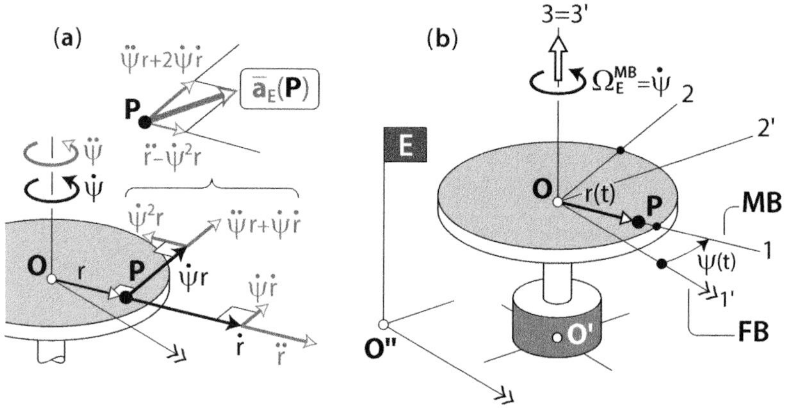

Fig. 2.5

variable. The change in value produces $\dot{r}$ in the $r(t)$ increasing direction. That of orientation produces a vector with value $r\dot{\psi}$, perpendicular to $\overline{\mathbf{OP}}$ and directed according to the sense of the platform rotation (Fig. 2.4b).

The time derivative of $\bar{\mathbf{v}}_E(\mathbf{P})$ produces the four vectors shown in Fig. 2.5a. Their addition yields the two components of $\bar{\mathbf{a}}_E(\mathbf{P})$:

- component with value $(\ddot{r} - r\dot{\psi}^2)$, in the $r(t)$ increasing direction
- component with value $(r\ddot{\psi} + 2\dot{r}\dot{\psi})$, perpendicular to $\overline{\mathbf{OP}}$, directed according to the sense of the platform rotation

The same results can be obtained through a vector basis. The projection of $\overline{\mathbf{OP}}$ is straightforward in the platform-fixed basis MB with axes $(1, 2, 3)$ (moving in E, Fig. 2.5b):

$$\{\overline{\mathbf{OP}}\}_{MB} = \begin{Bmatrix} r \\ 0 \\ 0 \end{Bmatrix}. \tag{2.5}$$

The MB basis rotation relative to E is a simple rotation. Its angular velocity $\bar{\mathbf{\Omega}}_E^{MB}$ is $\bar{\dot{\boldsymbol{\psi}}}$ (as seen in Section 1.8). The velocity $\bar{\mathbf{v}}_E(\mathbf{P})$ is:

$$\{\bar{\mathbf{v}}_E(\mathbf{P})\}_{MB} = \frac{d}{dt}\{\overline{\mathbf{OP}}\}_{MB} + \left\{\bar{\boldsymbol{\Omega}}_E^{MB}\right\}_{MB} \times \{\overline{\mathbf{OP}}\}_{MB} = \begin{Bmatrix} \dot{r} \\ 0 \\ 0 \end{Bmatrix} + \begin{Bmatrix} 0 \\ 0 \\ \dot{\psi} \end{Bmatrix} \times \begin{Bmatrix} r \\ 0 \\ 0 \end{Bmatrix} = \begin{Bmatrix} \dot{r} \\ r\dot{\psi} \\ 0 \end{Bmatrix}. \quad (2.6)$$

The interpretation of this result is simple: the radial movement generates the $\dot{r}$ component on axis 1, and the platform rotation generates the $r\dot{\psi}$ component on axis 2 according to the sense of the platform rotation.

The time derivative of $\bar{\mathbf{v}}_E(\mathbf{P})$ in E yields the acceleration:

$$\{\bar{\mathbf{a}}_E(\mathbf{P})\}_{MB} = \begin{Bmatrix} \ddot{r} \\ \dot{r}\dot{\psi} + r\ddot{\psi} \\ 0 \end{Bmatrix} + \begin{Bmatrix} 0 \\ 0 \\ \dot{\psi} \end{Bmatrix} \times \begin{Bmatrix} \dot{r} \\ r\dot{\psi} \\ 0 \end{Bmatrix} = \begin{Bmatrix} \ddot{r} - r\dot{\psi}^2 \\ 2\dot{r}\dot{\psi} + r\ddot{\psi} \\ 0 \end{Bmatrix}. \quad (2.7)$$

The interpretation of these components is not as simple as those of the velocity.

If we use the ground-fixed basis FB with axes $(1', 2', 3')$, the projection of the position vector is more complicated:

$$\{\overline{\mathbf{OP}}\}_{FB} = \begin{Bmatrix} r\cos\psi \\ r\sin\psi \\ 0 \end{Bmatrix}, \quad (2.8)$$

and those of $\bar{\mathbf{v}}_E(\mathbf{P})$ and $\bar{\mathbf{a}}_E(\mathbf{P})$ even more. As FB is a ground-fixed basis, those components are the time derivative of $\{\overline{\mathbf{OP}}\}_{FB}$ and $\{\bar{\mathbf{v}}_T(\mathbf{P})\}_{FB}$ components, respectively:

$$\{\bar{\mathbf{v}}_E(\mathbf{P})\}_{FB} = \frac{d}{dt}\{\overline{\mathbf{OP}}\}_{FB} = \begin{Bmatrix} \dot{r}\cos\psi - r\dot{\psi}\sin\psi \\ \dot{r}\sin\psi + r\dot{\psi}\cos\psi \\ 0 \end{Bmatrix},$$
$$\{\bar{\mathbf{a}}_E(\mathbf{P})\}_{FB} = \frac{d}{dt}\{\bar{\mathbf{v}}_E(\mathbf{P})\}_{FB} = \begin{Bmatrix} \ddot{r}\cos\psi - r\dot{\psi}^2\cos\psi - 2\dot{r}\dot{\psi}\sin\psi - r\ddot{\psi}\sin\psi \\ \ddot{r}\sin\psi - r\dot{\psi}^2\sin\psi + 2\dot{r}\dot{\psi}\cos\psi + r\ddot{\psi}\cos\psi \\ 0 \end{Bmatrix}. \quad (2.9)$$

It is easy to check that Eq. (2.9) corresponds to the same vectors as those in Eqs. (2.6, 2.7) through a basis transformation. The direct comparison of the different expressions shows the advantage of the moving basis MB over the fixed basis FB (despite the apparent simplicity of the latter because of its zero angular velocity relative to E). ◀

2.3 Intrinsic Components of the Velocity and the Acceleration: Curvature Radius

There are important relationships between some geometric characteristics of the trajectory of a point **P** in a reference frame R and the corresponding velocity and acceleration vectors. They become evident when you project those vectors on the **intrinsic** or **Frenet basis** associated to the geometry of that trajectory. The versors of the intrinsic basis are (Fig. 2.6):

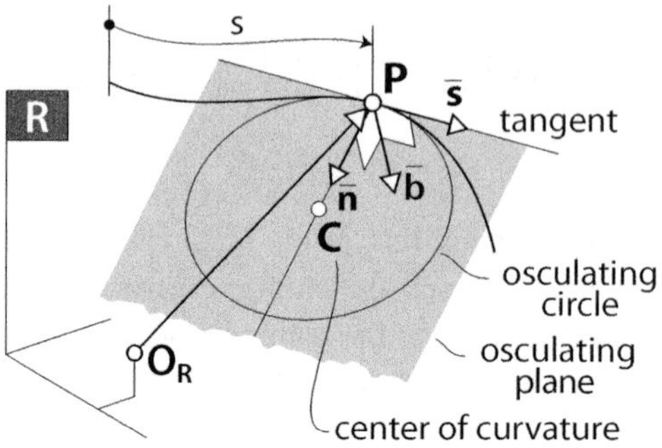

Fig. 2.6

- tangential : $$\bar{\mathbf{s}} \equiv \left.\frac{\mathrm{d}\overline{\mathbf{O}_\mathrm{R}\mathbf{P}}}{\mathrm{ds}}\right]_\mathrm{R} \qquad (2.10)$$
- normal : $$\bar{\mathbf{n}} \equiv \Re_\mathrm{R}(\mathbf{P}) \left.\frac{\mathrm{d}\bar{\mathrm{s}}}{\mathrm{ds}}\right]_\mathrm{R} \qquad (2.11)$$
- binormal : $$\bar{\mathbf{b}} \equiv \bar{\mathbf{s}} \times \bar{\mathbf{n}} \qquad (2.12)$$

where s is the **arc length along the trajectory** and $\Re_\mathrm{R}(\mathbf{P})$ is the **radius of curvature** of the trajectory at point **P**. The versor $\bar{\mathbf{n}}$ is directed toward the **center of curvature** of the trajectory at point **P**. In Eqs. (2.10, 2.11), it is necessary to take into account that the s-derivative of vectors is R-dependent (as the time derivative).

The plane defined by versors $\bar{\mathbf{s}}$ and $\bar{\mathbf{n}}$ is the **osculating plane** at point **P**, and the circle on that plane, with center **C** and radius $\Re_\mathrm{R}(\mathbf{P})$, is the **osculating circle** at point **P**. The trajectory first-order approximation is the tangent line, defined by $\bar{\mathbf{s}}$, and the second-order approximation is the osculating circle.

The velocity and acceleration components according to this basis are the **intrinsic components**:

$$\bar{\mathbf{v}}_\mathrm{R}(\mathbf{P}) = \dot{\mathrm{s}}(\mathbf{P})\bar{\mathbf{s}}, \ \bar{\mathbf{a}}_\mathrm{R}(\mathbf{P}) = \ddot{\mathrm{s}}(\mathbf{P})\bar{\mathbf{s}} + \frac{\dot{\mathrm{s}}^2(\mathbf{P})}{\Re_\mathrm{R}(\mathbf{P})}\bar{\mathbf{n}}, \qquad (2.13)$$

where $\dot{\mathrm{s}}(\mathbf{P})$ is the **speed** (velocity module) of **P** relative to R.

The velocity is strictly tangential in the sense of the advance, and the acceleration is on the osculating plane with components:

- ***tangential acceleration*** : $$\bar{\mathbf{a}}^\mathrm{s}_\mathrm{R}(\mathbf{P}) = \ddot{\mathrm{s}}(\mathbf{P})\bar{\mathbf{s}} \qquad (2.14)$$
- ***normal acceleration*** : $$\bar{\mathbf{a}}^\mathrm{n}_\mathrm{R}(\mathbf{P}) = \frac{\dot{\mathrm{s}}^2}{\Re_\mathrm{R}(\mathbf{P})}\bar{\mathbf{n}} = \frac{\mathrm{v}^2_\mathrm{R}(\mathbf{P})}{\Re_\mathrm{R}(\mathbf{P})} \qquad (2.15)$$

The tangential component is associated with the variation of the velocity value (it is what you read on a velocity indicator of a car and a chronometer: positive when “accelerating,” negative when “braking”). The normal component is associated with the change of the velocity direction. It is always directed to the curvature center. As the speed $\dot{\mathrm{s}}$ is squared, it can be substituted by the squared value of the velocity module.

When using the intrinsic basis, one has to bear in mind that the tangent versor $\bar{\mathrm{s}}$ reverses its sign whenever **P** reverses its traveling direction (as $\bar{\mathrm{s}}$ must always point to the advance direction of point **P** on the trajectory).

♣ *Proof*

The definitions of velocity and acceleration (Eq. (2.4)) together with Eqs. (2.10, 2.11) yield:

$$\bar{\mathbf{v}}_{\mathrm{R}}(\mathbf{P}) = \left.\frac{\mathrm{d}\overline{\mathbf{O}_{\mathrm{R}}\mathbf{P}}}{\mathrm{dt}}\right]_{\mathrm{R}} = \left.\frac{\mathrm{d}\overline{\mathbf{O}_{\mathrm{R}}\mathbf{P}}}{\mathrm{ds}}\right]_{\mathrm{R}} \frac{\mathrm{ds}}{\mathrm{dt}} = \dot{\mathrm{s}}\bar{\mathrm{s}}, \tag{2.16}$$

$$\bar{\mathbf{a}}_{\mathrm{R}}(\mathbf{P}) = \left.\frac{\mathrm{d}\bar{\mathbf{v}}_{\mathrm{R}}(\mathbf{P})}{\mathrm{dt}}\right]_{\mathrm{R}} = \ddot{\mathrm{s}}\bar{\mathrm{s}} + \dot{\mathrm{s}}^2 \left.\frac{\mathrm{d}\bar{\mathrm{s}}}{\mathrm{ds}}\right]_{\mathrm{R}} = \ddot{\mathrm{s}}\bar{\mathrm{s}} + \frac{\dot{\mathrm{s}}^2}{\mathfrak{R}_{\mathrm{R}}(\mathbf{P})}\bar{\mathbf{n}}. \tag{2.17}$$

♣

The expression of the normal acceleration may be very useful to determine the radius of curvature $\mathfrak{R}_{\mathrm{R}}(\mathbf{P})$ of a curve without having to resort to the analytical expression that comes from the trajectory equation. You just have to consider a point **P** following that curve and calculate $\mathrm{v}_{\mathrm{R}}(\mathbf{P})$ and $|\bar{\mathbf{a}}_{\mathrm{R}}^{\mathrm{n}}(\mathbf{P})|$. The movement of **P** on the curve can be chosen freely. It is advisable to choose it so that the kinematic study is as simple as possible.

► **Example 2.2** Point **P** follows a helical trajectory of constant radius r and pitch h (axial advance in a turn) with variable speed relative to the reference frame R (Fig. 2.7a). We want to find the general expressions of the intrinsic components of $\bar{\mathbf{v}}_{\mathrm{R}}(\mathbf{P})$ and $\bar{\mathbf{a}}_{\mathrm{R}}(\mathbf{P})$, and the center of curvature.

Vector $\bar{\mathbf{v}}_{\mathrm{R}}(\mathbf{P})$ can be found as the time derivative of $\overline{\mathbf{O}_{\mathrm{R}}\mathbf{P}}$. As it is a movement in space, the geometric derivation can be complicated, and it is safer to proceed through a vector basis. Figure 2.7a proposes a suitable basis that moves according to a simple rotation relative to R $\left(\bar{\mathbf{\Omega}}_{\mathrm{R}}^{\mathrm{B}} = \bar{\dot{\psi}}\right)$:

$$\{\overline{\mathbf{OP}}\}_{\mathrm{B}} = \begin{Bmatrix} \mathrm{r} \\ 0 \\ \mathrm{h}\psi/2\pi \end{Bmatrix} = \begin{Bmatrix} \mathrm{r} \\ 0 \\ \mathrm{r}\psi\tan\beta_0 \end{Bmatrix}, \quad \left\{\bar{\mathbf{\Omega}}_{\mathrm{R}}^{\mathrm{B}}\right\}_{\mathrm{B}} = \begin{Bmatrix} 0 \\ 0 \\ \dot{\psi} \end{Bmatrix}, \tag{2.18}$$

$$\{\bar{\mathbf{v}}_{\mathrm{R}}(\mathbf{P})\}_{\mathrm{B}} = \begin{Bmatrix} 0 \\ 0 \\ \mathrm{r}\dot{\psi}\tan\beta_0 \end{Bmatrix} + \begin{Bmatrix} 0 \\ 0 \\ \dot{\psi} \end{Bmatrix} \times \begin{Bmatrix} \mathrm{r} \\ 0 \\ \mathrm{r}\psi\tan\beta_0 \end{Bmatrix} = \begin{Bmatrix} 0 \\ \mathrm{r}\dot{\psi} \\ \mathrm{r}\dot{\psi}\tan\beta_0 \end{Bmatrix}, \tag{2.19}$$

$$\{\bar{\mathbf{a}}_{\mathrm{R}}(\mathbf{P})\}_{\mathrm{B}} = \begin{Bmatrix} 0 \\ \mathrm{r}\ddot{\psi} \\ \mathrm{r}\ddot{\psi}\tan\beta_0 \end{Bmatrix} + \begin{Bmatrix} 0 \\ 0 \\ \dot{\psi} \end{Bmatrix} \times \begin{Bmatrix} 0 \\ \mathrm{r}\dot{\psi} \\ \mathrm{r}\dot{\psi}\tan\beta_0 \end{Bmatrix} = \begin{Bmatrix} -\mathrm{r}\dot{\psi}^2 \\ \mathrm{r}\ddot{\psi} \\ \mathrm{r}\ddot{\psi}\tan\beta_0 \end{Bmatrix}. \tag{2.20}$$

The speed is $|\bar{\mathbf{v}}_{\mathrm{R}}(\mathbf{P})| = \mathrm{r}\dot{\psi}\sqrt{1+\tan^2\beta_0} = \mathrm{r}\dot{\psi}/\cos\beta_0$. The tangential versor $\bar{\mathrm{s}}$ is the normalized velocity vector:

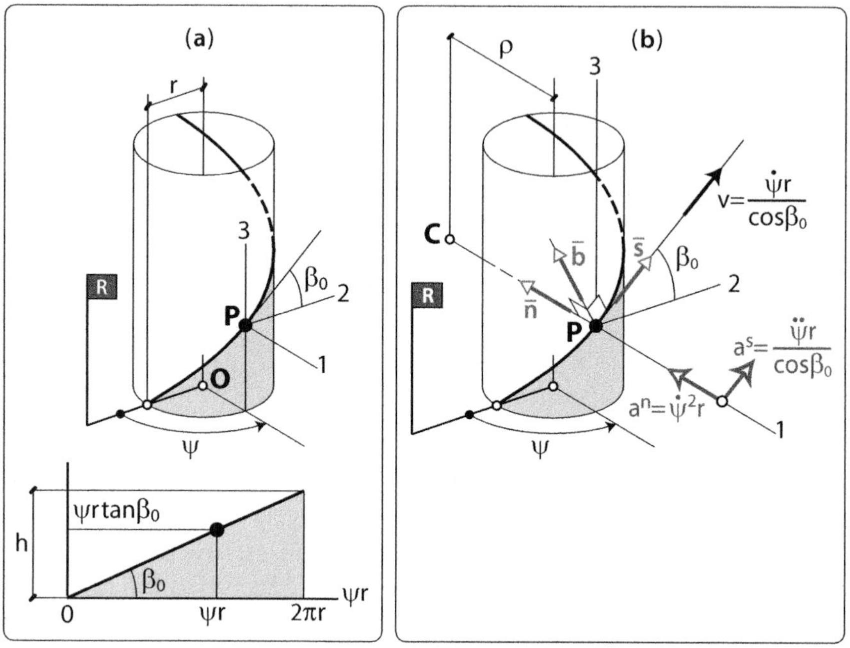

Fig. 2.7

$$\{\bar{\mathbf{s}}\}_B = \frac{1}{|\bar{\mathbf{v}}_R(\mathbf{P})|}\{\bar{\mathbf{v}}_R(\mathbf{P})\}_B = \frac{1}{r\dot{\psi}\sqrt{1+\tan^2\beta_0}}\left\{\begin{array}{c} 0 \\ r\dot{\psi} \\ r\dot{\psi}\tan\beta_0 \end{array}\right\} = \left\{\begin{array}{c} 0 \\ \cos\beta_0 \\ \sin\beta_0 \end{array}\right\}. \tag{2.21}$$

The tangential acceleration is:

$$a_R^s(\mathbf{P}) = \dot{v}_R(\mathbf{P}) = \frac{r\ddot{\psi}}{\cos\beta_0} \Rightarrow \left\{\bar{\mathbf{a}}_R^s(\mathbf{P})\right\}_B = a_R^s(\mathbf{P})\{\bar{\mathbf{s}}\}_B = \left\{\begin{array}{c} 0 \\ r\ddot{\psi} \\ r\ddot{\psi}\tan\beta_0 \end{array}\right\}. \tag{2.22}$$

The normal acceleration vector can be obtained by subtracting the tangential acceleration $\bar{\mathbf{a}}_R^s(\mathbf{P})$ from the total acceleration $\bar{\mathbf{a}}_R(\mathbf{P})$:

$$\left\{\bar{\mathbf{a}}_R^n(\mathbf{P})\right\}_B = \{\bar{\mathbf{a}}_R(\mathbf{P})\}_B - \left\{\bar{\mathbf{a}}_R^s(\mathbf{P})\right\}_B = \left\{\begin{array}{c} -r\dot{\psi}^2 \\ 0 \\ 0 \end{array}\right\}. \tag{2.23}$$

This expression indicates that the normal versor is on the negative semiaxis 1 (Fig. 2.7b).

The curvature radius is:

$$\mathfrak{R}_R(\mathbf{P}) = \frac{v_R^2(\mathbf{P})}{|a_R^n(\mathbf{P})|} = \frac{r}{\cos^2\beta_0} > r. \tag{2.24}$$

The curvature center is located on the straight line through **P** perpendicular to the helix axis, opposite to **P**, and at a distance $\rho = \mathfrak{R}_R(\mathbf{P}) - r = r\tan^2\beta_0$ from that axis.

The locus of the curvature centers is a helix with the same pitch and sense but with radius $\rho = r\tan^2\beta_0$. ◀

2.4 Composition of Velocities: Transportation Velocity

There are many situations where it is interesting to relate movements relative to two different references frames. Take the case of an observer on a ship measuring the motion of other ships to prevent collisions. Those measurements are relative to the ship (or to a trihedral with an origin fixed to the ship and the north–south and east–west axes), while the hypotheses about the trajectory of the other vessels are formulated relative to the sea (considered at rest relative to the earth).

In automobile kinematics, the ground frame is suitable to study the wheel–ground contact conditions (whether there is longitudinal/transversal sliding), while the movement of the internal elements of the propulsion system (motor and transmission) is more easily analyzed in the chassis reference frame.

Relating the movement of a same point relative to two different reference frames is called **composition of movements**. The two frames are usually called absolute frame (AB) and relative frame (REL). In general, the AB frame is that from which the movement of the other frame (REL) has a simpler description. For the examples mentioned in the previous paragraphs, the AB frame is the ground (AB = E), and the relative one is the ship (REL = S) or the vehicle chassis (REL = V).

The relationship between the velocity in the two frames AB and REL is:

$$\bar{\mathbf{v}}_{AB}(\mathbf{P}) = \bar{\mathbf{v}}_{REL}(\mathbf{P}) + \bar{\mathbf{v}}_{tr}(\mathbf{P}), \tag{2.25}$$

where $\bar{\mathbf{v}}_{tr}(\mathbf{P})$ is the **transportation velocity** of point **P**. It describes the AB velocity that **P** would have if there were no REL motion. Alternatively, it is the **P** velocity relative to AB if **P** were fixed to REL in the precise location where it is at every time instant:

$$\bar{\mathbf{v}}_{tr}(\mathbf{P}) = \bar{\mathbf{v}}_{AB}(\mathbf{P} \in \mathrm{REL}) = \bar{\mathbf{v}}_{AB}(\mathbf{P}_{REL}). \tag{2.26}$$

The transportation velocity $\bar{\mathbf{v}}_{tr}(\mathbf{P})$ can be found, according to its meaning, by means of the usual kinematic resources taking into account that all the variables describing the relative movement are kept constant. Alternatively, it can be calculated through the specific expression:

$$\bar{\mathbf{v}}_{tr}(\mathbf{P}) = \bar{\mathbf{v}}_{AB}(\mathbf{O}_{REL}) + \bar{\boldsymbol{\Omega}}_{tr} \times \overline{\mathbf{O}_{REL}\mathbf{P}}. \tag{2.27}$$

Vector $\bar{\boldsymbol{\Omega}}_{tr}$ is the **transportation angular velocity**. It is the angular velocity of frame REL relative to frame AB: $\bar{\boldsymbol{\Omega}}_{tr} = \bar{\boldsymbol{\Omega}}_{AB}^{REL}$ (Fig. 2.8). Point $\mathbf{O}_{REL}$ is any point in frame REL.

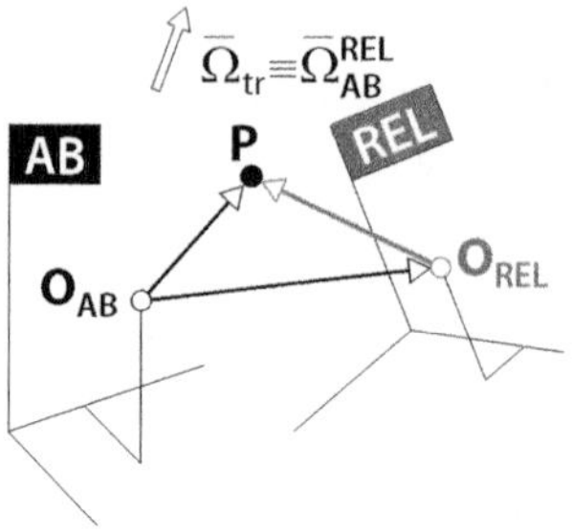

Fig. 2.8

♣ *Proof*

The **P** absolute velocity may be obtained as the time derivative of an absolute position vector $\overline{\mathbf{O}_{AB}\mathbf{P}}$ described as $\overline{\mathbf{O}_{AB}\mathbf{P}} = \overline{\mathbf{O}_{AB}\mathbf{O}_{REL}} + \overline{\mathbf{O}_{REL}\mathbf{P}}$ (Fig. 2.8):

$$\bar{\mathbf{v}}_{AB}(\mathbf{P}) = \left.\frac{d\overline{\mathbf{O}_{AB}\mathbf{P}}}{dt}\right]_{AB} = \left.\frac{d\overline{\mathbf{O}_{AB}\mathbf{O}_{REL}}}{dt}\right]_{AB} + \left.\frac{d\overline{\mathbf{O}_{REL}\mathbf{P}}}{dt}\right]_{AB} = \bar{\mathbf{v}}_{AB}(\mathbf{O}_{REL}) + \left.\frac{d\overline{\mathbf{O}_{REL}\mathbf{P}}}{dt}\right]_{AB}. \tag{2.28}$$

The last vector in Eq. (2.28) has the right velocity dimensions, but it cannot be physically interpreted as a velocity since it is the time derivative of a *relative* position vector in the *absolute* frame. Applying Eq. (1.25) of Section 1.6 (which relates the time derivative of a same vector in two different frames), Eq. (2.28) can be rewritten as:

$$\bar{\mathbf{v}}_{AB}(\mathbf{P}) = \bar{\mathbf{v}}_{AB}(\mathbf{O}_{REL}) + \bar{\mathbf{v}}_{REL}(\mathbf{P}) + \bar{\mathbf{\Omega}}_{AB}^{REL} \times \overline{\mathbf{O}_{REL}\mathbf{P}}. \tag{2.29}$$

In this equation, the sum of the first and the third terms of the right-hand side corresponds to the transportation velocity because it is what is left when you impose $\bar{\mathbf{v}}_{REL}(\mathbf{P}) = \bar{\mathbf{0}}$. ♣

► **Example 2.3** The motion studied in Example 2.1, where a point **P** moves radially on a rotating platform, can be considered as a composition of:

- the relative movement of **P** on the platform (REL), which is a nonuniform rectilinear motion
- the transportation movement associated with the platform rotation relative to the ground (AB), which is a nonuniform simple rotation

From these motions (Fig. 2.9), simple enough to not require an analytical manipulation of vectors, we can obtain $\bar{\mathbf{v}}_{AB}(\mathbf{P})$.

Projected on the platform-fixed basis B with axes $(1,2,3)$: $\{\bar{\mathbf{v}}_{AB}(\mathbf{P})\}_B = \begin{Bmatrix} \dot{r} \\ r\dot{\psi} \\ 0 \end{Bmatrix}$. ◄

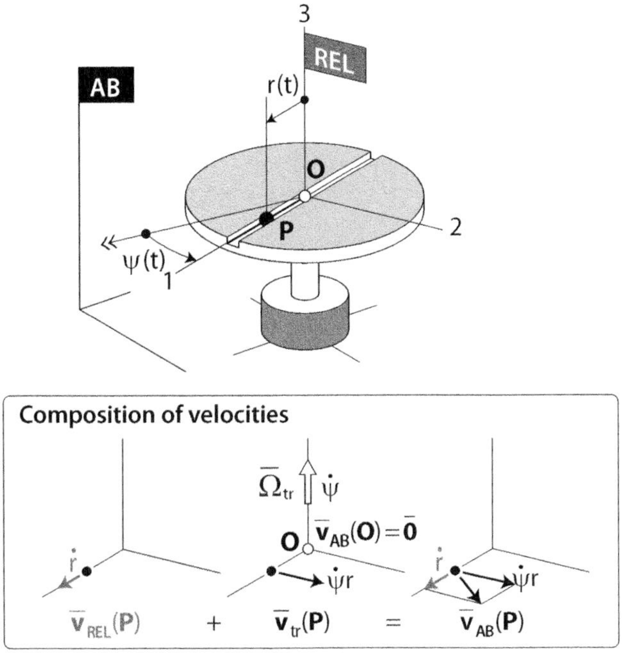

Fig. 2.9

► **Example 2.4** The ring of the Ferris wheel, whose radius is R, rotates in counterclockwise direction with angular velocity Ω_0 relative to the ground (E). Point **Q** is fixed to the ground, so $\bar{v}_E(\mathbf{Q}) = \bar{0}$. Its velocity relative to the ring can be obtained through a composition of velocities.

Taking the ground as AB and the ring as REL, the transportation motion is a circular motion about the ring center **O** (Fig. 2.10). As $\bar{v}_{AB}(\mathbf{Q}) = \bar{0}$, $\bar{v}_{REL}(\mathbf{Q}) = -\bar{v}_{tr}(\mathbf{Q})$.

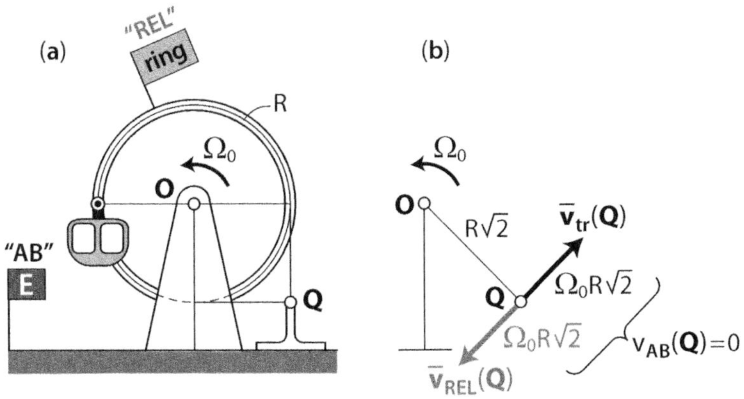

Fig. 2.10

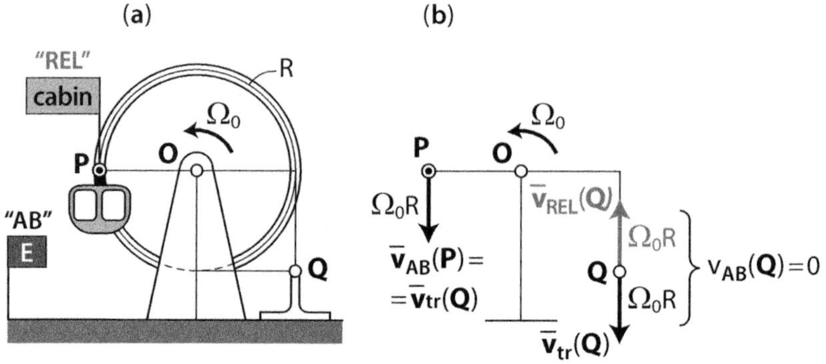

Fig. 2.11

If the REL reference frame were the cabin, which is articulated to the ring and consequently does not share the ring rotation Ω_0, the transportation motion would be totally different. If we neglect the small oscillations of the cabin, its angular velocity relative to the ground is zero. As a result, all its points describe exactly the same circular trajectory with radius R relative to the ground (though their centers of curvature do not coincide). Hence, $\bar{\mathbf{v}}_{tr}(\mathbf{Q}) = \bar{\mathbf{v}}_{AB}(\mathbf{P})$ at all time instants (Fig. 2.11).

Again, as $\bar{\mathbf{v}}_{AB}(\mathbf{Q}) = \bar{0}$, $\bar{\mathbf{v}}_{REL}(\mathbf{Q}) = -\bar{\mathbf{v}}_{tr}(\mathbf{Q})$. ◀

The composition of the relative motion with the transportation motion is a simple superposition when we are dealing with velocity. This agrees with the most direct intuition and places the composition of velocities within the framework of linear algebra. This is not the case when it comes to composing accelerations.

2.5 Composition of Accelerations: Transportation and Coriolis Accelerations

The composition of accelerations relating the absolute and the relative acceleration of point **P** is:

$$\bar{\mathbf{a}}_{AB}(\mathbf{P}) = \bar{\mathbf{a}}_{REL}(\mathbf{P}) + \bar{\mathbf{a}}_{tr}(\mathbf{P}) + \bar{\mathbf{a}}_{Cor}(\mathbf{P}), \tag{2.30}$$

where $\bar{\mathbf{a}}_{tr}(\mathbf{P})$ is the **transportation acceleration**:

$$\bar{\mathbf{a}}_{tr}(\mathbf{P}) = \bar{\mathbf{a}}_{AB}(\mathbf{P} \in \mathrm{REL}) = \bar{\mathbf{a}}_{AB}(\mathbf{P}_{REL}). \tag{2.31}$$

The term $\bar{\mathbf{a}}_{Cor}(\mathbf{P})$ in Eq. (2.30) is the **complementary** or **Coriolis acceleration.**[1] Unlike the transportation acceleration, this term has no simple and useful physical interpretation.

[1] The acceleration of Coriolis gets its name from G. G. Coriolis, a French scientist who introduced it in an article published in 1835 entitled "Sur les équations du mouvement relatif des systèmes de corps."

The transportation acceleration $\bar{\mathbf{a}}_{\text{tr}}(\mathbf{P})$ can be found by means of the usual kinematic resources (taking into account that the relative movement is blocked) or through the specific expression:

$$\bar{\mathbf{a}}_{\text{tr}}(\mathbf{P}) = \bar{\mathbf{a}}_{\text{AB}}(\mathbf{O}_{\text{REL}}) + \bar{\boldsymbol{\alpha}}_{\text{tr}} \times \overline{\mathbf{O}_{\text{REL}}\mathbf{P}} + \bar{\boldsymbol{\Omega}}_{\text{tr}} \times \left(\bar{\boldsymbol{\Omega}}_{\text{tr}} \times \overline{\mathbf{O}_{\text{REL}}\mathbf{P}}\right), \tag{2.32}$$

where $\bar{\boldsymbol{\alpha}}_{\text{tr}}$ is the transportation angular acceleration:

$$\bar{\boldsymbol{\alpha}}_{\text{tr}} = \bar{\boldsymbol{\alpha}}_{\text{AB}}^{\text{REL}} = \left.\frac{\mathrm{d}\bar{\boldsymbol{\Omega}}_{\text{AB}}^{\text{REL}}}{\mathrm{dt}}\right]_{\text{AB}}. \tag{2.33}$$

The Coriolis acceleration is:

$$\bar{\mathbf{a}}_{\text{Cor}}(\mathbf{P}) = 2\bar{\boldsymbol{\Omega}}_{\text{tr}} \times \bar{\mathbf{v}}_{\text{REL}}(\mathbf{P}). \tag{2.34}$$

♣ *Proof*

The absolute acceleration $\bar{\mathbf{a}}_{\text{AB}}(\mathbf{P})$ can be obtained as the time derivative of $\bar{\mathbf{v}}_{\text{AB}}(\mathbf{P})$:

$$\bar{\mathbf{a}}_{\text{AB}}(\mathbf{P}) = \left.\frac{\mathrm{d}\bar{\mathbf{v}}_{\text{AB}}(\mathbf{P})}{\mathrm{dt}}\right]_{\text{AB}} = \left.\frac{\mathrm{d}\bar{\mathbf{v}}_{\text{REL}}(\mathbf{P})}{\mathrm{dt}}\right]_{\text{AB}} + \left.\frac{\mathrm{d}\bar{\mathbf{v}}_{\text{tr}}(\mathbf{P})}{\mathrm{dt}}\right]_{\text{AB}}. \tag{2.35}$$

Taking into account that $\bar{\mathbf{v}}_{\text{tr}}(\mathbf{P}) = \bar{\mathbf{v}}_{\text{AB}}(\mathbf{O}_{\text{REL}}) + \bar{\boldsymbol{\Omega}}_{\text{tr}} \times \overline{\mathbf{O}_{\text{REL}}\mathbf{P}}$:

$$\bar{\mathbf{a}}_{\text{AB}}(\mathbf{P}) = \left.\frac{\mathrm{d}\bar{\mathbf{v}}_{\text{REL}}(\mathbf{P})}{\mathrm{dt}}\right]_{\text{AB}} + \left.\frac{\mathrm{d}\bar{\mathbf{v}}_{\text{AB}}(\mathbf{O}_{\text{REL}})}{\mathrm{dt}}\right]_{\text{AB}} + \left.\frac{\mathrm{d}\bar{\boldsymbol{\Omega}}_{\text{tr}}}{\mathrm{dt}}\right]_{\text{AB}} \times \overline{\mathbf{O}_{\text{REL}}\mathbf{P}} + \bar{\boldsymbol{\Omega}}_{\text{tr}} \times \left.\frac{\mathrm{d}\overline{\mathbf{O}_{\text{REL}}\mathbf{P}}}{\mathrm{dt}}\right]_{\text{AB}}$$

$$= \left.\frac{\mathrm{d}\bar{\mathbf{v}}_{\text{REL}}(\mathbf{P})}{\mathrm{dt}}\right]_{\text{AB}} + \bar{\mathbf{a}}_{\text{AB}}(\mathbf{O}_{\text{REL}}) + \bar{\alpha}_{\text{tr}} \times \overline{\mathbf{O}_{\text{REL}}\mathbf{P}} + \bar{\boldsymbol{\Omega}}_{\text{tr}} \times \left.\frac{\mathrm{d}\overline{\mathbf{O}_{\text{REL}}\mathbf{P}}}{\mathrm{dt}}\right]_{\text{AB}}.$$

Applying Eq. (1.25) of Section 1.6:

$$\left.\frac{\mathrm{d}\bar{\mathbf{v}}_{\text{REL}}(\mathbf{P})}{\mathrm{dt}}\right]_{\text{AB}} = \left.\frac{\mathrm{d}\bar{\mathbf{v}}_{\text{REL}}(\mathbf{P})}{\mathrm{dt}}\right]_{\text{REL}} + \bar{\boldsymbol{\Omega}}_{\text{tr}} \times \bar{\mathbf{v}}_{\text{REL}}(\mathbf{P}) = \bar{\mathbf{a}}_{\text{REL}}(\mathbf{P}) + \bar{\boldsymbol{\Omega}}_{\text{tr}} \times \bar{\mathbf{v}}_{\text{REL}}(\mathbf{P}),$$

$$\left.\frac{\mathrm{d}\overline{\mathbf{O}_{\text{REL}}\mathbf{P}}}{\mathrm{dt}}\right]_{\text{AB}} = \left.\frac{\mathrm{d}\overline{\mathbf{O}_{\text{REL}}\mathbf{P}}}{\mathrm{dt}}\right]_{\text{REL}} + \bar{\boldsymbol{\Omega}}_{\text{tr}} \times \overline{\mathbf{O}_{\text{REL}}\mathbf{P}} = \bar{\mathbf{v}}_{\text{REL}}(\mathbf{P}) + \bar{\boldsymbol{\Omega}}_{\text{tr}} \times \overline{\mathbf{O}_{\text{REL}}\mathbf{P}}, \tag{2.36}$$

and Eq. (2.35) becomes:

$$\bar{\mathbf{a}}_{\text{AB}}(\mathbf{P}) = \bar{\mathbf{a}}_{\text{REL}}(\mathbf{P}) + \bar{\boldsymbol{\Omega}}_{\text{tr}} \times \bar{\mathbf{v}}_{\text{REL}}(\mathbf{P})$$
$$+ \bar{\mathbf{a}}_{\text{AB}}(\mathbf{O}_{\text{REL}}) + \bar{\boldsymbol{\alpha}}_{\text{tr}} \times \overline{\mathbf{O}_{\text{REL}}\mathbf{P}} + \bar{\boldsymbol{\Omega}}_{\text{tr}} \times \bar{\mathbf{v}}_{\text{REL}}(\mathbf{P}) + \bar{\boldsymbol{\Omega}}_{\text{tr}} \times \left(\bar{\boldsymbol{\Omega}}_{\text{tr}} \times \overline{\mathbf{O}_{\text{REL}}\mathbf{P}}\right). \tag{2.37}$$

Blocking the relative motion in Eq. (2.37) yields:

$$\bar{\mathbf{a}}_{\text{tr}}(\mathbf{P}) = \bar{\mathbf{a}}_{\text{AB}}(\mathbf{O}_{\text{REL}}) + \bar{\boldsymbol{\alpha}}_{\text{tr}} \times \overline{\mathbf{O}_{\text{REL}}\mathbf{P}} + \bar{\boldsymbol{\Omega}}_{\text{tr}} \times \left(\bar{\boldsymbol{\Omega}}_{\text{tr}} \times \overline{\mathbf{O}_{\text{REL}}\mathbf{P}}\right). \tag{2.38}$$

Finally, introducing Eq. (2.38) in Eq. (2.37) and reorganizing the terms yields $\bar{\mathbf{a}}_{\text{Cor}}(\mathbf{P}) = 2\bar{\boldsymbol{\Omega}}_{\text{tr}} \times \bar{\mathbf{v}}_{\text{REL}}(\mathbf{P})$. ♣

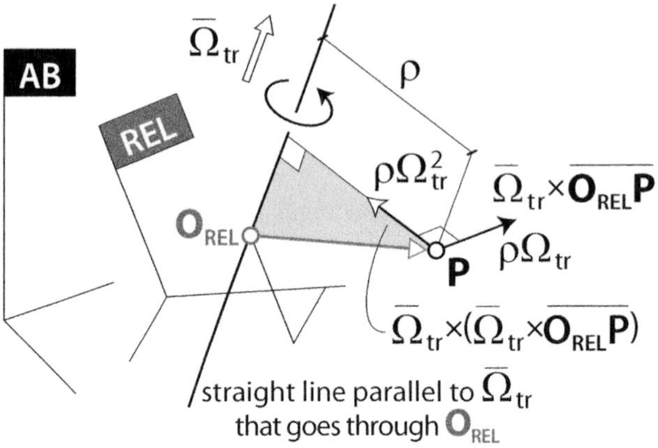

Fig. 2.12

Figure 2.12 illustrates the geometry of the term $\bar{\Omega}_{tr} \times \left(\bar{\Omega}_{tr} \times \overline{\mathbf{O}_{REL}\mathbf{P}}\right)$: it is directed perpendicularly from **P** to the line parallel to $\bar{\Omega}_{tr}$ through $\mathbf{O}_{REL}$, and its module is Ω_{tr}^2 times the distance ρ from **P** to that line.

► **Example 2.5** The composition of accelerations applied to point **P** in Example 2.3 yields the **P** acceleration relative to the ground ($\bar{\mathbf{a}}_{AB}(\mathbf{P})$) from:

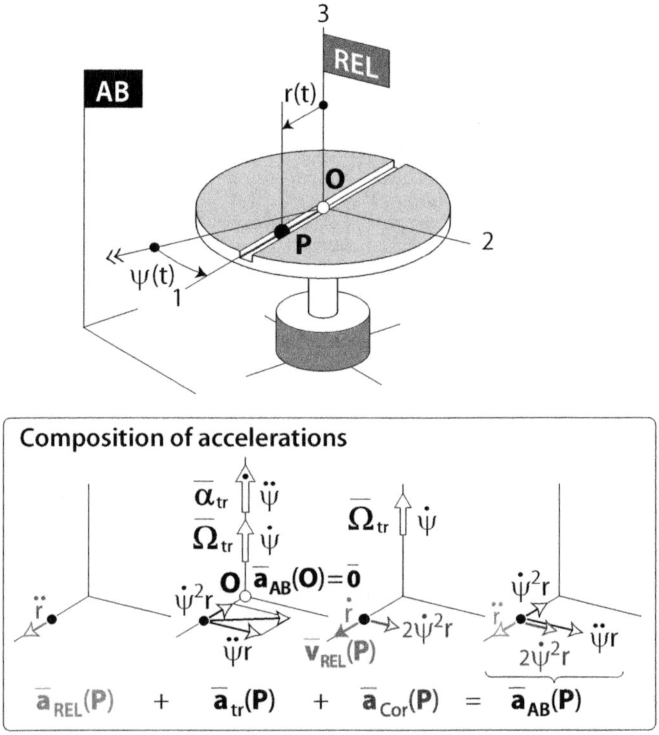

Fig. 2.13

- the tangential acceleration associated with the nonuniform rectilinear motion of **P** on the platform (taken as REL reference frame)
- the tangential and normal components of the transportation acceleration associated with the platform rotation relative to the ground (AB), which is a nonuniform simple rotation
- the Coriolis acceleration, obtained as $\bar{\mathbf{a}}_{\mathrm{Cor}}(\mathbf{P}) = 2\bar{\boldsymbol{\Omega}}_{\mathrm{tr}} \times \bar{\mathbf{v}}_{\mathrm{REL}}(\mathbf{P})$

The superposition of those terms is shown in Fig. 2.13.

Projected on the platform-fixed basis B with axes $(1, 2, 3)$: $\{\bar{\mathbf{a}}_{\mathrm{AB}}(\mathbf{P})\}_{\mathrm{B}} = \begin{Bmatrix} \ddot{r} - r\dot{\psi}^2 \\ r\ddot{\psi} + 2\dot{r}\dot{\psi} \\ 0 \end{Bmatrix}$. ◀

► **Example 2.6** Let's consider again the Ferris wheel and the ground-fixed point **Q** of Example 2.4. The composition of accelerations can be applied to obtain $\bar{\mathbf{a}}_{\mathrm{ring}}(\mathbf{Q})$ and $\bar{\mathbf{a}}_{\mathrm{cabin}}(\mathbf{Q})$.

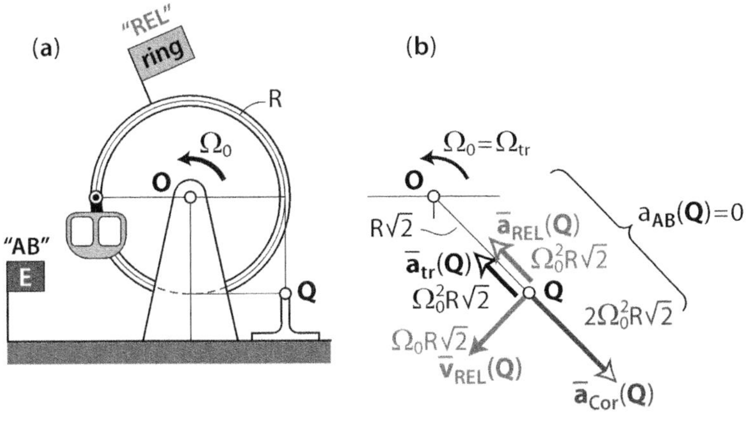

Fig. 2.14

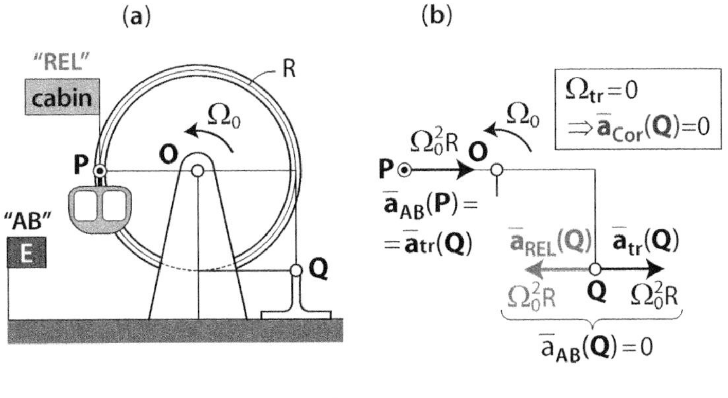

Fig. 2.15

Taking the ground as AB and the ring as REL, the transportation acceleration has just a normal component, directed to the ring center **O**, associated with the circular motion about **O**. As $\bar{\mathbf{a}}_{AB}(\mathbf{Q}) = \bar{0}$, $\bar{\mathbf{a}}_{REL}(\mathbf{Q}) = -\bar{\mathbf{a}}_{tr}(\mathbf{Q}) - \bar{\mathbf{a}}_{Cor}(\mathbf{Q})$. The Coriolis acceleration is nonzero as $\bar{\mathbf{\Omega}}_{tr} = \bar{\mathbf{\Omega}}_0$ (Fig. 2.14).

If the REL reference frame is the cabin, $\bar{\mathbf{\Omega}}_{tr} = \bar{0}$. Hence, $\bar{\mathbf{a}}_{Cor}(\mathbf{Q}) = \bar{0}$ and $\bar{\mathbf{a}}_{REL}(\mathbf{Q}) = -\bar{\mathbf{a}}_{tr}(\mathbf{Q})$. The transportation acceleration has just one normal component whose value and direction are identical to those of $\bar{\mathbf{a}}_{AB}(\mathbf{P})$ (Fig. 2.15). ◄

2.6 Additional Interesting Points regarding the Composition of Movements

Besides the immediate interest associated with problems that refer directly to it, the composition of movements is interesting for many other reasons:

- The concept of transportation movement is the basis of rigid body kinematics (Chapter 3)
- The relationship between accelerations in two different frames allows the extension of the formulation of the dynamics, which is initially formulated for Galilean or inertial frames, to all reference frames
- It is an alternative resource to calculate velocities and accelerations involving only one single reference frame. It is particularly useful when the relative and transportation movements are simple enough to dispense with analytic manipulation (and represent vectors their arrows and their corresponding values)

The vector operations in the expressions of the composition of movements are performed in one single time instant. Consequently, the vectors appearing in them may be particularized to a time instant: as they do not contain explicit time derivatives, the vectors' general expressions (those that describe them at any moment) are not needed.

► **Example 2.7** On a smooth (frictionless) horizontal platform with radius R, which rotates relative to the ground about the vertical axis passing through its center **O** (fixed to the ground) with constant angular velocity $\bar{\mathbf{\Omega}}_0$, a person located at **Q** launches a particle **P** with initial velocity $\bar{\mathbf{v}}_0$ relative to the platform (Fig. 2.16a). The dynamics of the problem is responsible for a *uniform rectilinear movement of* **P** *relative to the ground.*

Different orientations of the launching velocity $\bar{\mathbf{v}}_0$ relative to the platform (different values of the angle β_0 defined by $\bar{\mathbf{v}}_0$ and the radial direction) yield different relative motions. The conditions on β_0 to achieve relative trajectories passing through particular points can be studied by means of the composition of movements.

Trajectory through the Platform Center

As point **O** is both fixed to the platform and the ground, imposing a trajectory relative to the platform (REL) through **O** is equivalent to imposing that the trajectory relative to the

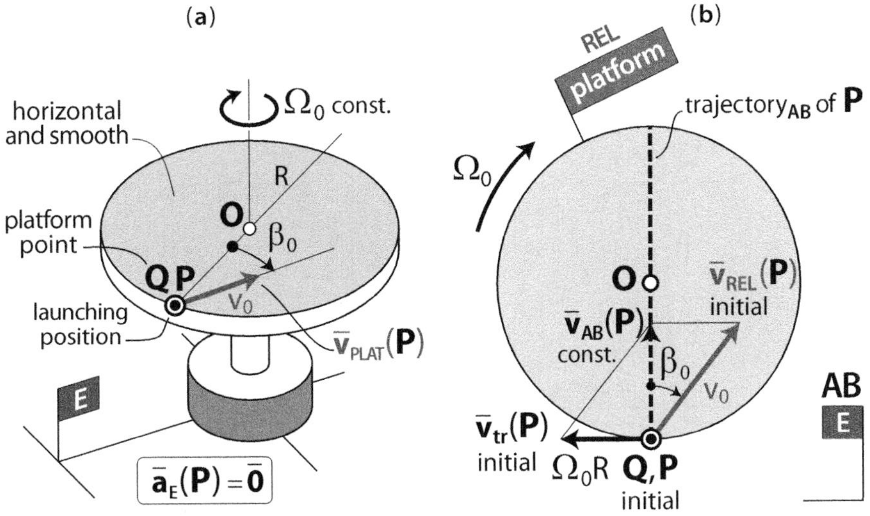

Fig. 2.16

ground (AB) goes through that same point. This calls for a $\bar{v}_{AB}(\mathbf{P})$ radially oriented (Fig. 2.16b). This condition is fulfilled if $\sin\beta_0 = R\Omega_0/v_0$.

Trajectory through the Point Diametrically Opposite to the Launching Position (Diametrical Launching)

In a diametrical launching, the particle **P** is caught by a person **Q′** located on the platform diametrically opposite to the launching person **Q**. The condition on β_0 is found in a simple way from the absolute motions of **P** and **Q′**: their absolute trajectories have to intersect simultaneously at the same point (Fig. 2.17). It follows that the time interval needed by **P** to cover the rectilinear path with length $\rho (= 2R\cos\delta_0)$ has to be the same as that needed by **Q′** to describe the arc associated with the angle $2\delta_0$ (plus any integer number n of complete turns). As the absolute speeds of **P** and **Q′** are constant, that condition is:

$$\left.\begin{array}{l}\rho = 2R\cos\delta_0 = |\bar{v}_{AB}(\mathbf{P})|t\\ 2\delta_0 + 2\pi n = \Omega_0 t\end{array}\right\} \Rightarrow |\bar{v}_{AB}(\mathbf{P})| = \frac{\Omega_0 R\cos\delta_0}{\delta_0 + \pi n}. \tag{2.39}$$

The absolute velocity of **P** can be calculated composing the REL and the transportation motion at the initial time instant: $\bar{v}_{AB}(\mathbf{P}) = \bar{v}_{REL}(\mathbf{P}) + \bar{v}_{tr}(\mathbf{P})$. Equating the radial and tangential components we obtain:

$$|\bar{v}_{AB}(\mathbf{P})|\cos\delta_0 = v_0\cos\beta_0, \quad |\bar{v}_{AB}(\mathbf{P})|\sin\delta_0 + \Omega_0 R = v_0\sin\beta_0. \tag{2.40}$$

Combining Eqs. (2.38) and (2.40) yields the condition for a diametrical launching:

$$\tan\beta_0 = \frac{\sin\delta_0\cos\delta_0 + \delta_0 + \pi n}{\cos^2\delta_0}, \quad \frac{v_0}{\Omega_0 R} = \frac{\cos^2\delta_0}{(\delta_0 + \pi n)\cos\beta_0}. \tag{2.41}$$

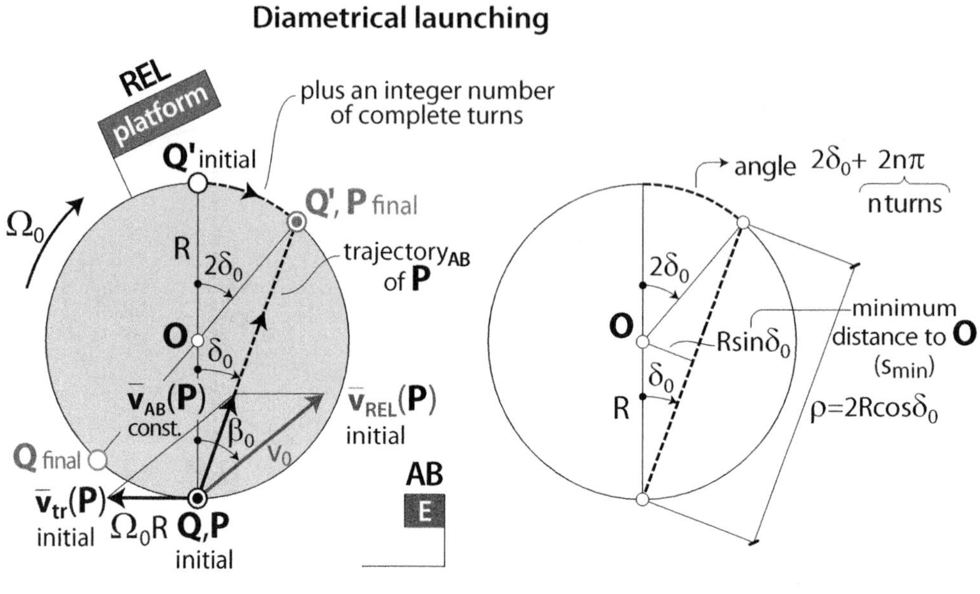

Fig. 2.17

The launching parameters are $(v_0, \delta_0, \beta_0, n)$. As we have two relationships, only two of them can be chosen freely. We can give values to (δ_0, n) and calculate (v_0, β_0) (which is what Eq. (2.41) suggests). A more visual choice would be to give values to (s_{min}, n), where s_{min} is the minimum distance between the absolute trajectory and the platform center. In that case, δ_0 is calculated as $\delta_0 = \arcsin(s_{min}/R)$(Fig. 2.17).

Figure 2.18 shows a collection of trajectories relative to the platform, for different values of s_{min} and n.

Return to the Launching Position (*Boomerang* Launching)

In a *boomerang* launching, the same person launches and catches the particle **P**. The condition on β_0 is formulated as in the previous case: the time interval needed by **P** to cover the straight path ρ has to be the same as that needed by **Q** to cover the arc corresponding to a half-turn plus that associated with angle $2\delta_0$ (plus any integer number n of complete turns, Fig. 2.19):

$$\tan\beta_0 = \frac{\sin\delta_0 + \delta_0 + \pi\left(n + \dfrac{1}{2}\right)}{\cos\delta_0}, \quad \frac{v_0}{\Omega_0 R} = \frac{\cos^2\delta_0}{\left(\delta_0 + \pi\left(n + \dfrac{1}{2}\right)\right)\cos\beta_0}. \tag{2.42}$$

Figure 2.20 shows a collection of boomerang trajectories relative to the platform, for different values of s_{min} and n. ◀

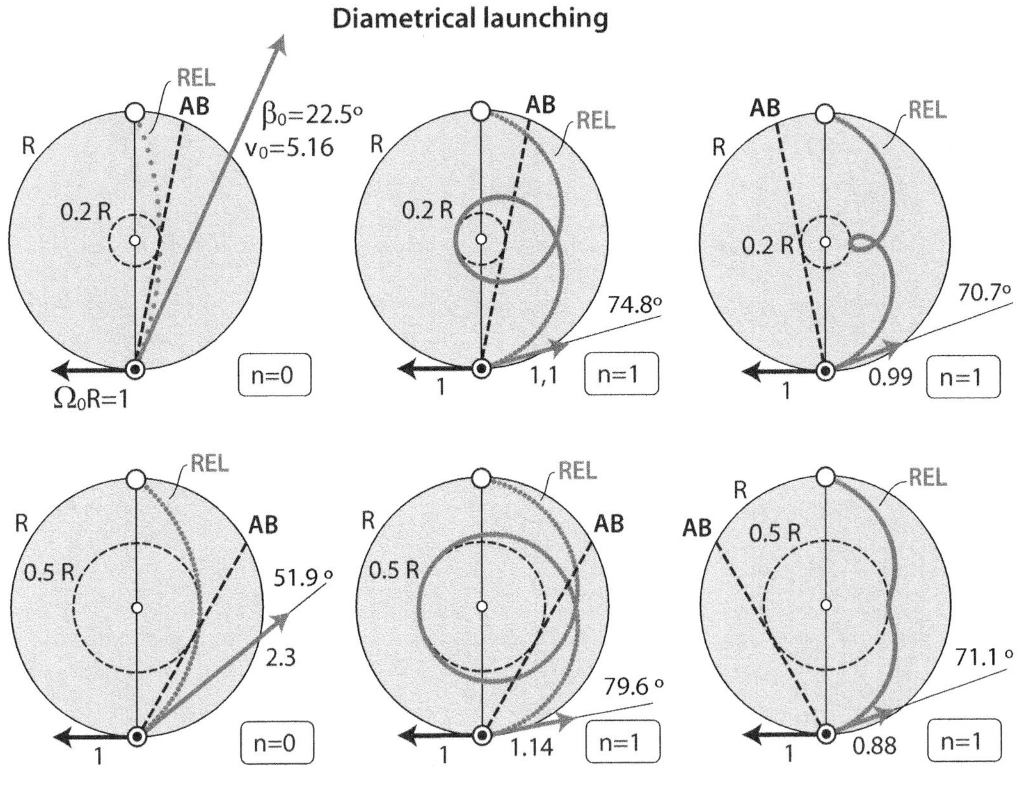

Fig. 2.18

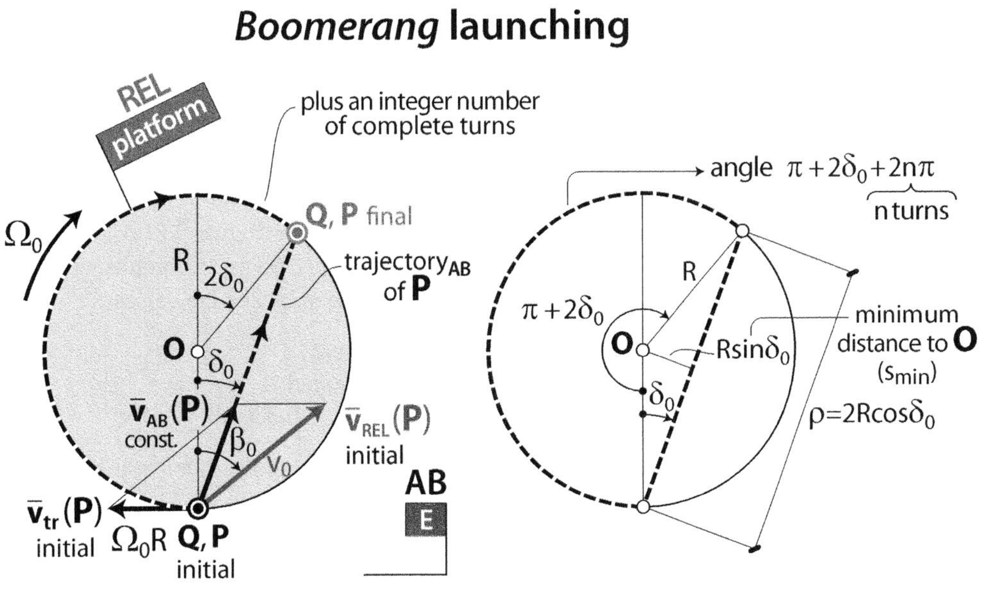

Fig. 2.19

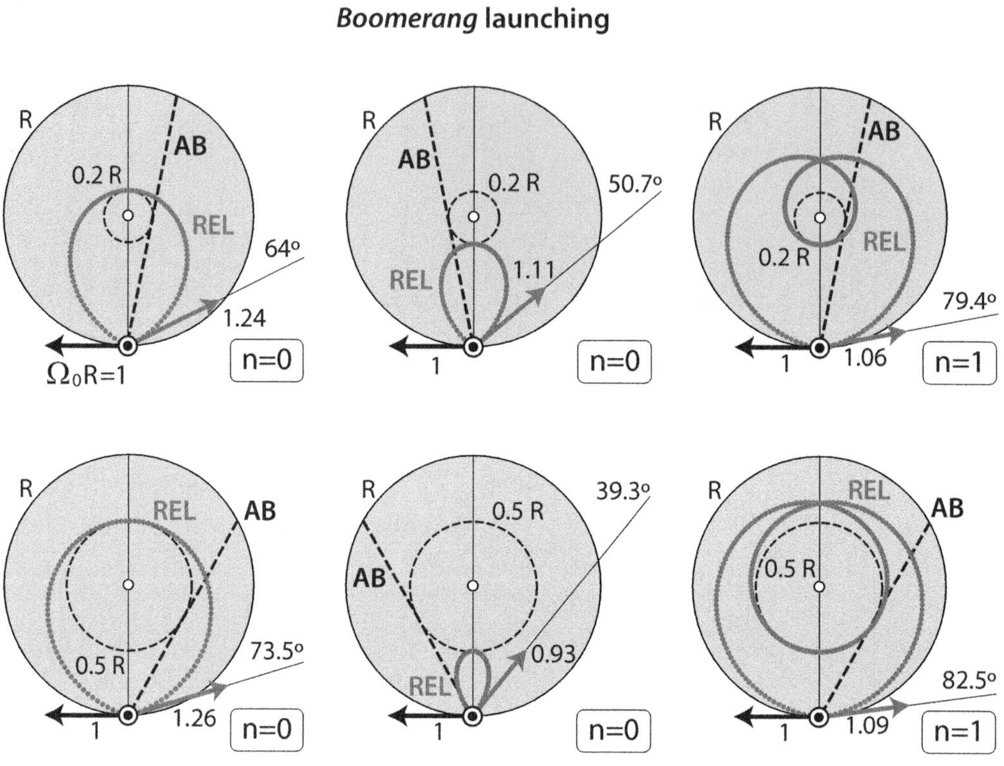

Fig. 2.20

▶ **Example 2.8** A point **O** of a ship describes a circular trajectory to starboard (right side when looking in the forward direction) relative to the ground (E). The radius is 1,000 m and the speed v is variable. The ship axis is always tangent to the **O** trajectory.

An observer in the ship measures the polar coordinates r and θ, plus their two first time derivatives, of a motorboat (modeled as a mass point **P**, Fig. 2.21a).

At a certain time instant, $v_E(\mathbf{O}) = 10\ \text{m/s}$, $\dot{v}_E(\mathbf{O}) = 0.05\ \text{m/s}^2$. Simultaneously, the variables associated with the **P** motion relative to the ship take the values:

$$\begin{array}{ll} r = 1,500\ \text{m} & \theta = 36.87^\circ = 0.006\ \text{rad} \\ \dot{r} = -16\ \text{m/s} & \dot{\theta} = 0.344^\circ/\text{s} = 0.006\ \text{rad/s} \\ \ddot{r} = 0.044\ \text{m/s}^2 & \ddot{\theta} = 0.2177 \cdot 10^{-2}\ {}^\circ/\text{s}^2 = 0.380 \cdot 10^{-4}\ \text{rad/s}^2 \end{array}$$

We want to calculate the velocity and the acceleration of the motorboat **P** relative to the sea – which we assume to be at rest relative to the ground.

As the measurements correspond to the movement of **P** relative to the ship (REL), we have to compose movements to obtain the movement of **P** relative to the sea (AB). First, we have to understand perfectly the ship motion relative to the sea (the motion of REL relative to AB).

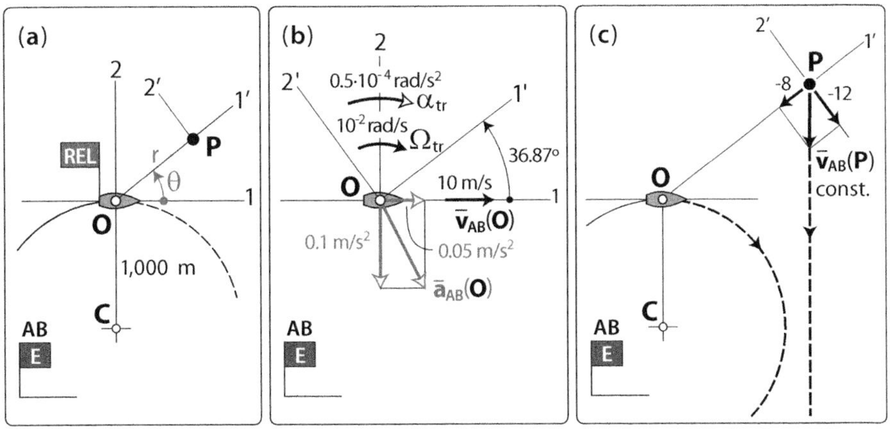

Fig. 2.21

Absolute motion of the relative reference frame: it is described through $\bar{\mathbf{v}}_{\mathrm{AB}}(\mathbf{O})$, $\bar{\mathbf{a}}_{\mathrm{AB}}(\mathbf{O})$, $\bar{\mathbf{\Omega}}^{\mathrm{REL}}_{\mathrm{AB}}$ and $\bar{\boldsymbol{\alpha}}^{\mathrm{REL}}_{\mathrm{AB}}$. At this time instant, the projection of these vectors on the vector basis B with axes $(1,2,3)$ (associated with the absolute motion of $\mathbf{O}$, Fig. 2.21b) yields:

$$\{\bar{\mathbf{v}}_{\mathrm{AB}}(\mathbf{O})\}_{\mathrm{B}} = \begin{Bmatrix} 10 \\ 0 \\ 0 \end{Bmatrix} \mathrm{m/s},$$

$$\left.\begin{array}{l} \mathrm{a}^{\mathrm{s}}_{\mathrm{AB}}(\mathbf{O}) = 0.05\,\mathrm{m/s^2} \\ \mathrm{a}^{\mathrm{n}}_{\mathrm{AB}}(\mathbf{O}) = \mathrm{v}^2_{\mathrm{AB}}(\mathbf{O})/\mathfrak{R}_{\mathrm{AB}}(\mathbf{O}) = 0.1\,\mathrm{m/s^2} \end{array}\right] \Rightarrow \{\bar{\mathbf{a}}_{\mathrm{AB}}(\mathbf{O})\}_{\mathrm{B}} = \begin{Bmatrix} 0.05 \\ -0.1 \\ 0 \end{Bmatrix} \mathrm{m/s^2},$$

$$\left|\bar{\mathbf{\Omega}}^{\mathrm{REL}}_{\mathrm{AB}}\right| = \left|\bar{\mathbf{\Omega}}_{\mathrm{tr}}\right| = \frac{\mathrm{v}_{\mathrm{AB}}(\mathbf{O})}{\mathfrak{R}_{\mathrm{AB}}(\mathbf{O})} = 0.01\,\mathrm{rad/s} \Rightarrow \{\bar{\mathbf{\Omega}}_{\mathrm{tr}}\} = \begin{Bmatrix} 0 \\ 0 \\ -0.01 \end{Bmatrix} \mathrm{rad/s},$$

$$\left|\bar{\boldsymbol{\alpha}}^{\mathrm{REL}}_{\mathrm{AB}}\right| = \left|\bar{\boldsymbol{\alpha}}_{\mathrm{tr}}\right| = \frac{\mathrm{a}^{\mathrm{s}}_{\mathrm{AB}}(\mathbf{O})}{\mathfrak{R}_{\mathrm{AB}}(\mathbf{O})} = 0.5\cdot 10^{-4}\mathrm{rad/s^2} \Rightarrow \{\bar{\boldsymbol{\alpha}}_{\mathrm{tr}}\} = \begin{Bmatrix} 0 \\ 0 \\ -0.5\cdot 10^{-4} \end{Bmatrix} \mathrm{rad/s^2}\ .$$

The next calculations are simpler in the vector basis B′ with axes $(1',2',3')$ (Fig. 2.21b):

$$\{\bar{\mathbf{v}}_{\mathrm{AB}}(\mathbf{O})\}_{\mathrm{B'}} = \begin{Bmatrix} 8 \\ -6 \\ 0 \end{Bmatrix} \mathrm{m/s},\ \ \{\bar{\mathbf{a}}_{\mathrm{AB}}(\mathbf{O})\}_{\mathrm{B'}} = \begin{Bmatrix} -0.02 \\ -0.11 \\ 0 \end{Bmatrix} \mathrm{m/s^2},$$

$$\{\bar{\mathbf{\Omega}}_{\mathrm{tr}}\}_{\mathrm{B'}} = \begin{Bmatrix} 0 \\ 0 \\ -0.01 \end{Bmatrix} \mathrm{rad/s},\ \ \{\bar{\boldsymbol{\alpha}}_{\mathrm{tr}}\}_{\mathrm{B'}} = \begin{Bmatrix} 0 \\ 0 \\ -0.5\cdot 10^{-4} \end{Bmatrix} \mathrm{rad/s^2}.$$

Relative motion of P: it is a planar motion described through polar coordinates. The results in the previous example lead to:

$$\{\bar{\mathbf{v}}_{\mathrm{REL}}(\mathbf{P})\}_{\mathrm{B}'}=\begin{Bmatrix}\dot{r}\\ r\dot{\theta}\\ 0\end{Bmatrix}=\begin{Bmatrix}-16\\ 9\\ 0\end{Bmatrix}\mathrm{m/s};\ \{\bar{\mathbf{a}}_{\mathrm{REL}}(\mathbf{P})\}_{\mathrm{B}'}=\begin{Bmatrix}\ddot{r}-r\dot{\theta}^2\\ r\ddot{\theta}+2\dot{r}\dot{\theta}\\ 0\end{Bmatrix}=\begin{Bmatrix}-0.01\\ -0.135\\ 0\end{Bmatrix}\mathrm{m/s^2}.$$

Transportation motion of P: as the absolute motion of the ship is a simple rotation around an axis through **C**, the transportation motion of **P** is circular with curvature center at **C**. In this case, the projection of the transportation velocity and acceleration in the B′ basis is not straightforward. It is easier to calculate them from Eqs. (2.27) and (2.33), respectively:

$$\{\bar{\mathbf{v}}_{\mathrm{tr}}(\mathbf{P})\}_{\mathrm{B}'}=\{\bar{\mathbf{v}}_{\mathrm{AB}}(\mathbf{O})+\bar{\boldsymbol{\Omega}}_{\mathrm{tr}}\times\overline{\mathbf{OP}}\}_{\mathrm{B}'}=\begin{Bmatrix}8\\ -6\\ 0\end{Bmatrix}+\begin{Bmatrix}0\\ 0\\ -0.01\end{Bmatrix}\times\begin{Bmatrix}1,500\\ 0\\ 0\end{Bmatrix}$$

$$=\begin{Bmatrix}8\\ -21\\ 0\end{Bmatrix}\mathrm{m/s},$$

$$\{\bar{\mathbf{a}}_{\mathrm{tr}}(\mathbf{P})\}_{\mathrm{B}'}=\left\{\bar{\mathbf{a}}_{\mathrm{AB}}(\mathbf{O})+\bar{\boldsymbol{\alpha}}_{\mathrm{tr}}\times\overline{\mathbf{OP}}+\bar{\boldsymbol{\Omega}}_{\mathrm{tr}}\times\left(\bar{\boldsymbol{\Omega}}_{\mathrm{tr}}\times\overline{\mathbf{OP}}\right)\right\}_{\mathrm{B}'}$$

$$=\begin{Bmatrix}-0.02\\ -0.11\\ 0\end{Bmatrix}+\begin{Bmatrix}0\\ 0\\ -0.5\cdot10^{-4}\end{Bmatrix}\times\begin{Bmatrix}1,500\\ 0\\ 0\end{Bmatrix}+\begin{Bmatrix}0\\ 0\\ -0.01\end{Bmatrix}$$

$$\times\left(\begin{Bmatrix}0\\ 0\\ -0.01\end{Bmatrix}\times\begin{Bmatrix}1,500\\ 0\\ 0\end{Bmatrix}\right)=\begin{Bmatrix}-0.170\\ -0.185\\ 0\end{Bmatrix}\mathrm{m/s^2}.$$

Coriolis acceleration of P: from Eq. (2.34):

$$\{\bar{\mathbf{a}}_{\mathrm{Cor}}(\mathbf{P})\}_{\mathrm{B}'}=2\begin{Bmatrix}0\\ 0\\ -0.01\end{Bmatrix}\times\begin{Bmatrix}-16\\ 9\\ 0\end{Bmatrix}=\begin{Bmatrix}0.18\\ 0.32\\ 0\end{Bmatrix}\mathrm{m/s^2}$$

Finally:

$$\{\bar{\mathbf{v}}_{\mathrm{AB}}(\mathbf{P})\}_{\mathrm{B}'}=\{\bar{\mathbf{v}}_{\mathrm{REL}}(\mathbf{P})+\bar{\mathbf{v}}_{\mathrm{tr}}(\mathbf{P})\}_{\mathrm{B}'}=\begin{Bmatrix}-8\\ -12\\ 0\end{Bmatrix}\mathrm{m/s}\ \Rightarrow\ |\bar{\mathbf{v}}_{\mathrm{AB}}(\mathbf{P})|=14.42\ \mathrm{m/s},$$

$$\{\bar{\mathbf{a}}_{\mathrm{AB}}(\mathbf{P})\}_{\mathrm{B}'}=\{\bar{\mathbf{a}}_{\mathrm{REL}}(\mathbf{P})+\bar{\mathbf{a}}_{\mathrm{tr}}(\mathbf{P})+\bar{\mathbf{a}}_{\mathrm{Cor}}(\mathbf{P})\}_{\mathrm{B}'}=\begin{Bmatrix}0\\ 0\\ 0\end{Bmatrix}\mathrm{m/s^2}.$$

These results show that **P** is following a linear trajectory with constant speed (Fig. 2.21c). As that trajectory does not intersect the ship's trajectory, there would be no collision risk if it were maintained. ◀

Appendix 2A Angular Velocity of the Intrinsic (or Frenet) Vector Basis

As seen in Section 2.3, the intrinsic basis IB associated with a point **P** moving relative to a reference frame R (Fig. 2A.1) is defined by the versors:

$$\bar{\mathbf{s}} = \left.\frac{\mathrm{d}\overline{\mathbf{OP}}}{\mathrm{ds}}\right]_{\mathrm{R}}, \quad \bar{\mathbf{n}} = \mathfrak{R}_{\mathrm{R}}(\mathbf{P})\left.\frac{\mathrm{d}\bar{\mathbf{s}}}{\mathrm{ds}}\right]_{\mathrm{R}}, \quad \bar{\mathbf{b}} = \bar{\mathbf{s}} \times \bar{\mathbf{n}}. \tag{2A.1}$$

This basis changes its orientation in R with angular velocity $\bar{\boldsymbol{\Omega}}_{\mathrm{R}}^{\mathrm{IB}}$. Its projection on this same basis $(\bar{\mathbf{s}}, \bar{\mathbf{n}}, \bar{\mathbf{b}})$ yields the components:

$$\left\{\bar{\boldsymbol{\Omega}}_{\mathrm{R}}^{\mathrm{IB}}\right\}_{\mathrm{IB}} = \left\{\begin{matrix} \mathrm{v}_{\mathrm{R}}(\mathbf{P})/\varsigma_{\mathrm{R}}(\mathbf{P}) \\ 0 \\ \mathrm{v}_{\mathrm{R}}(\mathbf{P})/\mathfrak{R}_{\mathrm{R}}(\mathbf{P}) \end{matrix}\right\}, \tag{2A.2}$$

where $\varsigma_{\mathrm{R}}(\mathbf{P})$ is the torsional curvature radius of the trajectory,[2] and is variable in general.

♣ *Proof*

The s-derivatives of the IB versors are:

$$\left.\frac{\mathrm{d}\bar{\mathbf{s}}}{\mathrm{ds}}\right]_{\mathrm{R}} = \frac{\bar{\mathbf{n}}}{\mathfrak{R}_{\mathrm{R}}(\mathbf{P})}, \quad \left.\frac{\mathrm{d}\bar{\mathbf{n}}}{\mathrm{ds}}\right]_{\mathrm{R}} = -\frac{\bar{\mathbf{s}}}{\mathfrak{R}_{\mathrm{R}}(\mathbf{P})} + \frac{\bar{\mathbf{b}}}{\varsigma_{\mathrm{R}}(\mathbf{P})}, \quad \left.\frac{\mathrm{d}\bar{\mathbf{b}}}{\mathrm{ds}}\right]_{\mathrm{R}} = -\frac{\bar{\mathbf{n}}}{\varsigma_{\mathrm{R}}(\mathbf{P})}. \tag{2A.3}$$

Their time derivatives (Fig. 2A.2) can be obtained by multiplying Eq. (2A.3) and the speed of **P**:

$$\left.\frac{\mathrm{d}\bar{\mathbf{s}}}{\mathrm{dt}}\right]_{\mathrm{R}} = \bar{\mathbf{n}}\frac{\mathrm{v}_{\mathrm{R}}(\mathbf{P})}{\mathfrak{R}_{\mathrm{R}}(\mathbf{P})}, \quad \left.\frac{\mathrm{d}\bar{\mathbf{n}}}{\mathrm{dt}}\right]_{\mathrm{R}} = -\bar{\mathbf{s}}\frac{\mathrm{v}_{\mathrm{R}}(\mathbf{P})}{\mathfrak{R}_{\mathrm{R}}(\mathbf{P})} + \bar{\mathbf{b}}\frac{\mathrm{v}_{\mathrm{R}}(\mathbf{P})}{\varsigma_{\mathrm{R}}(\mathbf{P})}, \quad \left.\frac{\mathrm{d}\bar{\mathbf{b}}}{\mathrm{ds}}\right]_{\mathrm{R}} = -\bar{\mathbf{n}}\frac{\mathrm{v}_{\mathrm{R}}(\mathbf{P})}{\varsigma_{\mathrm{R}}(\mathbf{P})}. \tag{2A.4}$$

These expressions show that vector $\bar{\boldsymbol{\Omega}}_{\mathrm{R}}^{\mathrm{IB}}$ is:

$$\bar{\boldsymbol{\Omega}}_{\mathrm{R}}^{\mathrm{IB}} = \bar{\mathbf{s}}\frac{\mathrm{v}_{\mathrm{R}}(\mathbf{P})}{\varsigma_{\mathrm{R}}(\mathbf{P})} + \bar{\mathbf{b}}\frac{\mathrm{v}_{\mathrm{R}}(\mathbf{P})}{\mathfrak{R}_{\mathrm{R}}(\mathbf{P})}. \tag{2A.5}$$

♣

[2] $\varsigma_{\mathrm{R}}^{-1}(\mathbf{P})$ is the torsional curvature and gives the rate of change of the osculating plane along the trajectory.

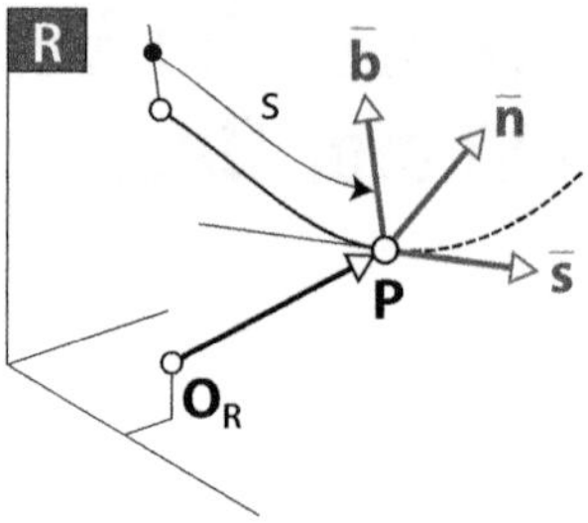

Fig. 2A.1

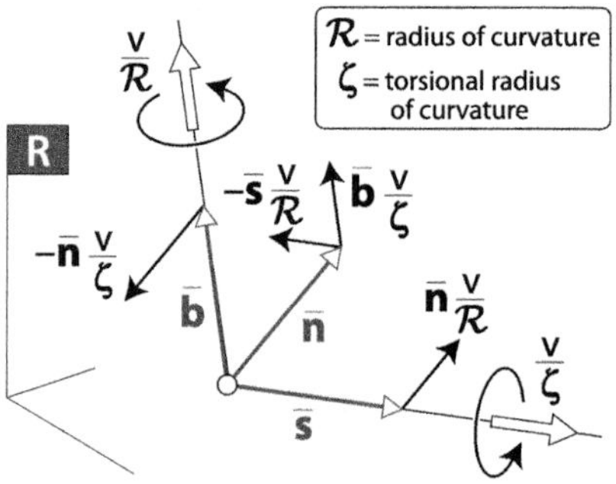

Fig. 2A.2

Appendix 2B The Inertial Guidance

The **inertial guidance** is used in many vehicles (planes, missiles, submarines, etc.). It provides the position and velocity of the vehicle relative to the earth – an information essential for the guidance maneuvers – with no need of external observations or the capture of guidance signals.

The vehicle position is associated with one of its points **P**, and its motion is determined from the measurement of vector:

$$\Delta\bar{\mathbf{a}}_{\mathrm{A}}(\mathbf{P}) = \bar{\mathbf{a}}_{\mathrm{A}}(\mathbf{P}) - \bar{\mathbf{g}}(\mathbf{P}), \tag{2B.1}$$

which is immune to external interferences. Vector $\bar{\mathbf{a}}_{\mathrm{A}}(\mathbf{P})$ is the vehicle acceleration in the astronomical reference frame A (trihedral with origin in the center of inertia of the solar system and fixed orientation relative to the distant galaxies), and vector $\bar{\mathbf{g}}(\mathbf{P})$ is the resulting field of all the gravitational attractions acting on the vehicle and coming from the earth, the moon, the sun, the planets, etc.

There are different options in terms of the results to obtain. A frequent choice is the vehicle's position and velocity relative to the earth (E), given as components in an earth-fixed basis B′:

$$\{\overline{\mathbf{G}_{\mathrm{E}}\mathbf{P}}\}_{\mathrm{B}'}, \quad \{\bar{\mathbf{v}}_{\mathrm{E}}(\mathbf{P})\}_{\mathrm{B}'}, \tag{2B.2}$$

where $\mathbf{G}_{\mathrm{E}}$ is the earth center of inertia.

The earth reference frame E moves relative to the astronomical reference frame A. The acceleration of its origin $\bar{\mathbf{a}}_{\mathrm{A}}(\mathbf{G}_{\mathrm{E}})$ comes from the gravitational attractions of the sun, the moon, and the planets, and it rotates with an angular velocity $\bar{\mathbf{\Omega}}_{\mathrm{A}}^{\mathrm{E}}$ about the earth axis (south–north direction). The $\bar{\mathbf{\Omega}}_{\mathrm{A}}^{\mathrm{E}}$ value (nearly constant) is one turn per sidereal day (the time it takes to rotate about its axis relative to the distant galaxies; its duration is studied in appendix 1B of *Rigid Body Dynamics* [Cambridge University Press, forthcoming]).

Another frequent choice is to give the measurement in Eq. (2B.1) projected on a vector basis B with constant orientation relative to the astronomical reference frame A. We have to solve the transformation:

$$\{\bar{\mathbf{a}}_{\mathrm{A}}(\mathbf{P}) - \bar{\mathbf{g}}(\mathbf{P})\}_{\mathrm{B}} \quad \rightarrow \quad \left(\{\overline{\mathbf{G}_{\mathrm{E}}\mathbf{P}}\}_{\mathrm{B}'}, \{\bar{\mathbf{v}}_{\mathrm{E}}(\mathbf{P})\}_{\mathrm{B}'}\right), \tag{2B.3}$$

which involves the vector bases B and B′ with axes $(1,2,3)$ and $(1',2',3')$ (Fig. 2B.1).

This calculation can be done in five steps:

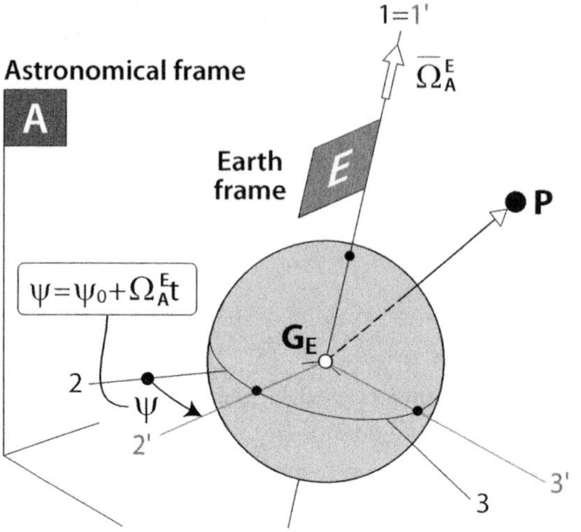

Fig. 2B.1

1. **Transformation to the earth-fixed basis** $\{\Delta\bar{\mathbf{a}}_A(\mathbf{P})\}_B \rightarrow \{\Delta\bar{\mathbf{a}}_A(\mathbf{P})\}_{B'}$

$$\{\Delta\bar{\mathbf{a}}_A(\mathbf{P})\}_{B'} = [S]\{\Delta\bar{\mathbf{a}}_A(\mathbf{P})\}_B \ , \quad [S] = \begin{bmatrix} 1 & 0 & 0 \\ 0 & \cos\left(\psi_0 + \Omega^E_A t\right) & \sin\left(\psi_0 + \Omega^E_A t\right) \\ 0 & -\sin\left(\psi_0 + \Omega^E_A t\right) & \cos\left(\psi_0 + \Omega^E_A t\right) \end{bmatrix}. \tag{2B.4}$$

As Ω^E_A is a well-known constant value, we only need the initial angle ψ_0 and the time interval given by a chronometer to determine the transformation matrix [S].

2. **Calculation of** $\{\bar{\mathbf{a}}_A(\mathbf{P})\}_{B'}$

$$\{\bar{\mathbf{a}}_A(\mathbf{P})\}_{B'} = \{\bar{\mathbf{a}}_A(\mathbf{P}) - \bar{\mathbf{g}}(\mathbf{P})\}_{B'} + \{\bar{\mathbf{g}}(\mathbf{P})\}_{B'}. \tag{2B.5}$$

Vector $\{\bar{\mathbf{g}}(\mathbf{P})\}_{B'}$ can be calculated at each time instant from the position of **P** relative to the earth – which is one of the results obtained – and the positions of the sun, moon and planets, which are perfectly predictable.

3. **Calculation of** $\{\bar{\mathbf{a}}_E(\mathbf{P})\}_{B'}$

If we take the astronomical reference frame as AB and the earth frame as REL:

$$\{\bar{\mathbf{a}}_E(\mathbf{P})\}_{B'} = \{\bar{\mathbf{a}}_A(\mathbf{P})\}_{B'} - \{\bar{\mathbf{a}}_{tr}(\mathbf{P})\}_{B'} - \{\bar{\mathbf{a}}_{Cor}(\mathbf{P})\}_{B'}. \tag{2B.6}$$

The acceleration $\bar{\mathbf{a}}_{tr}(\mathbf{P})$ depends on the **P** position relative to the Earth – one of the results obtained – and other known vectors. The $\bar{\mathbf{a}}_{Cor}(\mathbf{P})$ term depends on $\bar{\mathbf{v}}_E(\mathbf{P})$ – also one of the results obtained – and of $\bar{\mathbf{\Omega}}^E_A$(also known).

4. **Calculation of** $\{\bar{\mathbf{v}}_E(\mathbf{P})\}_{B'}$

As we are using an earth-fixed basis, the $\bar{\mathbf{v}}_E(\mathbf{P})$ components be found directly through the integration of the $\bar{\mathbf{a}}_E(\mathbf{P})$ components:

$$\{\bar{\mathbf{v}}_E(\mathbf{P})\}_{B'} = \left\{\bar{\mathbf{v}}_E(\mathbf{P})\right]_{t=0}\}_{B'} + \int_0^t \{\bar{\mathbf{a}}_E(\mathbf{P})\}_{B'} dt. \tag{2B.7}$$

5. **Calculation of $\left\{\overline{\mathbf{G_E P}}\right\}_{B'}$**

For the same reason as before:

$$\left\{\overline{\mathbf{G_E P}}\right\}_{B'} = \left\{\overline{\mathbf{G_E P}}\Big]_{t=0}\right\}_{B'} + \int_0^t \{\bar{\mathbf{v}}_E(\mathbf{P})\}_{B'} dt. \quad (2B.8)$$

Once the projection of the vectors $\overline{\mathbf{G_E P}}$ and $\bar{\mathbf{v}}_E(\mathbf{P})$ on the B′ basis are known, we can calculate $\bar{\mathbf{g}}(\mathbf{P})$, $\bar{\mathbf{a}}_{tr}(\mathbf{P})$ y $\bar{\mathbf{a}}_{Cor}(\mathbf{P})$. The calculation of $\bar{\mathbf{g}}(\mathbf{P})$ and $\bar{\mathbf{a}}_{tr}(\mathbf{P})$ calls for the knowledge over time of the position of the stars whose gravitational field has been taken into account.

As the results $\overline{\mathbf{G_E P}}$ and $\bar{\mathbf{v}}_E(\mathbf{P})$ (in the B′ basis) appear in the calculation, we need an approximation. It can be either a value known for a previous time instant $(t - \Delta t)$ or that value corrected through one of the usual numerical integration algorithms that improve the accuracy of the calculation.

Quiz Questions

For the sake of brevity, the following assumptions are made throughout the whole collection of questions unless stated otherwise:

- Threads, ropes, strings, and cables are inextensible
- The wheels of the vehicles do not slide on the ground
- Skis do not have a lateral translational motion

The ground reference frame is always denoted as E, and the calculations refer to the configuration shown in the figure. All data are declared in the figures, but only part of them are declared in the text.

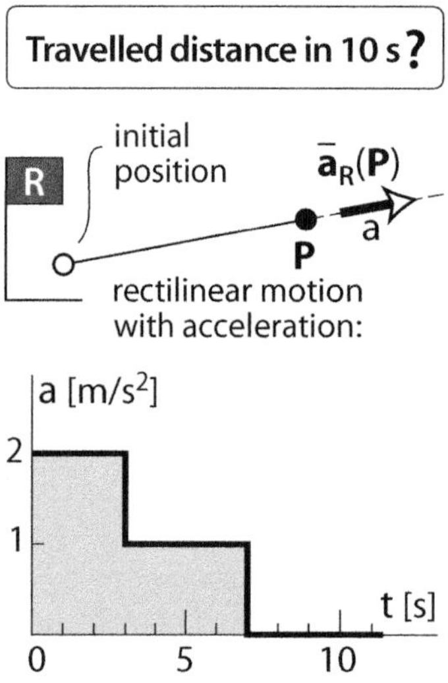

2.1 The acceleration of the rectilinear motion of point **P** in R is stepwise. What is the traveled distance in R after 10 s?

A 88 m
B 59 m
C 71 m
D 100 m
E 0 m

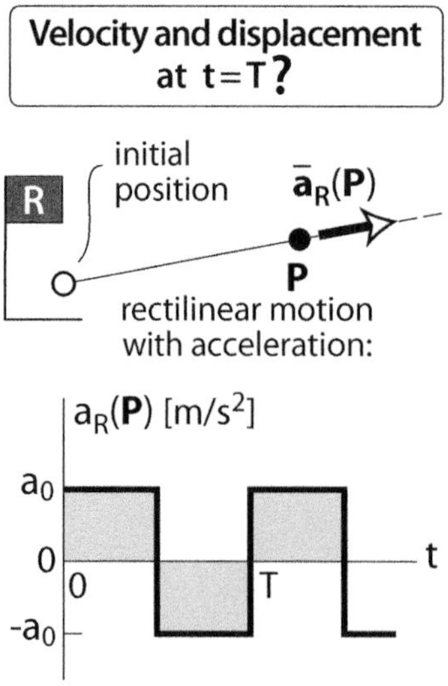

2.2 The acceleration of the rectilinear motion of point **P** in R is a square wave. What are the speed and traveled distance in R at t = T?

	speed	distance
A	0	0
B	a_0T	$(1/4)a_0T^2$
C	0	$(1/4)a_0T^2$
D	a_0T	0
E	0	$(1/2)a_0T^2$

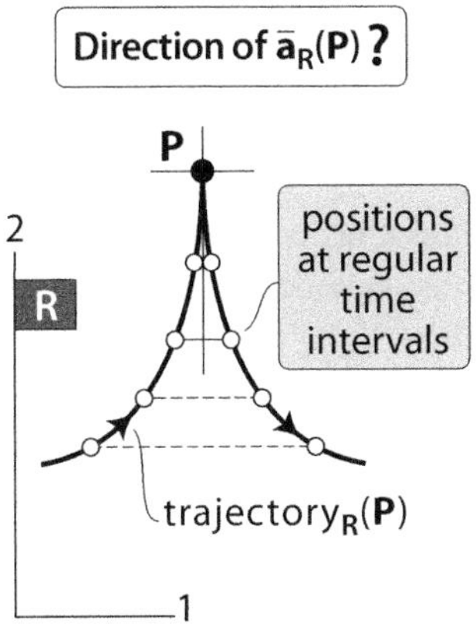

2.3 Point **P** describes a curved planar trajectory in R. Its locations at regular time intervals are marked with white dots. What is the direction of $\bar{\mathbf{a}}_R(\mathbf{P})$ at the shown location**?**

A It is indeterminate because the trajectory contains a discontinuity
B It is zero
C It has the direction of axis 2, opposite to the advance of **P** along the trajectory
D It has the direction of the positive semiaxis 2
E It has the direction of the negative semiaxis 2

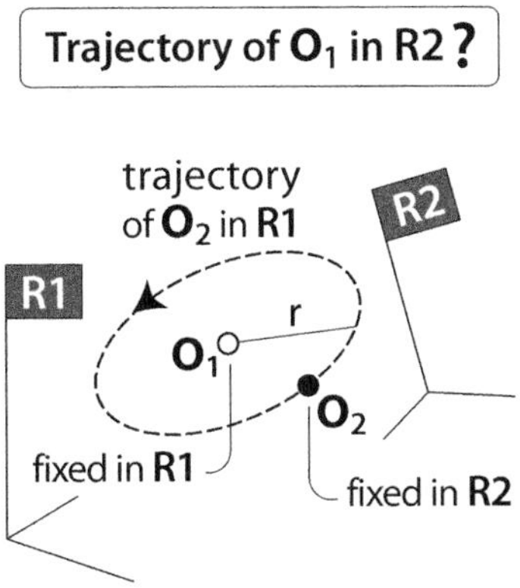

2.4 What is the shape of the $\mathbf{O_1}$ trajectory in R2?

A Any shape on a sphere with radius r
B Any shape except rectilinear
C Circular with radius $\neq$ r
D Circular with radius = r
E It depends on the vector basis used to describe the $\mathbf{O_1}$ position vector

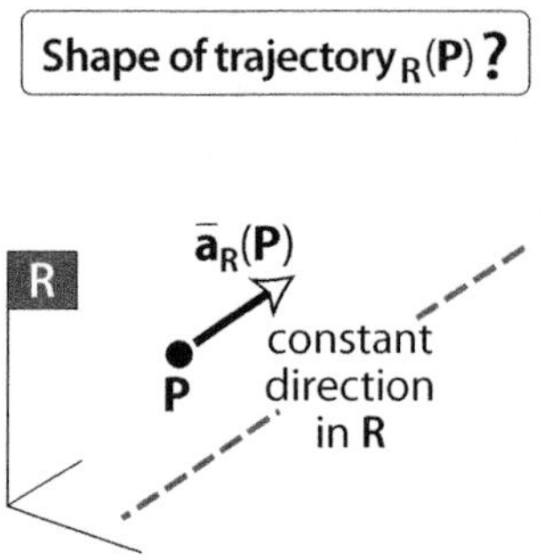

2.5 The acceleration $\bar{\mathbf{a}}_R(\mathbf{P})$ has a constant direction. What is the shape of the **P** trajectory in R?

A Rectilinear
B Parabolic
C Planar but not necessarily rectilinear or parabolic
D It tends asymptotically to the direction of the acceleration
E Any shape because it is not constrained

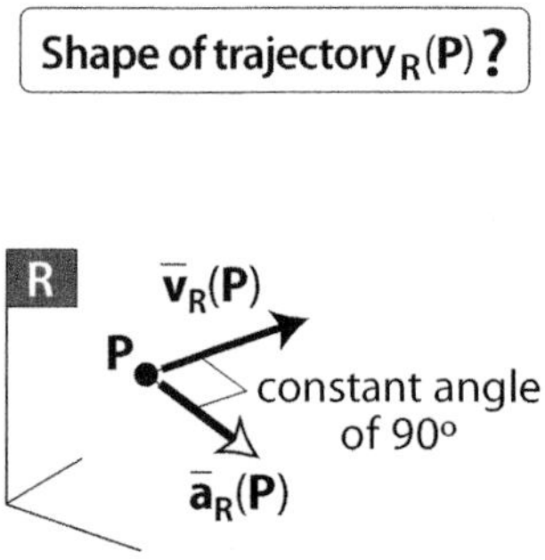

2.6 The vectors $\bar{\mathbf{v}}_R(\mathbf{P})$ and $\bar{\mathbf{a}}_R(\mathbf{P})$ define constantly a right angle. What is the shape of the **P** trajectory in R?

A Circular
B Planar but not necessarily circular or spiral
C Any curved shape on a spherical surface
D Any shape because it is not constrained
E Spiral

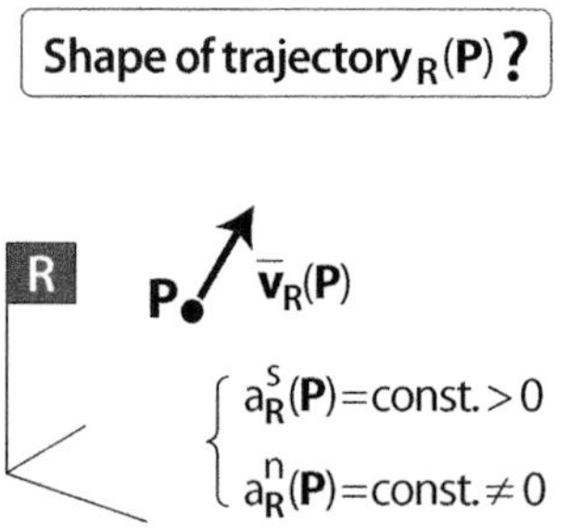

2.7 The intrinsic components $a_R^s(\mathbf{P})$ and $a_R^n(\mathbf{P})$ are constant and nonzero. If $a_R^s(\mathbf{P}) > 0$, what is the shape of the **P** trajectory in R?

A It is a planar trajectory but not necessarily circular or rectilinear
B A shape with increasing curvature radius along the trajectory
C It is circular
D It is rectilinear
E It is helical

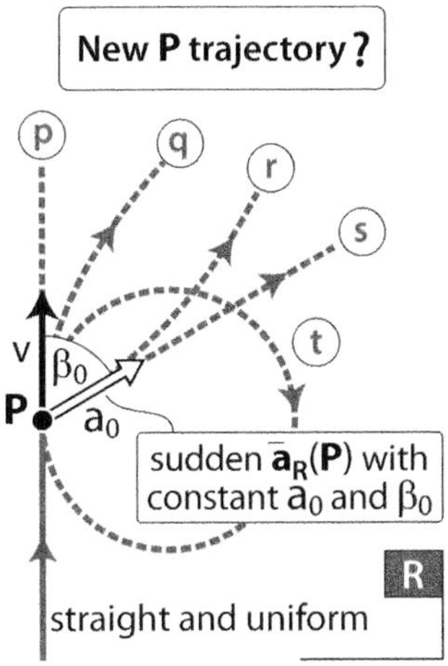

2.8 The initial motion of **P** in R is uniform and rectilinear. Suddenly it acquires an acceleration with constant value a_0 and defining a constant angle β_0 with the velocity $\bar{\mathbf{v}}_R(\mathbf{P})$. What will be the new trajectory of **P** in R?

A p
B q
C r
D s
E t

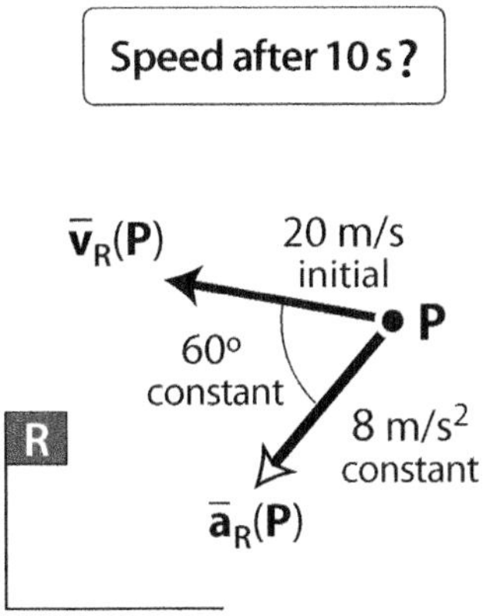

2.9 The initial speed of point **P** relative to R is 20 m/s. What will be the speed 10 s later?

A 100 m/s
B 26.93 m/s
C 20 m/s
D 60 m/s
E Not enough data to calculate it.

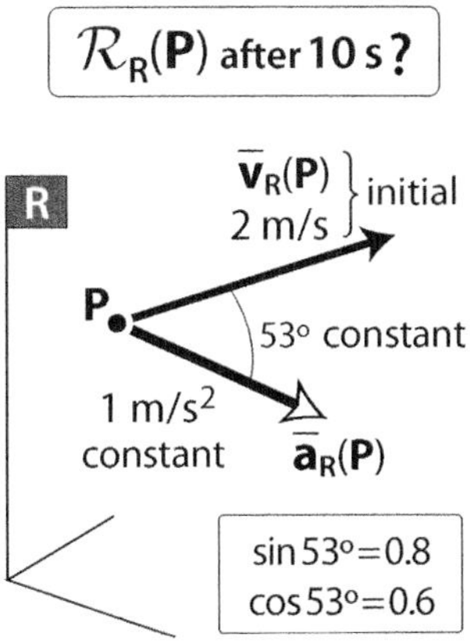

2.10 The initial speed of point **P** relative to R is 2 m/s. What will be $\mathfrak{R}_R(\mathbf{P})$ 10 s later?

A 167 m
B 100 m
C 80 m, provided that $\bar{\mathbf{a}}_R(\mathbf{P})$ is constrained to a planar surface
D 80 m, independent from the orientation changes of $\bar{\mathbf{a}}_R(\mathbf{P})$
E 60 m, provided that $\bar{\mathbf{a}}_R(\mathbf{P})$ is constrained to a planar surface

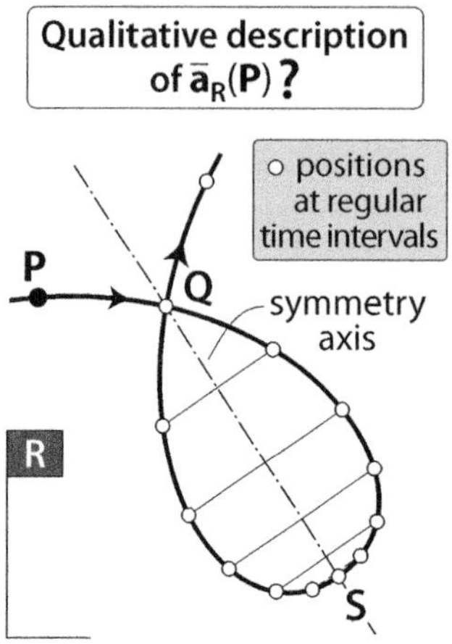

2.11 Point **P** describes a curved planar trajectory in R. Its locations at regular time intervals are marked with white dots. What can be concluded about $\bar{\mathbf{a}}_R(\mathbf{P})$?

A It is the same for symmetrical points
B It is symmetrical for symmetrical points
C It is opposite for symmetrical points
D It is the same only in locations **Q**
E It is zero at location **S**

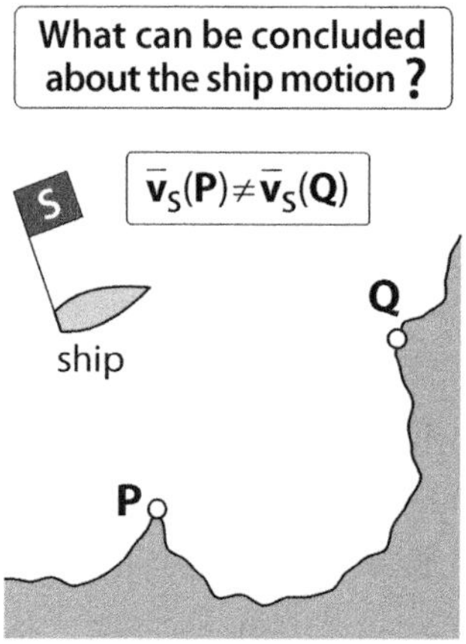

2.12 Two lighthouses **P** and **Q** on the coastline (modeled as particles) have different velocities relative to a ship S ($\bar{\mathbf{v}}_S(\mathbf{P}) \neq \bar{\mathbf{v}}_S(\mathbf{Q})$). We can conclude that:

A The ship maintains its orientation relative to the coast
B The ship is changing its orientation relative to the coast
C It is impossible because two points fixed to the coast will always have the same velocity relative to the ship
D It is always the case because the distances between each lighthouse and the ship are different
E It is always true for certain observation points

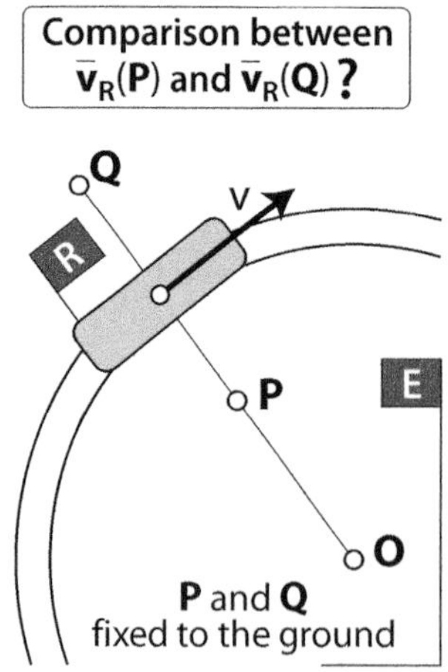

2.13 A train car (R) moves relative to the ground (E) on a circular railway with center **O** (fixed in E). **P** and **Q** are two points fixed to the ground and equally spaced from R. How do $\bar{\mathbf{v}}_R(\mathbf{P})$ and $\bar{\mathbf{v}}_R(\mathbf{Q})$ compare?

A $|v_R(\mathbf{P})| < |v_R(\mathbf{Q})|$
B $|v_R(\mathbf{P})| = |v_R(\mathbf{Q})| \neq 0$, but $\bar{\mathbf{v}}_R(\mathbf{P}) \neq \bar{\mathbf{v}}_R(\mathbf{Q})$
C $|v_R(\mathbf{P})| > |v_R(\mathbf{Q})|$
D $|v_R(\mathbf{P})| = |v_R(\mathbf{Q})| = 0$
E $\bar{\mathbf{v}}_R(\mathbf{P}) = \bar{\mathbf{v}}_R(\mathbf{Q}) \neq 0$

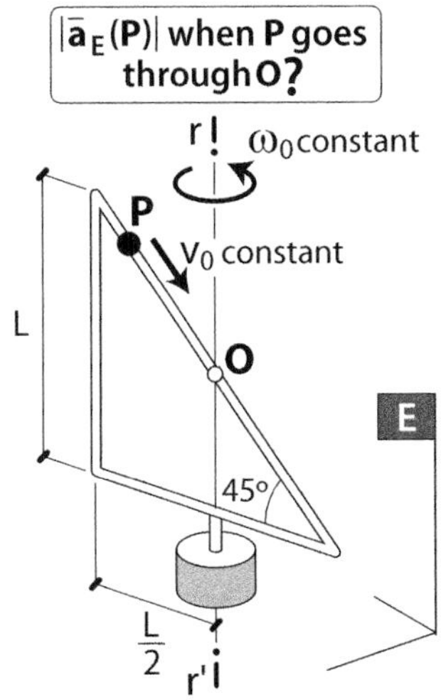

2.14 What is the acceleration modulus of point **P** relative to the ground (E) when it goes through **O**?

A 0
B $2v_0\omega_0$
C $\sqrt{2}v_0\omega_0$
D $2L\omega_0^2$
E $\sqrt{2}L\omega_0^2$

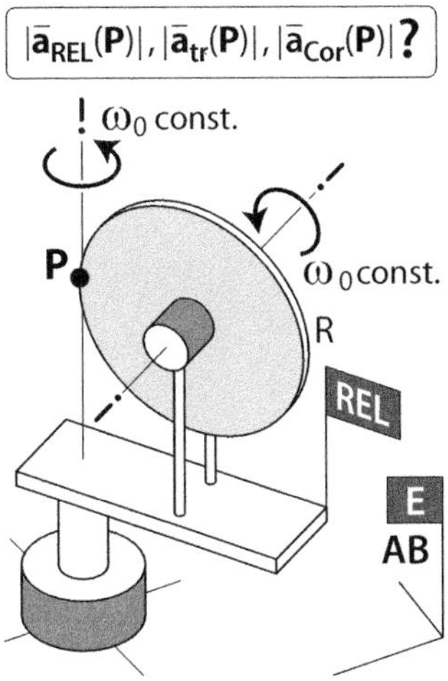

2.15 What are the modules of $\bar{\mathbf{a}}_{REL}, \bar{\mathbf{a}}_{tr}, \bar{\mathbf{a}}_{Cor}$ of point **P** of the disk just when it goes through the location shown in the figure?

	$\lvert\bar{\mathbf{a}}_{REL}\rvert$	$\lvert\bar{\mathbf{a}}_{tr}\rvert$	$\lvert\bar{\mathbf{a}}_{Cor}\rvert$
A	$R\omega_0^2$	$R\omega_0^2$	$2R\omega_0^2$
B	0	$R\omega_0^2$	$2R\omega_0^2$
C	$R\omega_0^2$	0	0
D	$R\omega_0^2$	$2R\omega_0^2$	0
E	$R\omega_0^2$	$R\omega_0^2$	0

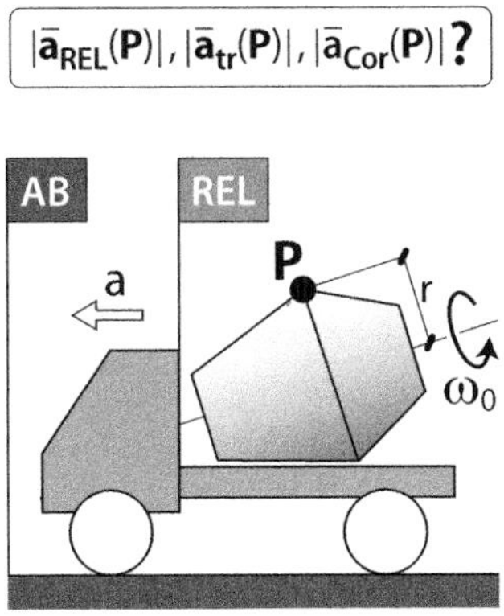

2.16 What are the modules of $\bar{\mathbf{a}}_{REL}, \bar{\mathbf{a}}_{tr}, \bar{\mathbf{a}}_{Cor}$ of point **P** of the concrete mixer tank when it goes through the highest position?

	$\lvert\bar{\mathbf{a}}_{REL}\rvert$	$\lvert\bar{\mathbf{a}}_{tr}\rvert$	$\lvert\bar{\mathbf{a}}_{Cor}\rvert$
A	$r\omega_0^2$	$a + r\omega_0^2$	$2r\omega_0^2$
B	$r\omega_0^2$	a	$2r\omega_0^2$
C	$r\omega_0^2$	a	0
D	$r\omega_0^2$	0	$2r\omega_0^2$
E	0	$a + r\omega_0^2$	$2r\omega_0^2$

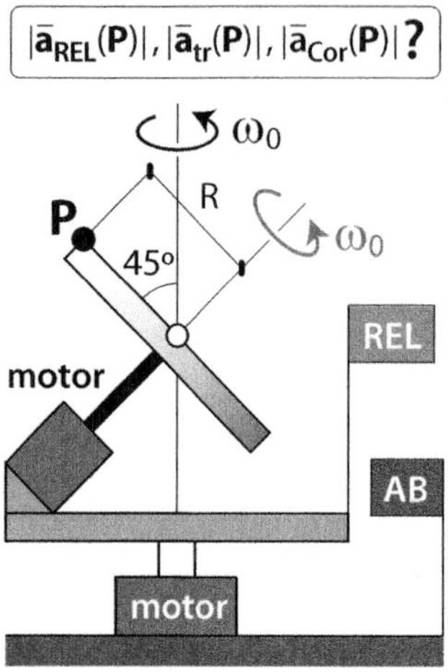

2.17 The disk with radius R rotates with constant angular velocity ω_0 relative to the support, which in turn rotates with constant angular velocity ω_0 relative to the ground. What are the modules of $\bar{\mathbf{a}}_{REL}$, $\bar{\mathbf{a}}_{tr}$, $\bar{\mathbf{a}}_{Cor}$ of point **P** of the disk when it goes through the highest position?

	$\lvert\bar{\mathbf{a}}_{REL}\rvert$	$\lvert\bar{\mathbf{a}}_{tr}\rvert$	$\lvert\bar{\mathbf{a}}_{Cor}\rvert$
A	$R\omega_0^2$	$R\omega_0^2/\sqrt{2}$	$2R\omega_0^2$
B	0	$R\omega_0^2/\sqrt{2}$	$R\omega_0^2/\sqrt{2}$
C	$R\omega_0^2$	0	$2R\omega_0^2$
D	$R\omega_0^2$	$\sqrt{2}R\omega_0^2$	$2\sqrt{2}R\omega_0^2$
E	$R\omega_0^2$	$R\omega_0^2$	$\sqrt{2}R\omega_0^2$

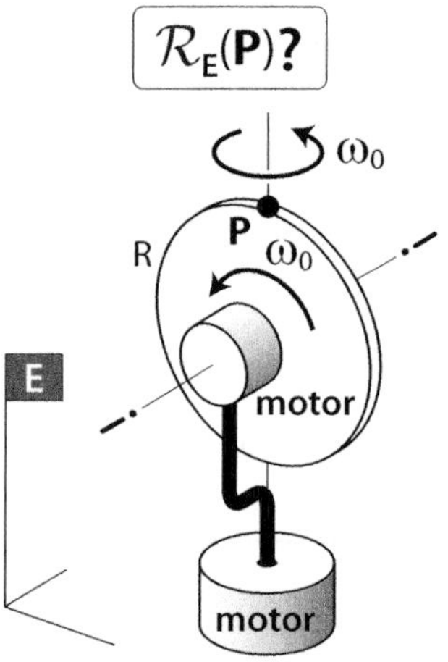

2.18 What is the value of the radius of curvature $\mathfrak{R}_E(\mathbf{P})$ when point **P** of the disk goes through the highest position?

A 0
B R
C R/2
D $R/\sqrt{2}$
E $R/\sqrt{5}$

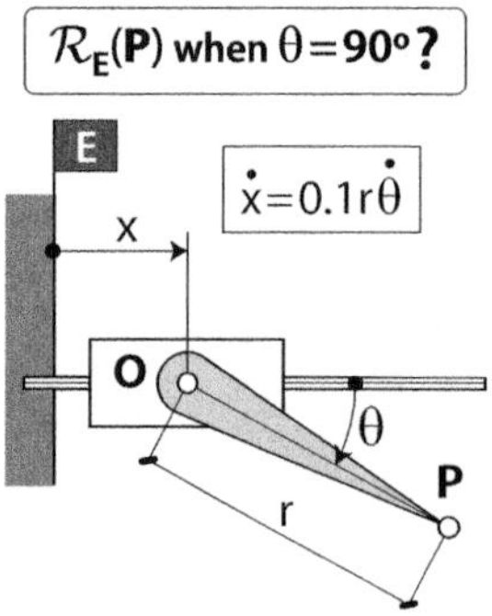

2.19 If $\dot{x} = 0.1r\dot{\theta}$, what is the value of the radius of curvature $\mathfrak{R}_E(\mathbf{P})$ when point **P** of the pendulum goes through the lowest position?

A 1.10 r
B r
C 1.21 r
D 0.81 r
E 0.90 r

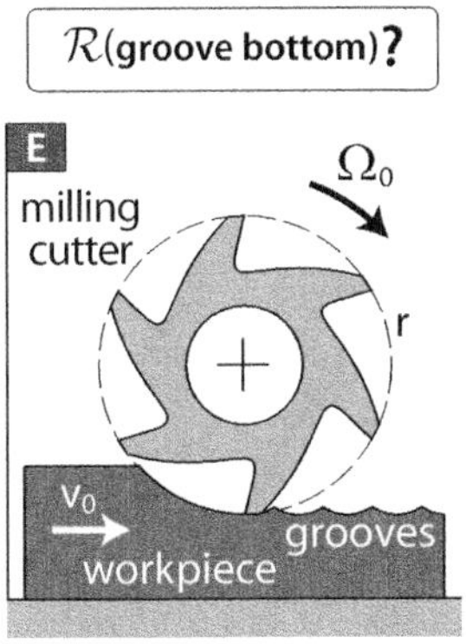

2.20 What is the value of the curvature radius of the lowest point in a groove?

A r
B $r(1 + v_0/(r\Omega_0))$
C $r(1 - v_0/(r\Omega_0))$
D $r(1 + v_0/(r\Omega_0))^2$
E $r(1 - v_0/(r\Omega_0))^2$

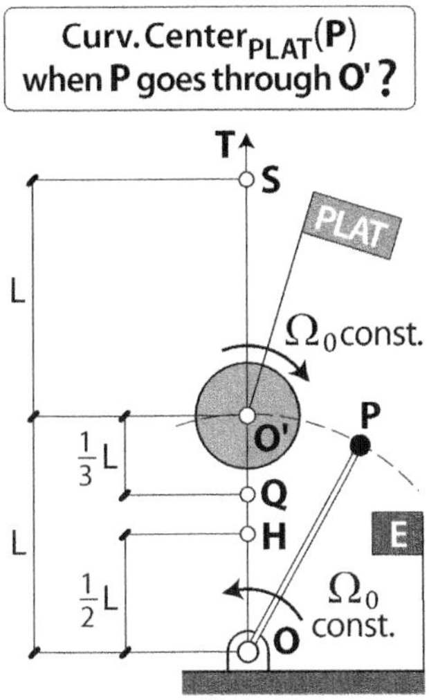

2.21 Where is the **P** curvature center relative to the platform (PLAT) when **P** goes through the platform center **O′** (fixed to the ground E)?

A **O**
B **H**
C **Q**
D **S**
E **T**

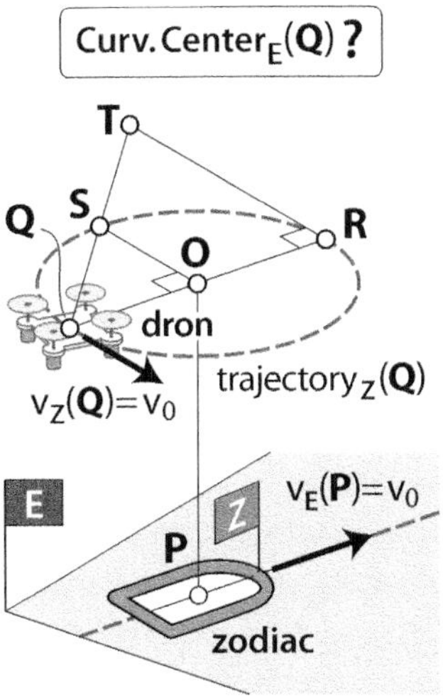

2.22 Point **Q** of a dron describes a circular trajectory relative to a zodiac (Z). Where is the **Q** curvature center relative to the ground (E) (points **O**, **Q**, **R**, **S**, **T** are coplanar)?

A **O**
B **P**
C **R**
D **S**
E **T**

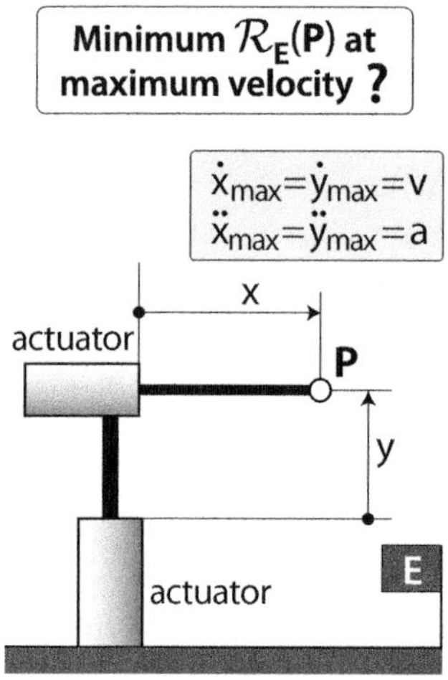

2.23 What is the minimum curvature radius of **P** relative to the ground (E) when it moves with maximum speed?

A v^2/a
B $\sqrt{2}v^2/a$
C $v^2/(\sqrt{2}a)$
D $2v^2/a$
E $v^2/(2a)$

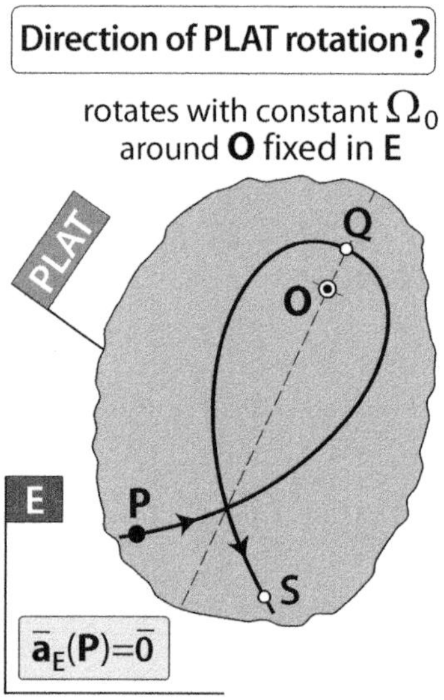

2.24 A particle **P** slides on the frictionless platform (which implies $\bar{\mathbf{a}}_E(\mathbf{P}) = \bar{\mathbf{0}}$) and describes a curved trajectory relative to it which goes through points **Q** and **S**. What is the direction of $\bar{\mathbf{\Omega}}_E^{PLAT}$?

A Clockwise
B Counterclockwise
C Not enough data
D Counterclockwise from **P** to **Q**, clockwise from **Q** to **S**
E Clockwise from **P** to **Q**, counterclockwise from **Q** to **S**

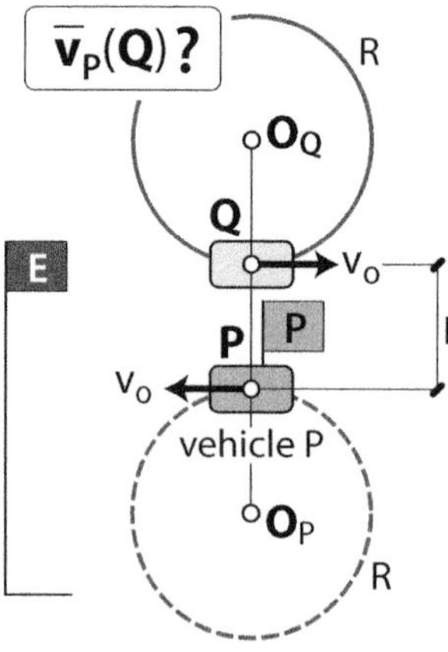

2.25 Points **P** and **Q** of the two vehicles describe a circular trajectory relative to the ground (E). What is the velocity $\bar{\mathrm{v}}_P(\mathbf{Q})$?

A $3v_0 \rightarrow$
B $2v_0 \leftarrow$
C $v_0 \leftarrow$
D $4v_0 \leftarrow$
E 0

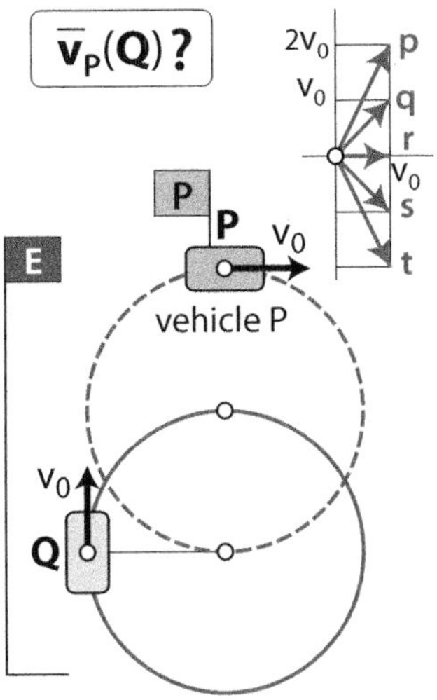

2.26 Points **P** and **Q** of the two vehicles describe a circular trajectory relative to the ground (E). What is the velocity $\bar{v}_P(\mathbf{Q})$?

A **p**
B **q**
C **r**
D **s**
E **t**

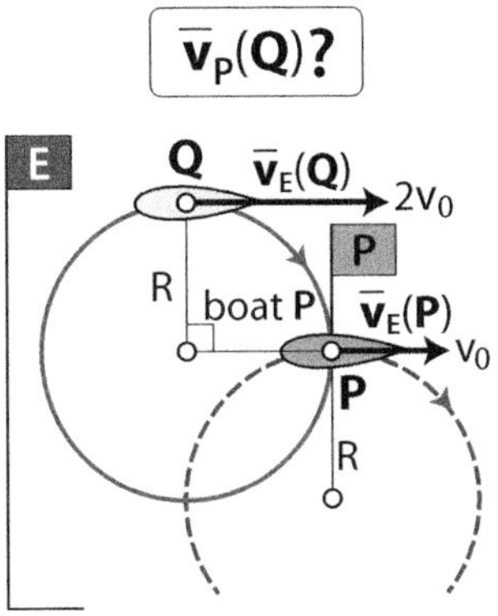

2.27 Points **P** and **Q** of the two boats describe a circular trajectory relative to the ground (E). What is the velocity $\bar{v}_P(\mathbf{Q})$?

A 0
B $\sqrt{2}v_0 \nearrow$
C $3v_0 \rightarrow$
D $v_0 \rightarrow$
E $v_0 \downarrow$

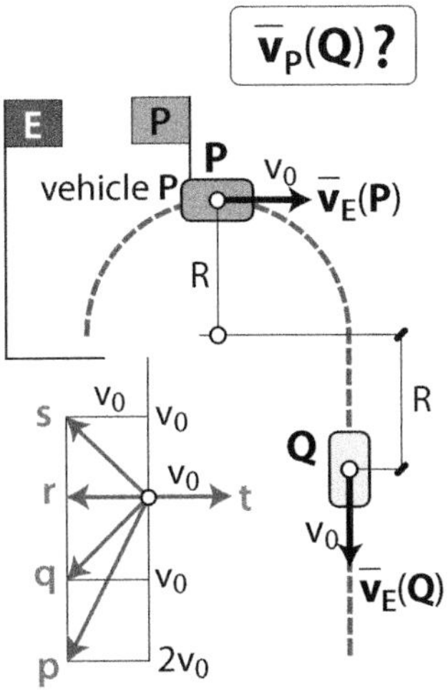

2.28 Points **P** and **Q** of the two vehicles move relative to the ground (E). What is the velocity $\bar{v}_P(\mathbf{Q})$?

A **p**
B **q**
C **r**
D **s**
E **t**

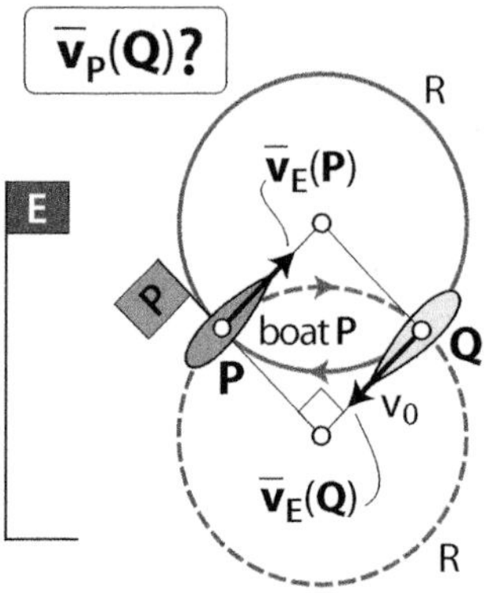

2.29 Points **P** and **Q** of the two boats describe a circular trajectory relative to the ground (E). What is the velocity $\bar{v}_P(\mathbf{Q})$?

A $\sqrt{2}v_0 \rightarrow$
B $\sqrt{2}v_0 \downarrow$
C $v_0 \swarrow$
D $2v_0 \swarrow$
E $\sqrt{2}v_0 \leftarrow$

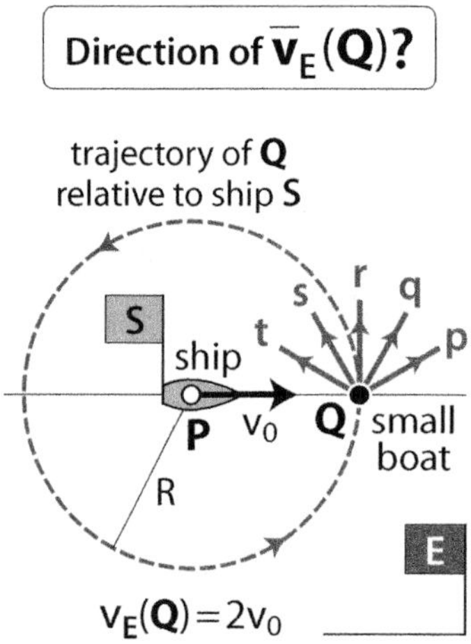

2.30 The ship S has a uniform rectilinear motion relative to the ground (E). Point **Q** has a circular motion with radius R relative to S and keeps a constant speed $2v_0$ relative to E. What is the direction of $\bar{v}_E(\mathbf{Q})$?

A **p**
B **q**
C **r**
D **s**
E **t**

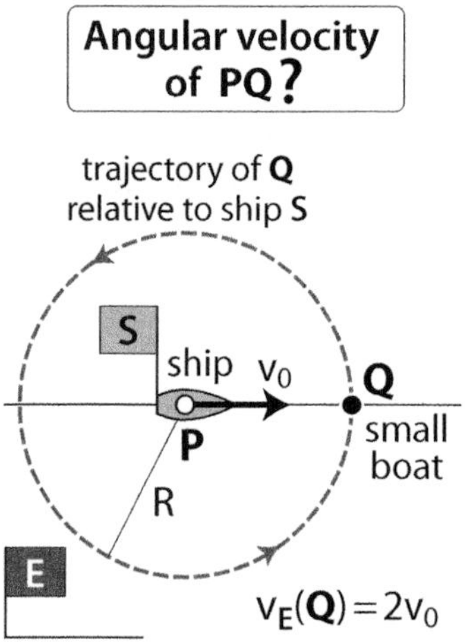

2.31 The ship S has a uniform rectilinear motion relative to the ground (E). Point **Q** has a circular motion with radius R relative to S and keeps a constant speed $2v_0$ relative to E. What is the module of the angular velocity $\bar{\Omega}_S^{PQ}$?

A v_0/R
B $2v_0/R$
C $v_0/(\sqrt{3}R)$
D $\sqrt{3}v_0/R$
E $2v_0/(\sqrt{3}R)$

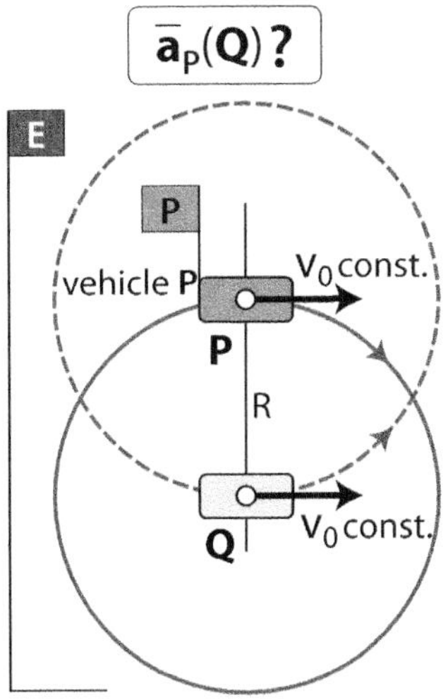

2.32 Points **P** and **Q** of the two vehicles describe a circular trajectory with constant speed v_0 relative to the ground (E). What is the acceleration $\bar{\mathbf{a}}_P(\mathbf{Q})$?

A 0
B $(v_0^2/R)\uparrow$
C $(v_0^2/R)\downarrow$
D $(2v_0^2/R)\uparrow$
E $(3v_0^2/R)\uparrow$

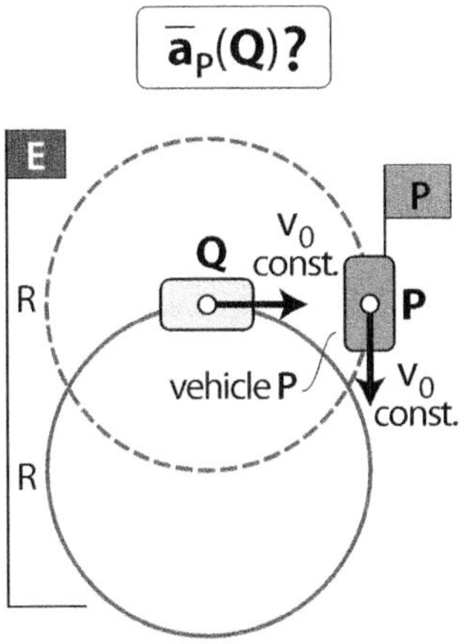

2.33 Points **P** and **Q** of the two vehicles describe a circular trajectory with constant speed v_0 relative to the ground (E). What is the acceleration $\bar{\mathbf{a}}_P(\mathbf{Q})$?

A $(v_0^2/R)\leftarrow$
B $(v_0^2/R)\uparrow$
C $(v_0^2/R)\rightarrow$
D $(v_0^2/R)\downarrow$
E $3(v_0^2/R)\downarrow$

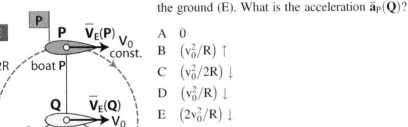

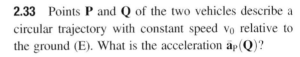

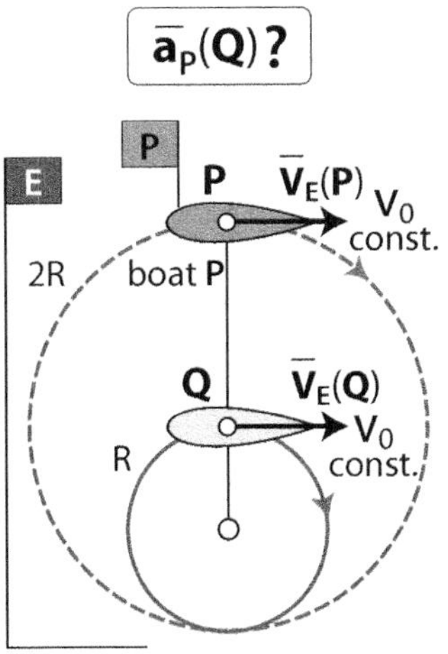

2.34 Points **P** and **Q** of the two boats describe a circular trajectory with constant speed v_0 relative to the ground (E). What is the acceleration $\bar{\mathbf{a}}_P(\mathbf{Q})$?

A 0
B $(v_0^2/R)\uparrow$
C $(v_0^2/2R)\downarrow$
D $(v_0^2/R)\downarrow$
E $(2v_0^2/R)\downarrow$

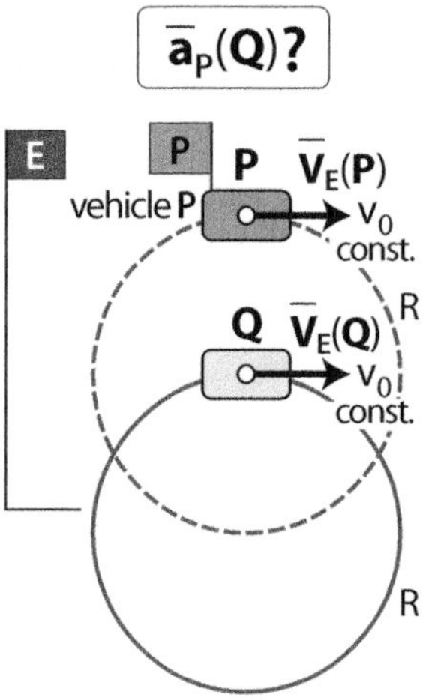

2.35 Points **P** and **Q** of the two vehicles describe a circular trajectory with constant speed v_0 relative to the ground (E). What is the acceleration $\bar{\mathbf{a}}_P(\mathbf{Q})$?

A 0
B $(2v_0^2/R)\uparrow$
C $(v_0^2/R)\uparrow$
D $(v_0^2/R)\downarrow$
E $(2v_0^2/R)\downarrow$

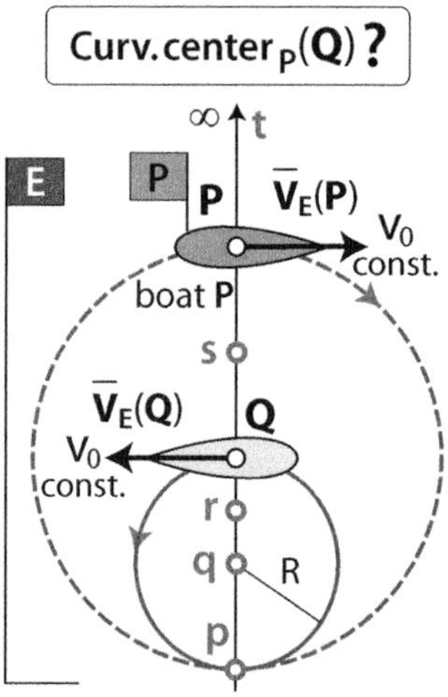

2.36 Points **P** and **Q** of the two boats describe a circular trajectory with constant speed v_0 relative to the ground (E). What is the center of curvature of **Q** relative to boat P?

A **p**
B **q**
C **r**
D **s**
E **t**

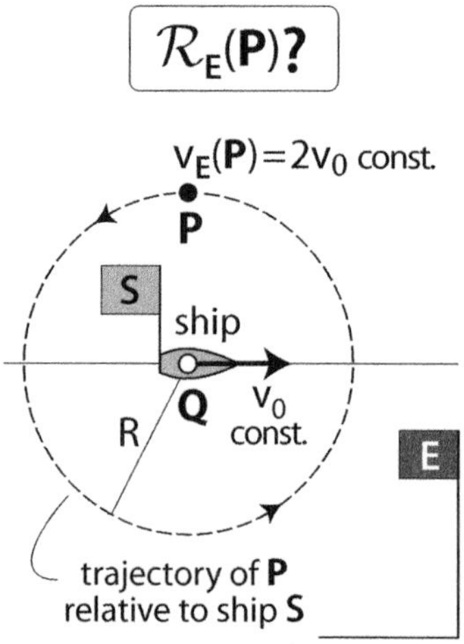

2.37 The ship S has a uniform rectilinear motion relative to the ground (E). Point **P** has a circular motion with radius R relative to S and keeps a constant speed $2v_0$ relative to E. What is the radius of curvature $\Re_E(\mathbf{P})$ when $v_S(\mathbf{P})$ goes through its maximum value?

A 0
B (1/2)R
C (2/3)R
D (4/3)R
E (4/9)R

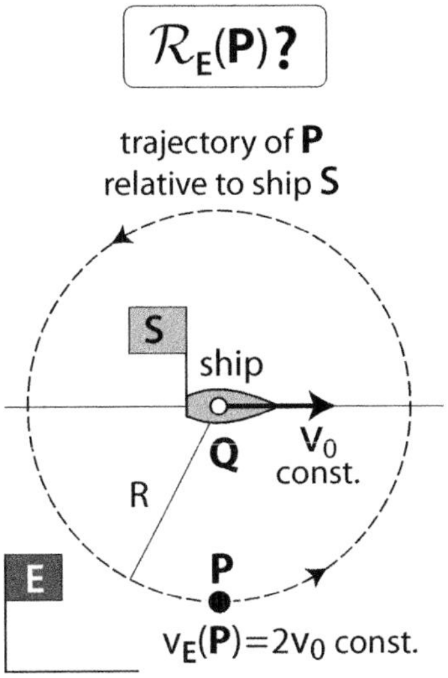

2.38 The ship S has a uniform rectilinear motion relative to the ground (E). Point **P** has a circular motion with radius R relative to S and keeps a constant speed $2v_0$ relative to E. What is the radius of curvature $\mathfrak{R}_E(\mathbf{P})$ at this particular configuration?

A R
B R/2
C 2R
D 3R
E 4R

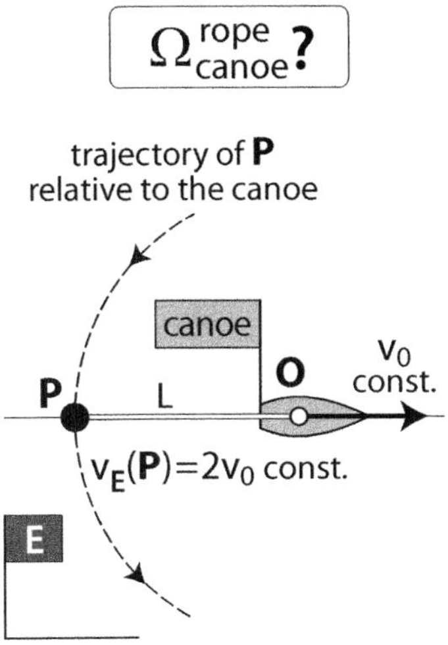

2.39 The canoe has a uniform rectilinear motion relative to the ground (E). A person **P** (modeled as a particle) holds the end point of a taut rope attached to the canoe and has a constant speed $2v_0$ relative to E. What is the value of the angular velocity $\bar{\mathbf{\Omega}}^{rope}_{canoe}$?

A v_0/L
B $2(v_0/L)$
C $(v_0/L)/\sqrt{3}$
D $\sqrt{3}(v_0/L)$
E $(2v_0/L)/\sqrt{3}$

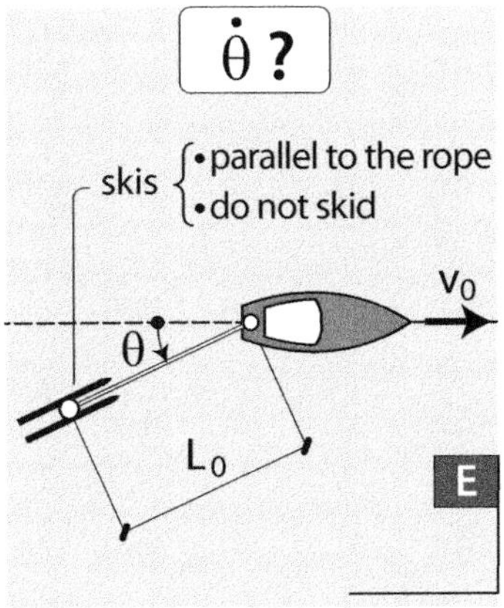

2.40 The canoe has a rectilinear motion relative to the ground (E). The skis are parallel to the taut rope attached to the canoe. What is the value of the change of orientation $\dot{\theta}$?

A $+(v_0/L)\sin\theta$
B $+(v_0/L)$
C $-(v_0/L)\sin\theta$
D $-(v_0/L)$
E This is an impossible motion

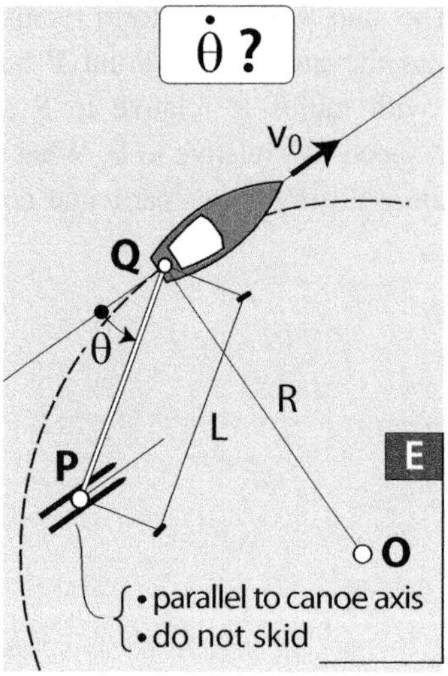

2.41 Point **O** of the canoe has a uniform circular motion relative to the ground (E). The canoe drags a skier **P** (modeled as a particle) through a taut rope. What is the rate of change $\dot{\theta}$ of angle θ?

A $\dot{\theta} = 0$
B $\dot{\theta} = v_0/R$
C $\dot{\theta} = -(v_0/R)$
D $\dot{\theta} = -(v_0/L)\sin\theta$
E $\dot{\theta} = (v_0/L)\sin\theta$

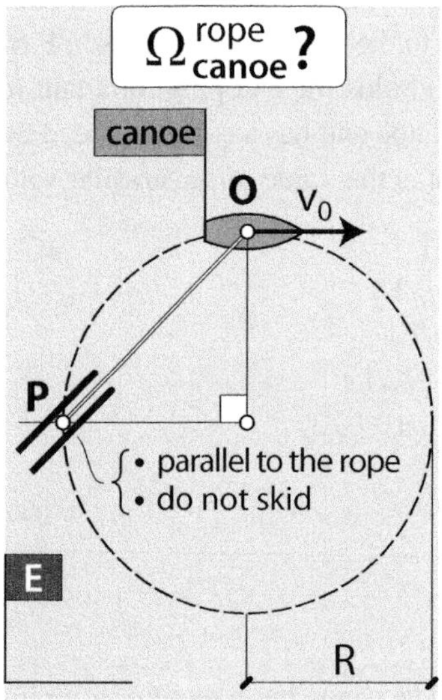

2.42 Point **O** of the canoe has a uniform circular motion relative to the ground (E). The canoe drags a skier **P** (modeled as a particle) through a taut rope. What is the rope angular velocity relative to the canoe $\left(\Omega^{rope}_{canoe}\right)$ at this particular configuration?

A 0
B $v_0/\left(\sqrt{2}R\right)$, clockwise direction
C $v_0/\left(\sqrt{2}R\right)$, counterclockwise direction
D $v_0/(2R)$, clockwise direction
E $v_0/(2R)$, counterclockwise direction

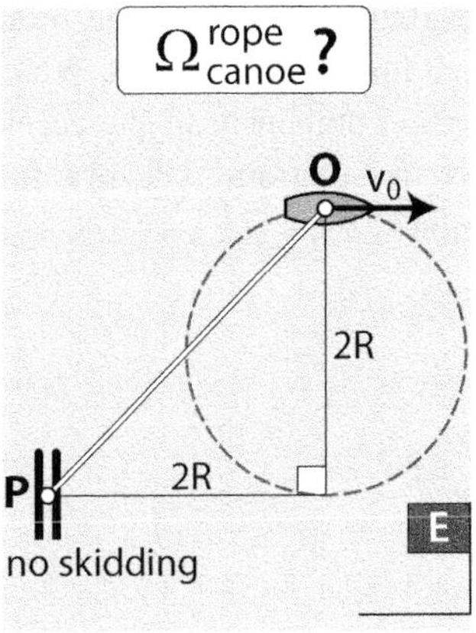

2.43 Point **O** of the canoe has a uniform circular motion relative to the ground (E). The canoe drags a skier **P** (modeled as a particle) through a taut rope. What is the rope angular velocity relative to the canoe (Ω^{rope}_{canoe}) at this particular configuration?

A 0 (no orientation change)
B v_O/R, clockwise direction
C v_O/R, counterclockwise direction
D $v_O/(2R)$, clockwise direction
E $v_O/(2R)$, counterclockwise direction

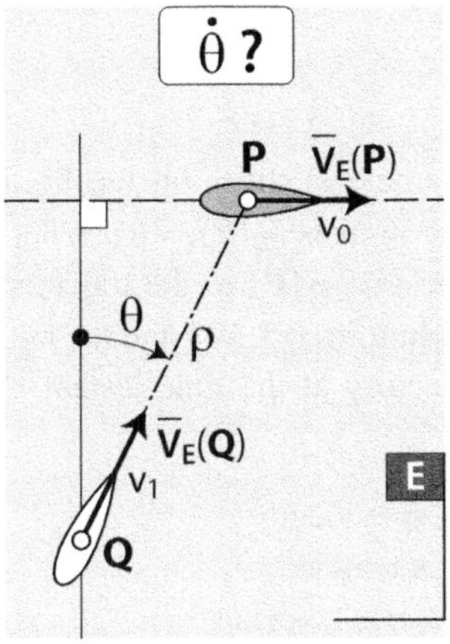

2.44 Point **P** of a ship has a rectilinear motion relative to the ground (E). Point **Q** of a second ship has a velocity relative to E permanently directed to **P**. What is the angular change of course $\dot{\theta}$ of the second ship relative to E?

A $v_1 \cos\theta/\rho$
B $v_1 \sin\theta/\rho$
C $(v_O - v_1 \sin\theta)/\rho$
D $v_O \cos\theta/\rho$
E $v_O \sin\theta/\rho$

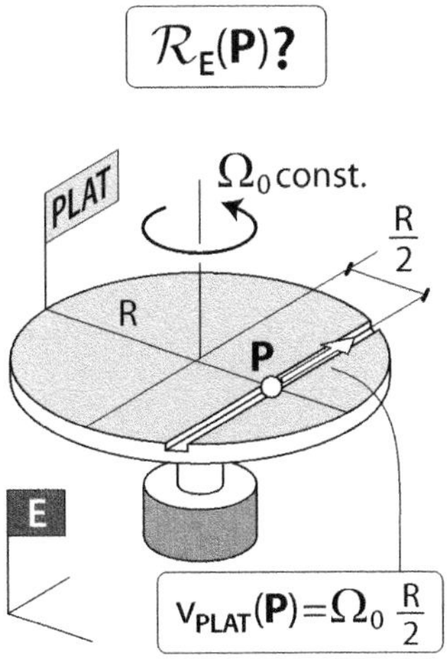

2.45 What is the radius of curvature of the trajectory of point **P** relative to the ground (E) for the given configuration?

A 0
B R
C R/2
D 2R/3
E ∞

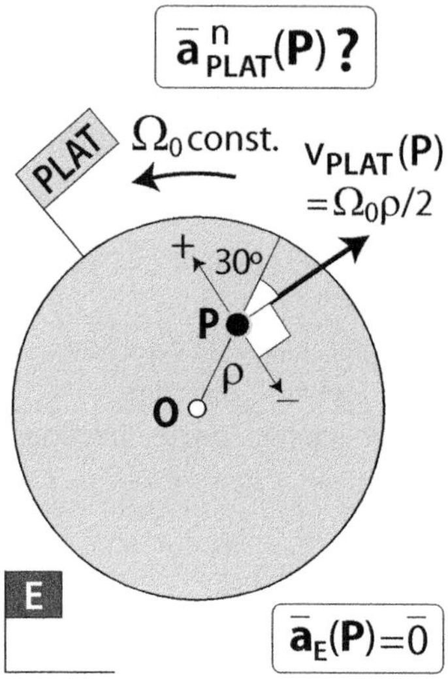

2.46 A particle **P** slides on the frictionless platform (which implies $\bar{\mathbf{a}}_E(\mathbf{P}) = \bar{\mathbf{0}}$). What is the value of the normal component of the acceleration of **P** relative to the platform (PLAT) for the given configuration?

A $+\Omega_0^2\rho/2$
B $-\Omega_0^2\rho/2$
C $+\Omega_0^2\rho$
D $-\Omega_0^2\rho$
E 0

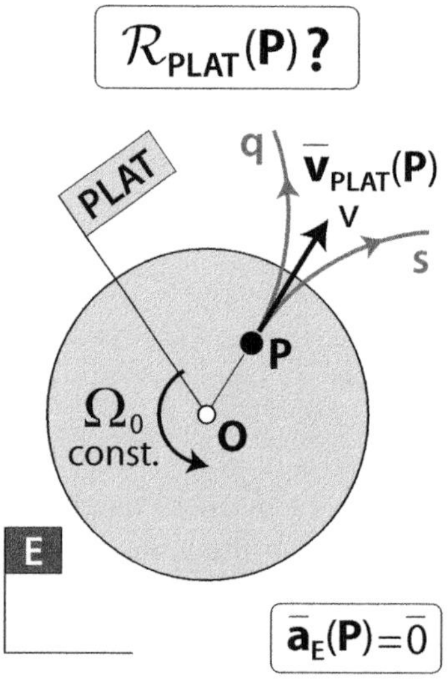

2.47 A particle **P** slides on the frictionless platform (which implies $\bar{\mathbf{a}}_E(\mathbf{P}) = \bar{\mathbf{0}}$). What is the radius of curvature $\mathfrak{R}_{PLAT}(\mathbf{P})$ of the trajectory of **P** relative to the platform (PLAT) and the initial tendency of that trajectory at the time instant shown in the figure?

A v_0/Ω_0, **q** tendency
B v_0/Ω_0, **s** tendency
C $v_0/(2\Omega_0)$, **q** tendency
D $v_0/(2\Omega_0)$, **s** tendency
E $2v_0/\Omega_0$, **q** tendency

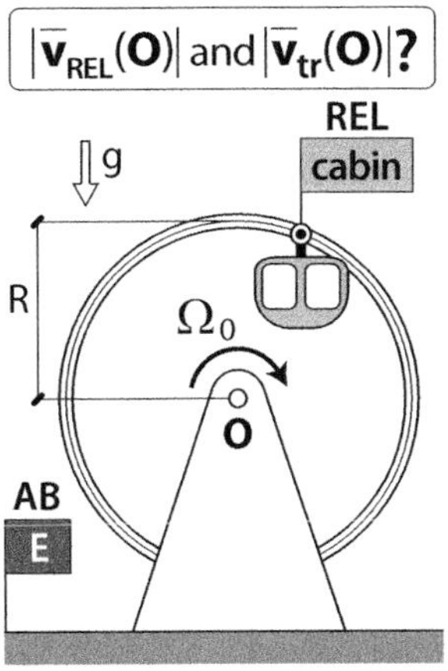

2.48 What are the modules of the relative and the transportation velocities of the center **O** of the Ferris wheel?

	$\lvert\bar{v}_{REL}\rvert$	$\lvert\bar{v}_{tr}\rvert$
A	0	0
B	0	$R\Omega_0$
C	$R\Omega_0$	0
D	$R\Omega_0$	$R\Omega_0$
E	$2R\Omega_0$	0

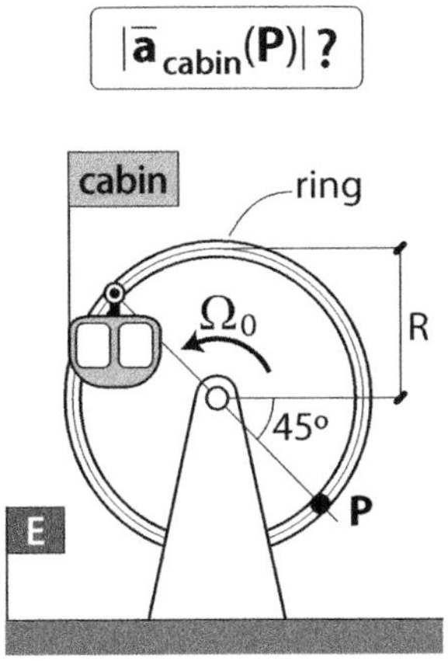

2.49 What is the modulus of the acceleration of point **P** of the ring relative to the cabin?

A 0
B $R\Omega_0^2$
C $2R\Omega_0^2$
D $3R\Omega_0^2$
E $4R\Omega_0^2$

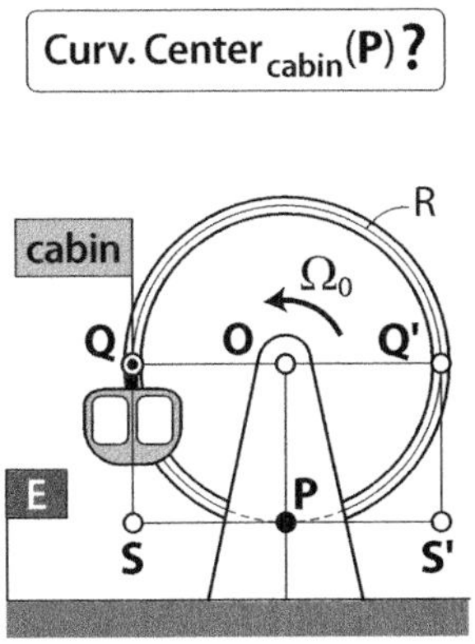

2.50 What is the center of curvature of the trajectory relative to the cabin of the ground-fixed point **P**?

A **O**
B **Q**
C **Q′**
D **S**
E **S′**

Problems

2.1 The mass point **P** slides on the block. The block rotates without sliding on the semicylindrical support fixed to the ground (E). for $\theta = 0$, **C** and **C′** are both on the vertical line through **O**. Find:

(a) $\bar{v}_E(\mathbf{P})$

(b) $\bar{a}_E(\mathbf{P})$

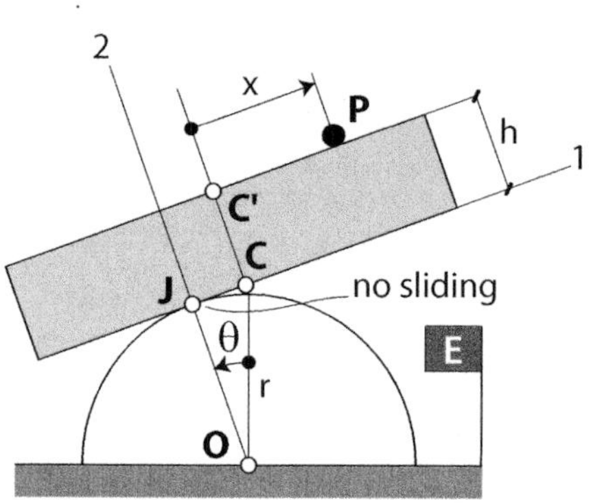

2.2 The mass point **P** is attached to one end of an inextensible thread. The other end is fixed to the pulley, whose radius is r. The pulley rotates with angular velocity $\dot{\varphi}(t)$ around point **O**, fixed to the ground. The thread-free portion (not touching the pulley) is taut and has a variable orientation $\theta(t)$, and its length is L_0 when $(\theta = \varphi = 0)$. Find the velocity and acceleration of **P** relative to:

(a) the ground (E)

(b) the reference frame R containing the ground point **O** and oriented as the vector basis

(c) the pulley (PL)

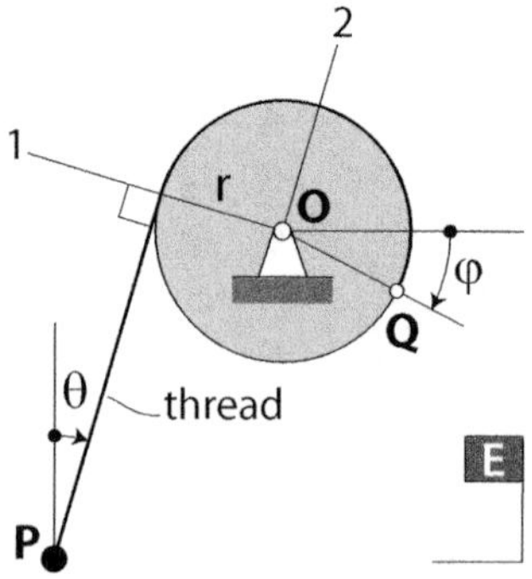

2.3 A telescopic antenna is oriented through the motorized rotations $(\psi(t), \theta(t))$. Another drive controls its length $\rho(t)$. Find the $\bar{\mathbf{v}}_E(\mathbf{P})$ components:

(a) in the B basis with axes $(1, 2, 3)$
(b) in the B′ basis with axes $(1', 2', 3')$

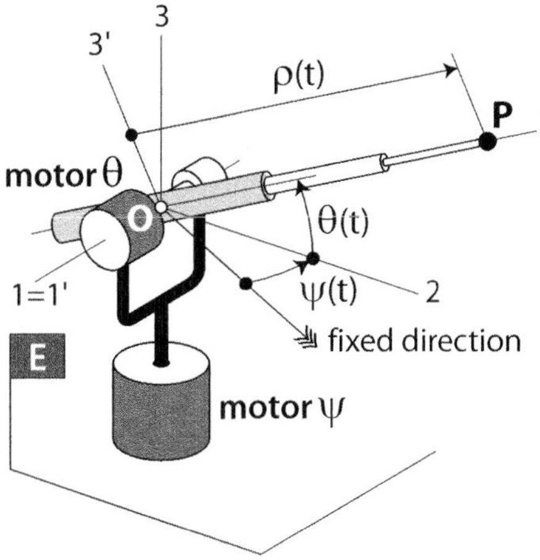

2.4 A spinning top has a general rotation motion while keeping its vertex **O** fixed to the ground (E). The top orientation is described through the three Euler angles defined in the figure. Find $\bar{\mathbf{v}}_E(\mathbf{P})$.

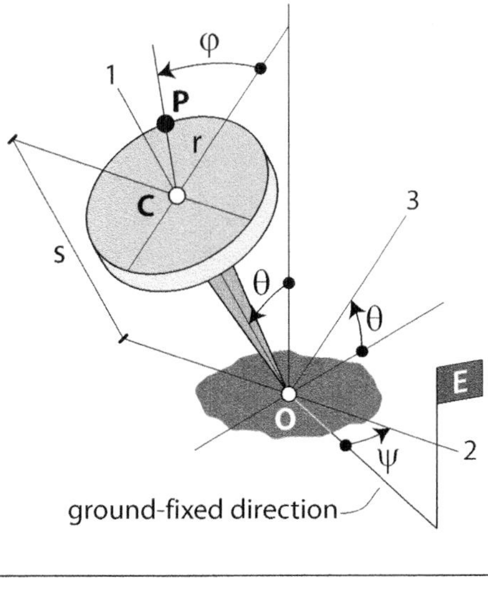

2.5 A canoe with a variable translational motion $v(t)$ relative to the water (considered at rest relative to the ground E) tows a skier. The horizontal rope **PO** is inextensible, and its length is L. The skier is modeled as a point mass **P**, and his velocity relative to the water has the skis direction (the skis do not skid). Find:

(a) the skier speed v' relative to the water and the angular velocity of the rope relative to the canoe $\dot{\theta}$, as a function of (v, θ, φ)

(b) the curvature radius $\Re_E(\mathbf{P})$, as a function of (v, θ, φ), when angle φ is constant

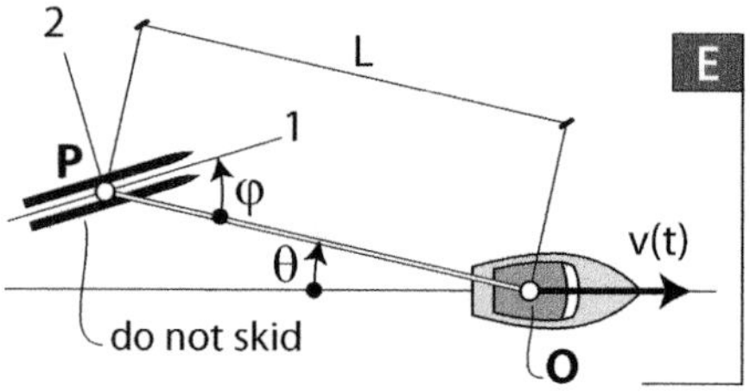

2.6 The mass point **P** is attached to one end of an inextensible thread, which in turn is rolled on a circular support fixed to the ground (E). When the taut thread unrolls, **P** describes a trajectory called *circumferential involute.* When $\theta = 0$, **P** is located at **P′** on the support. Find:

(a) the intrinsic components of $\bar{\mathbf{v}}_E(\mathbf{P})$

(b) the intrinsic components of $\bar{\mathbf{a}}_E(\mathbf{P})$

(c) the curvature radius $\Re_E(\mathbf{P})$

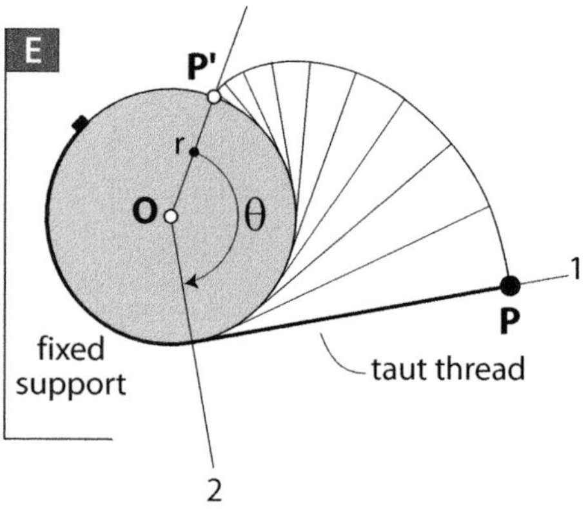

2.7 The mass point **P** moves freely on the circular support with radius r. The support rotates around point **O** fixed to the ground (E) with constant angular velocity $\bar{\boldsymbol{\Omega}}_0$ relative to the ground. Take the ground as the AB reference frame, and the support as the REL frame. Find:

(a) $\bar{\mathbf{v}}_{\text{REL}}(\mathbf{P}), \bar{\mathbf{v}}_{\text{tr}}(\mathbf{P})$
(b) $\bar{\mathbf{a}}_{\text{REL}}(\mathbf{P}), \bar{\mathbf{a}}_{\text{tr}}(\mathbf{P}), \bar{\mathbf{a}}_{\text{Cor}}(\mathbf{P})$
(c) the curvature radius $\mathfrak{R}_{\text{E}}(\mathbf{P})$

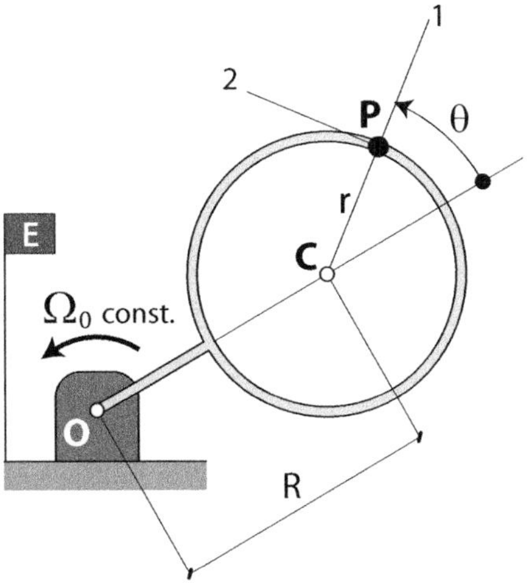

2.8 A plane has a simple rotation motion relative to the ground (E). Its propeller center **O** describes a circular trajectory in E, with center **C**, radius R, and constant speed v_0. The propeller rotates relative to the airplane with constant angular velocity $\dot{\theta}_0$ about its axis, which is tangent to the **O** trajectory. Take the ground as AB reference frame, and the plane as a REL frame. Find:

(a) $\bar{\mathbf{v}}_{\text{REL}}(\mathbf{P}), \bar{\mathbf{v}}_{\text{tr}}(\mathbf{P})$
(b) $\bar{\mathbf{a}}_{\text{REL}}(\mathbf{P}), \bar{\mathbf{a}}_{\text{tr}}(\mathbf{P}), \bar{\mathbf{a}}_{\text{Cor}}(\mathbf{P})$

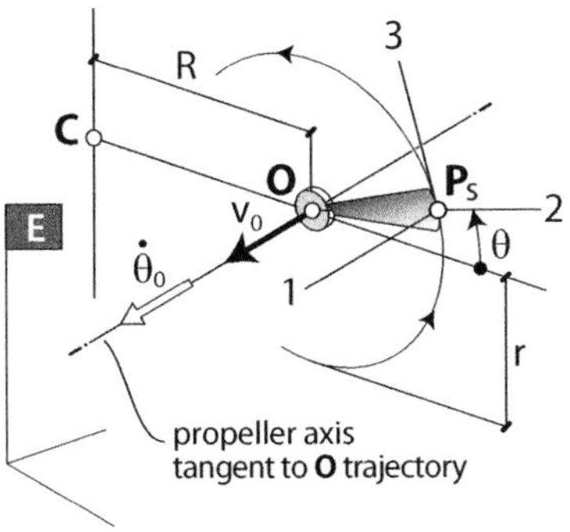

2.9 The mass point **P** impinges in radial direction on the entrance **A** of the slot **AB**. The slot is on an horizontal platform which rotates with constant angular velocity $\dot{\psi}_0$ about the vertical axis through its center **O** fixed to the ground (E).

(a) What has to be the **P** speed v_0 relative to the ground so that its velocity relative to the platform has the slot direction when impinging on **A**?

Once inside the slot, a driver guarantees a constant speed of **P** relative to the platform

(b) Find $\bar{\mathbf{v}}_E(\mathbf{P})$ and $\bar{\mathbf{a}}_E(\mathbf{P})$ for a general location x inside the slot

(c) What are the velocity and the location of **P** relative to E when it leaves the slot through **B**?

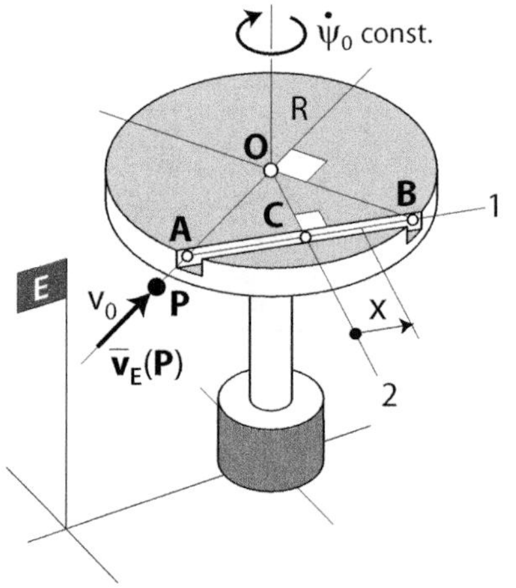

2.10 A point **Q** of a ship describes a circular trajectory to port (left side when looking in the forward direction) relative to the ground (E). The radius is 1,500 m and the speed is constant with value 15 m/s. The ship axis is always tangent to the **Q** trajectory. The ship's radar detects the presence of a vessel **P** (modeled as a particle), approaching from port. At a given time instant, the measured distance r and angle θ, plus their two first time derivatives, are:

$$\begin{array}{ll} r = 1,500\ \text{m} & \theta = 60^\circ \\ \dot{r} = -7.5\ \text{m/s} & \dot{\theta} = -0.3633^\circ/\text{s} \\ \ddot{r} = -0.1473\ \text{m/s}^2 & \ddot{\theta} = -0.7678\cdot 10^{-3}\ {}^\circ/\text{s}^2 \end{array}$$

(a) Find the velocity and acceleration of **P** relative to the sea (S) (assumed to be at rest relative to the ground)

(b) Make a forecast about the movement of **P** relative to the sea, and study the possibility of a collision between the two ships

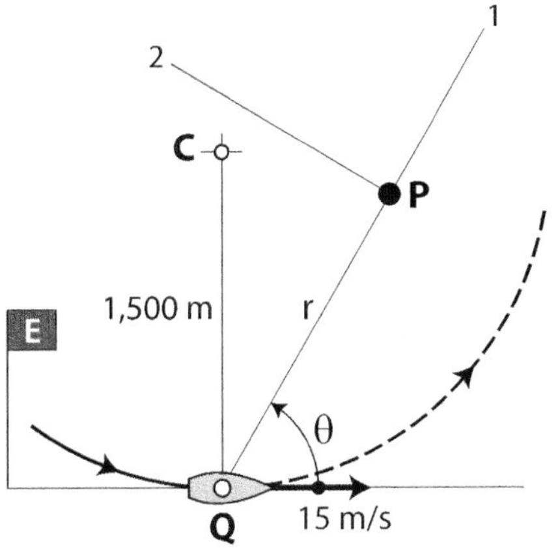

2.11 A canoe describes a simple rotation relative to the sea and tows a glider. The point **O** of the canoe describes a circular path of radius R with constant speed v_0 relative to the ground. An inextensible rope with length L is attached to **O**. The pilot of the glider – modeled as a particle **P** – maneuvers so that the angle β_0 remains constant and that the glider takes off with constant $\dot{\theta}_0$.

(a) Find the velocity and the acceleration of **P** relative to the sea S (assumed to be at rest relative to the ground)

Once reached a certain height, the pilot keeps it constant and slides down a ring (**Q**)

(b) Take the sea as AB frame, and the canoe as REL frame, and find $\bar{\mathbf{a}}_{tr}(\mathbf{Q})$ and $\bar{\mathbf{a}}_{Cor}(\mathbf{Q})$

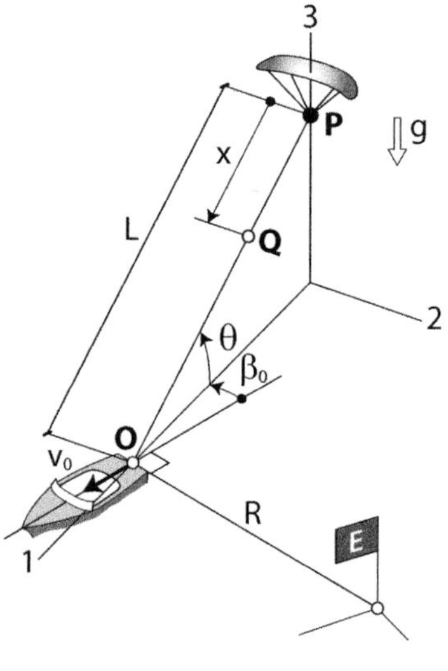

2.12 The "dog curve" problem. A dog **Q** runs toward his master **P** (both modeled as particles) so that its velocity relative to the ground (E) has a constant module v_Q and has always the $\overline{\mathbf{QP}}$ direction. If the master follows a rectilinear trajectory with constant speed v_P relative to the ground,

(a) What are the $\dot{\theta}$ and $\dot{\rho}$ values?
Tip: investigate the dog's speed in the reference frame that translates with **P** (RTP)

(b) Find $\Re_E(\mathbf{Q})$

(c) If $v_P = v_Q$, where is the curvature center of the **Q** trajectory relative to the ground?

(d) For the particular case $v_P = v_Q$, find $\rho(\theta)$ through integration of $d\rho/d\theta = \dot{\rho}/\dot{\theta}$ from the initial values (ρ_0, θ_0). What is the minimum distance between dog and master?

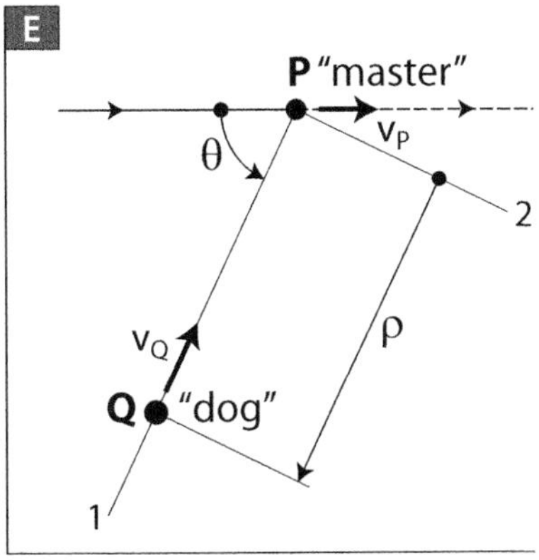

2.13 A fugitive **P** (modeled as a particle) is on a boat in the center **O** of a circular pond of radius r, while a pursuer **Q** (also modeled as a particle) runs along the pond shore with constant speed v_Q relative to the ground (E). **P** runs away keeping the pursuer in a diametrically opposite position (while he can) and maintaining its speed v_P constant relative to the ground. Find:

(a) $\bar{v}_E(\mathbf{P}), \bar{a}_E(\mathbf{P})$

(b) $\mathfrak{R}_E(\mathbf{P})$

(c) What is the minimum v_P value that allows **P** to leave the pond in a diametrically opposite position from the pursuer?

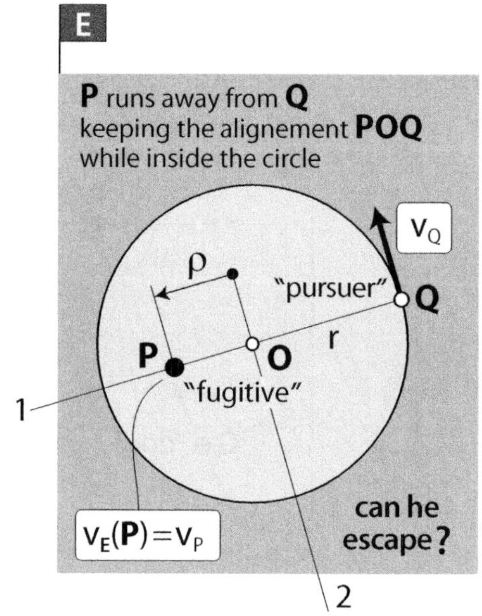

Puzzles

2.1 Determining the position change of a ground vehicle

The vehicle moves on a flat ground without sliding. The configuration of the steering wheel determines the curvature radius R of the trajectory of point **O** (which can be either positive or negative).

Is it possible to obtain the position change of the vehicle from the readings on the speedometer and the steering wheel configuration?

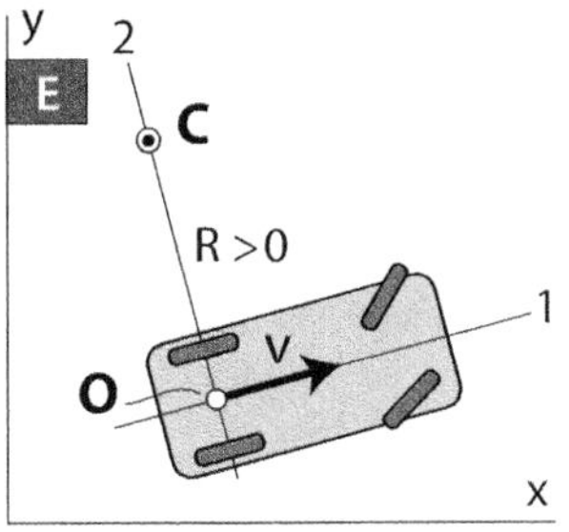

2.2 Determining the motion of a ship

The module $\rho(t)$ and the orientation angle $\theta(t)$ of the position vector $\overline{\mathbf{OP}}$ are constantly measured. The velocity of point **O** relative to the ground is always in the longitudinal direction of the ship.

Is it possible to obtain the velocity of point O and the angular velocity of the ship, relative to the ground, from $\rho(t)$ and $\theta(t)$?

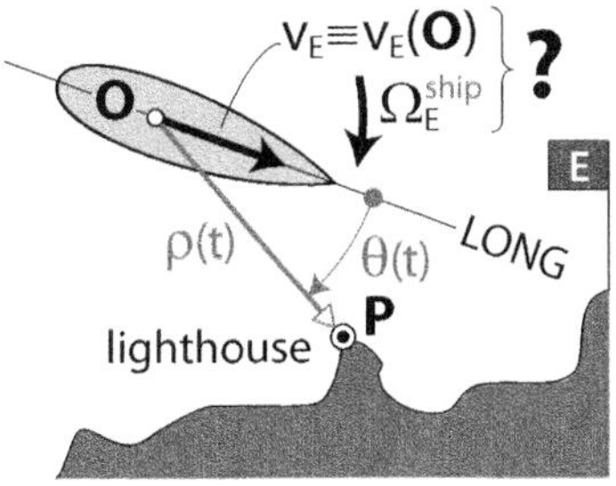

Quiz Questions: Answers

	1	2	3	4	5	6	7	8	9	10
	C	C	E	A	C	D	B	B	D	D
+10	B	B	A	C	C	C	A	E	D	D
+20	C	E	B	A	A	C	E	E	E	B
+30	D	E	B	A	C	C	E	E	D	C
+40	B	E	D	D	D	B	D	D	C	D

Problems: Answers

2.1 (a) $\{\bar{\mathbf{v}}_E(\mathbf{P})\}_B = \begin{Bmatrix} \dot{x} - h\dot{\theta} \\ (x + r\theta)\dot{\theta} \end{Bmatrix}$

(b) $\{\bar{\mathbf{a}}_E(\mathbf{P})\}_B = \begin{Bmatrix} \ddot{x} - h\ddot{\theta} - (x + r\theta)\dot{\theta}^2 \\ (x + r\theta)\ddot{\theta} + (r - h)\dot{\theta}^2 + 2\dot{x}\dot{\theta} \end{Bmatrix}$

2.2 (a) $\{\bar{\mathbf{v}}_E(\mathbf{P})\}_B = \begin{Bmatrix} [L_0 - r(\varphi - \theta)]\dot{\theta} \\ r\dot{\varphi} \end{Bmatrix},$

$\{\bar{\mathbf{a}}_E(\mathbf{P})\}_B = \begin{Bmatrix} [L_0 - r(\varphi - \theta)]\ddot{\theta} - 2r\dot{\varphi}\dot{\theta} + r\dot{\theta}^2 \\ r\ddot{\varphi} + [L_0 - r(\varphi - \theta)]\dot{\theta}^2 \end{Bmatrix}$

(b) $\{\bar{\mathbf{v}}_R(\mathbf{P})\}_B = \begin{Bmatrix} 0 \\ r(\dot{\varphi} - \dot{\theta}) \end{Bmatrix}, \{\bar{\mathbf{a}}_R(\mathbf{P})\}_B = \begin{Bmatrix} 0 \\ r(\ddot{\varphi} - \ddot{\theta})0 \end{Bmatrix}$

(c) $\{\bar{\mathbf{v}}_{PL}(\mathbf{P})\}_B = \begin{Bmatrix} -[L_0 - r(\varphi - \theta)](\dot{\varphi} - \dot{\theta}) \\ 0 \end{Bmatrix},$

$\{\bar{\mathbf{a}}_{PL}(\mathbf{P})\}_B = \begin{Bmatrix} -[L_0 - r(\varphi - \theta)](\ddot{\varphi} - \ddot{\theta}) + r(\dot{\varphi} - \dot{\theta})^2 \\ [L_0 - r(\varphi - \theta)](\dot{\varphi} - \dot{\theta})^2 \end{Bmatrix}$

2.3 (a) $\{\bar{\mathbf{v}}_E(\mathbf{P})\}_B = \begin{Bmatrix} -\rho\dot{\psi}\cos\theta \\ \dot{\rho}\cos\theta - \rho\dot{\theta}\sin\theta \\ \dot{\rho}\sin\theta + \rho\dot{\theta}\cos\theta \end{Bmatrix}$

(b) $\{\bar{\mathbf{v}}_E(\mathbf{P})\}_{B'} = \begin{Bmatrix} -\rho\dot{\psi}\cos\theta \\ \dot{\rho} \\ \rho\dot{\theta} \end{Bmatrix}$

2.4 $\{\bar{\mathbf{v}}_E(\mathbf{P})\}_B = \begin{Bmatrix} r\dot{\psi}\sin\theta\sin\varphi + r\dot{\theta}\cos\varphi \\ s\dot{\psi}\sin\theta - r\dot{\psi}\cos\theta\cos\varphi - r\dot{\varphi}\cos\varphi \\ -r\dot{\psi}\cos\theta\sin\varphi - s\dot{\theta} - r\dot{\varphi}\sin\varphi \end{Bmatrix}$

2.5 (a) $v' = v\dfrac{\cos\theta}{\cos\varphi}, \quad \dot{\theta} = \dfrac{v}{L}\dfrac{\sin(\varphi - \theta)}{\cos\varphi}$

(b) $\mathfrak{R}_E(\mathbf{P}) = L\dfrac{\cos\theta}{\sin(\varphi - \theta)}$

2.6 (a) $v_E(\mathbf{P}) = r\theta\dot{\theta}$

(b) $a_E^s(\mathbf{P}) = r\theta\ddot{\theta} + r\dot{\theta}^2, \quad a_E^n(\mathbf{P}) = r\theta\dot{\theta}^2$ (axis 1, negative)

(c) $\mathfrak{R}_E(\mathbf{P}) = r\theta$

2.7 (a) $\{\bar{\mathbf{v}}_{REL}(\mathbf{P})\}_B = \begin{Bmatrix} 0 \\ r\dot{\theta} \end{Bmatrix}, \quad \{\bar{\mathbf{v}}_{tr}(\mathbf{P})\}_B = \begin{Bmatrix} R\Omega_0\sin\theta \\ r\Omega_0 + R\Omega_0\cos\theta \end{Bmatrix}$

(b) $\{\bar{\mathbf{a}}_{\mathrm{REL}}(\mathbf{P})\}_B = \left\{ \begin{matrix} -r\dot{\theta}^2 \\ r\ddot{\theta} \end{matrix} \right\}$, $\{\bar{\mathbf{a}}_{\mathrm{tr}}(\mathbf{P})\}_B = \left\{ \begin{matrix} -\Omega_0^2(r + R\cos\theta) \\ R\Omega_0^2 \sin\theta \end{matrix} \right\}$,

$\{\bar{\mathbf{a}}_{\mathrm{Cor}}(\mathbf{P})\}_B = \left\{ \begin{matrix} -2r\Omega_0\dot{\theta} \\ 0 \end{matrix} \right\}$

(c) $\Re_E(\mathbf{P}) = \dfrac{r + R}{1 + \lambda}$, $\lambda = rR\left(\dfrac{\dot{\theta}}{r\dot{\theta} + (r + R)\Omega_0}\right)^2$

2.8 (a) $\{\bar{\mathbf{v}}_{\mathrm{REL}}(\mathbf{P})\}_B = \left\{ \begin{matrix} 0 \\ 0 \\ r\dot{\theta}_0 \end{matrix} \right\}$, $\{\bar{\mathbf{v}}_{\mathrm{tr}}(\mathbf{P})\}_B = \left\{ \begin{matrix} v_0 + \dfrac{r}{R} v_0 \cos\theta \\ 0 \\ 0 \end{matrix} \right\}$

(b) $\{\bar{\mathbf{a}}_{\mathrm{REL}}(\mathbf{P})\}_B = \left\{ \begin{matrix} 0 \\ -r\dot{\theta}_0^2 \\ 0 \end{matrix} \right\}$, $\{\bar{\mathbf{a}}_{\mathrm{tr}}(\mathbf{P})\}_B = \left\{ \begin{matrix} 0 \\ -(1 + (r/R)\cos\theta)v_0^2\cos\theta \\ (1 + (r/R)\cos\theta)v_0^2 \sin\theta \end{matrix} \right\}$,

$\{\bar{\mathbf{a}}_{\mathrm{Cor}}(\mathbf{P})\}_B = \left\{ \begin{matrix} -2\dfrac{r}{R} v_0\dot{\theta}_0 \sin\theta \\ 0 \\ 0 \end{matrix} \right\}$

2.9 (a) $v_0 = R\dot{\psi}_0$

(b) $\{\bar{\mathbf{v}}_E(\mathbf{P})\}_B = \left\{ \begin{matrix} R\dot{\psi}_0/\sqrt{2} \\ x\dot{\psi}_0 \end{matrix} \right\}$, $\{\bar{\mathbf{a}}_E(\mathbf{P})\}_B = \left\{ \begin{matrix} -x\dot{\psi}_0^2 \\ 3R\dot{\psi}_0^2/\sqrt{2} \end{matrix} \right\}$

(c) Direction **OB**, value $R\dot{\psi}_0$; location 32, 7° from **A** counterclockwise

2.10 (a) $\{\bar{\mathbf{v}}_S(\mathbf{P})\}_B = \left\{ \begin{matrix} 0 \\ -7.5\ \mathrm{m/s} \end{matrix} \right\}$, $\{\bar{\mathbf{a}}_S(\mathbf{P})\}_B = \left\{ \begin{matrix} -0.0375\ \mathrm{m/s^2} \\ 0 \end{matrix} \right\}$

(b) At that time instant, **P** describes a circular trajectory around **Q** with radius 1,500 m and constant speed. If this motion is maintained, they collide 104.7 s later

2.11 (a) $\{\bar{\mathbf{v}}_S(\mathbf{P})\}_B = \left\{ \begin{matrix} L\dot{\theta}_0 \sin\theta + v_0 \cos\beta_0 \\ -v_0 \sin\beta_0 - v_0(L/R)\cos\theta \\ L\dot{\theta}_0 \cos\theta \end{matrix} \right\}$,

$\{\bar{\mathbf{a}}_S(\mathbf{P})\}_B = \left\{ \begin{matrix} L\dot{\theta}_0^2 \cos\theta + \left(v_0^2/R\right)\left(\sin\beta_0 + (L/R)\cos\theta\right) \\ \left(v_0^2/R\right)\cos\beta_0 + 2v_0\dot{\theta}_0(L/R)\sin\theta \\ -L\dot{\theta}_0^2 \sin\theta \end{matrix} \right\}$

(b) $\{\bar{\mathbf{a}}_{\mathrm{tr}}(\mathbf{Q})\}_B = \dfrac{v_0^2}{R}\left\{ \begin{matrix} \sin\beta_0 + ((L - x)/R)\cos\theta \\ \cos\beta_0 \\ 0 \end{matrix} \right\}$,

$\{\bar{\mathbf{a}}_{\mathrm{Cor}}(\mathbf{Q})\}_B = \left\{ \begin{matrix} 0 \\ 2\dot{x}(v_0/R)\cos\theta \\ 0 \end{matrix} \right\}$

2.12 (a) $\dot{\theta} = -\dfrac{v_P}{\rho}\sin\theta, \quad \dot{\rho} = -v_Q + v_P\cos\theta$

(b) $\Re_E(\mathbf{Q}) = \dfrac{\rho}{\sin\theta}\dfrac{v_Q}{v_P}$

(c) It is the intersection of the directions through **P** and **Q** perpendicular to v_P y v_Q, respectively

(d) $\rho = \rho_0\left(\dfrac{\cos(\theta_0/2)}{\cos(\theta/2)}\right)^2, \quad \rho_{min} = \rho_0\cos^2(\theta_0/2)$

2.13 (a) $\{\bar{\mathbf{v}}_E(\mathbf{P})\}_B = \left\{\begin{matrix} v \\ v_Q(\rho/r) \end{matrix}\right\}, \quad \{\bar{\mathbf{a}}_E(\mathbf{P})\}_B = \left\{\begin{matrix} -2\rho(v_Q/r)^2 \\ 2v'' v_Q/r \end{matrix}\right\},$

$$v = \sqrt{v_P^2 - (v_Q\rho/r)^2}$$

(b) $\Re_E(\mathbf{P}) = \dfrac{r}{2}\dfrac{v_P}{v_Q}$

(c) $[v_P]_{min} = v_Q$

Puzzles: Solutions

2.1 Determining the position change of a ground vehicle

Yes, it is possible.

The change of orientation ψ can be obtained from the curvature radius R and the readings on the speedometer (v) through an integration:

$$\dot{\psi} = \frac{v}{R} \Rightarrow \psi = \psi_0 + \int_0^t \frac{v}{R}dt.$$

Then, the change of coordinates (x, y) of point **O** can be obtained as:

$$\dot{x} = v\cos\psi \Rightarrow \Delta x = \int_0^t v\cos\psi dt \ , \quad \dot{y} = v\sin\psi \Rightarrow \Delta y = \int_0^t v\sin\psi dt.$$

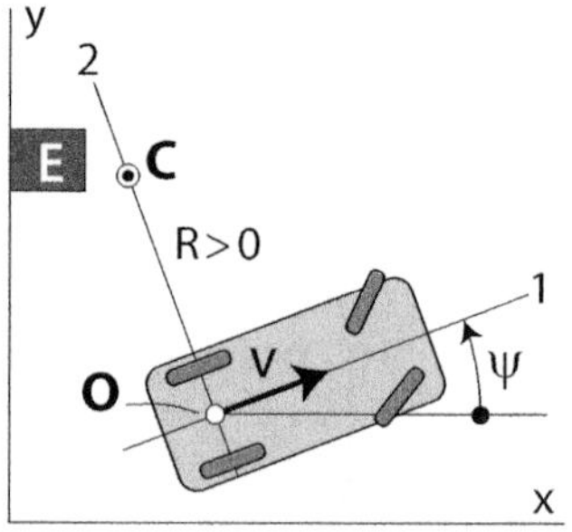

2.2 Determining the motion of a ship

It is possible except for configurations where OP is parallel to the ship transverse direction.

As the velocity of the lighthouse **P** relative to the ground (AB) is zero, its velocity relative to the ship (REL) is equal to the transportation motion of **P** but with reversed sign:

$$\bar{\mathbf{v}}_{\mathrm{REL}}(\mathbf{P}) = -\bar{\mathbf{v}}_{\mathrm{tr}}(\mathbf{P})$$

$$\left.\begin{array}{l} \bar{\mathbf{v}}_{\mathrm{REL}}(\mathbf{P})]_{\mathrm{LONG}} = \dot{\rho}\cos\theta - \rho\dot{\theta}\sin\theta \\ \bar{\mathbf{v}}_{\mathrm{REL}}(\mathbf{P})]_{\mathrm{TRANS}} = \dot{\rho}\sin\theta + \rho\dot{\theta}\cos\theta \\ \bar{\mathbf{v}}_{\mathrm{tr}}(\mathbf{P})]_{\mathrm{LONG}} = -\mathrm{v} + \dot{\psi}\rho\sin\theta \\ \bar{\mathbf{v}}_{\mathrm{tr}}(\mathbf{P})]_{\mathrm{TRANS}} = -\dot{\psi}\rho\cos\theta \end{array}\right\} \Rightarrow \left\{\begin{array}{l} \dot{\psi} = -\dot{\theta} - \dfrac{\dot{\rho}}{\rho}\tan\theta \\ \mathrm{v} = \rho\left(\dot{\psi} + \dot{\theta}\right)\sin\theta - \dot{\rho}\cos\theta = -\dfrac{\dot{\rho}}{\cos\theta} \end{array}\right.$$

These expressions are not defined when $\theta = 90^0$.

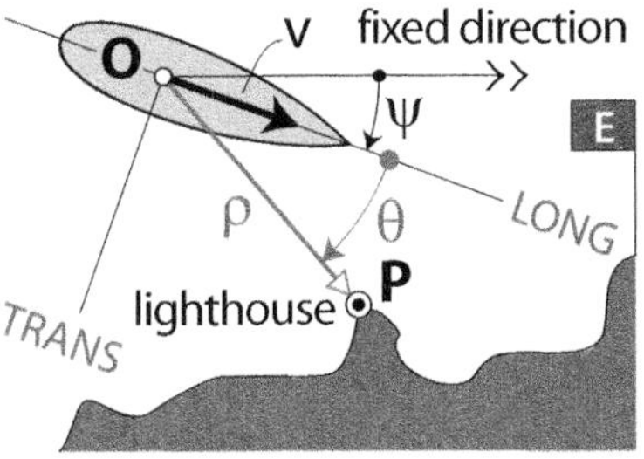

3 Rigid Body Kinematics

The mass point model is not suitable when the orientation is relevant to the problem under study. The simplest model incorporating the orientation is the *rigid body*, which is defined as a set of material points with constant mutual distances. This definition is so close to that of the reference frame that both concepts are taken as synonyms in a kinematics context. The reference frame of points fixed with respect to the rigid body is called *body reference frame*. The kinematics of the rigid body corresponds to the transportation motion (Chapter 2) associated with that reference frame.

The most complex aspect in rigid body kinematics (from a conceptual point of view) is the study of the body change of orientation, but this has already been addressed in Chapter 1.

Many interesting geometric properties can be deduced from the analytical structure of the expressions of velocity and acceleration of the points of a rigid body. When the instrumental resources of calculation were scarce, those properties were extensively studied because they allowed the treatment of many kinematic aspects through geometric constructions. Nowadays, computers bypass those procedures. In this book, the only geometric property that has been retained (because of its conceptual and practical interest in many cases) is the description of the velocity distribution as a rotation and a sliding motion about and along a same axis: the *screw axis*.

The *impenetrability of the solid state* gives rise to the phenomenon of *contact between bodies* – rigid or deformable – which constitutes a restriction in their relative movement. When two bodies are in contact, the *surface roughness* can cause a new restriction: the *nonslip* due to friction. These restrictions constitute the *basic constraint conditions*.

In the case of rigid bodies, the composition of these basic restrictions applied to different points originates the *constraint restrictions* between them. Their mathematical formulation is known as *constraint equations*.

3.1 Position and Orientation of a Rigid Body

The **rigid body** S is defined as "set of material points with constant mutual distances." Those points can be seen as points of a reference frame: the **body frame** SR. The situation of a rigid body S in a reference frame R is determined by:

- the position of a point of the rigid body S – or of the SR – which can be described by three position coordinates

- the orientation of the rigid body S – or of the SR – which can be described by three orientation coordinates (for example, three Euler angles)

Those six coordinates define the **configuration** (position and orientation) of S in R. From this information, one can determine the position of any point of the rigid body in R.

3.2 Velocity and Acceleration Distributions in a Rigid Body

The constant distances between points of S imply that their motion (velocity and acceleration) is related (but not equal, in principle).

Since the points $\mathbf{P}_\mathrm{S}$ of the rigid body S are fixed with respect to the body reference frame SR, their movement with respect to a frame R can be described as a transportation movement associated with SR if R is taken as *absolute* frame (R = AB). Equations (2.27) and (2.32), adapted to the notation in Fig. 3.1, take the form:

$$\mathrm{R}=\mathrm{AB}, \mathrm{SR}=\mathrm{REL} \Rightarrow \bar{\mathbf{v}}_\mathrm{R}(\mathbf{P}_\mathrm{S})=\bar{\mathbf{v}}_\mathrm{R}(\mathbf{Q}_\mathrm{S})+\bar{\boldsymbol{\Omega}}_\mathrm{R}^\mathrm{S}\times\overline{\mathbf{Q}_\mathrm{S}\mathbf{P}_\mathrm{S}}, \tag{3.1}$$

$$\bar{\mathbf{a}}_\mathrm{R}(\mathbf{P}_\mathrm{S})=\bar{\mathbf{a}}_\mathrm{R}\left(\mathbf{Q}'_\mathrm{S}\right)+\bar{\boldsymbol{\alpha}}_\mathrm{R}^\mathrm{S}\times\overline{\mathbf{Q}'_\mathrm{S}\mathbf{P}_\mathrm{S}}+\bar{\boldsymbol{\Omega}}_\mathrm{R}^\mathrm{S}\times\left(\bar{\boldsymbol{\Omega}}_\mathrm{R}^\mathrm{S}\times\overline{\mathbf{Q}'_\mathrm{S}\mathbf{P}_\mathrm{S}}\right). \tag{3.2}$$

These equations summarize the rigid body kinematics (RBK). Points $\mathbf{Q}_\mathrm{S}$ and $\mathbf{Q}'_\mathrm{S}$, suitable to express the velocity and acceleration of $\mathbf{P}_\mathrm{S}$, are usually different.

Equation (3.1) shows that as far as velocities are concerned, the most general movement of the rigid body depends only on two vectors (or six velocity scalar variables): the velocity of a point $\mathbf{Q}_\mathrm{S}$ of the body and the body angular velocity (alternatively, three scalar variables to define the velocity of one of its points and three to define the velocity of change of orientation – usually the time derivatives of the three Euler angles). Those two vectors form the body's **kinematic torsor** at point $\mathbf{Q}_\mathrm{S}$:

$$\text{kinematic torsor at } \mathbf{Q}_\mathrm{S}:\ \left\{\bar{\mathbf{v}}_\mathrm{R}(\mathbf{Q}_\mathrm{S}),\bar{\boldsymbol{\Omega}}_\mathrm{R}^\mathrm{S}\right\}^\mathrm{T}.$$

The velocity components of any point $\mathbf{P}_\mathrm{S}$ of the body are a linear superposition of these six variables; in principle, the coefficients of the linear superposition are a function of the position and orientation of the body (body's configuration).

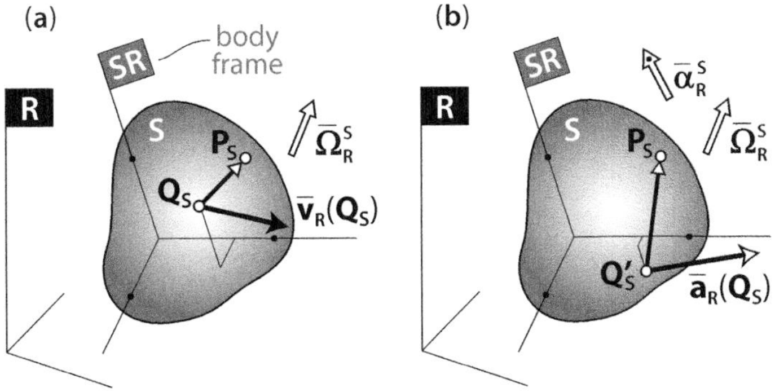

Fig. 3.1

A consequence of Eqs. (3.1) and (3.2) is that when a rigid body has a **translational motion** (no rotation), all points have the same velocity and the same acceleration, and they all describe equal trajectories. When that trajectory is curved, we say that the body has a curvilinear translational motion; if it is circular, it is a circular translational motion.[1]

▶ **Example 3.1** The configuration of a helicopter relative to the ground (E) is described through (Fig. 3.2):

- three Euler angles (ψ, θ, φ), defining its orientation
- three coordinates (x, y, z), defining the position of its point **O**

The velocity $\bar{\mathbf{v}}_{\mathrm{E}}(\mathbf{P})$ can be obtained from that of point **O**:

$$\bar{\mathbf{v}}_{\mathrm{E}}(\mathbf{P}) = \bar{\mathbf{v}}_{\mathrm{E}}(\mathbf{O}) + \bar{\boldsymbol{\Omega}}_{\mathrm{E}}^{\mathrm{S}} \times \overline{\mathbf{OP}},$$

$$\{\bar{\mathbf{v}}_{\mathrm{E}}(\mathbf{P})\}_{\mathrm{B}} = \begin{Bmatrix} \dot{x}\cos\psi + \dot{y}\sin\psi \\ \dot{y}\cos\psi - \dot{x}\sin\psi \\ \dot{z} \end{Bmatrix} + \begin{Bmatrix} \dot{\varphi}\cos\theta \\ \dot{\theta} \\ \dot{\psi} - \dot{\varphi}\sin\theta \end{Bmatrix} \times \begin{Bmatrix} \mathrm{L}\cos\theta \\ 0 \\ -\mathrm{L}\sin\theta \end{Bmatrix}$$

$$= \begin{Bmatrix} \dot{x}\cos\psi + \dot{y}\sin\psi - \dot{\theta}\mathrm{L}\sin\theta \\ -\dot{x}\sin\psi + \dot{y}\cos\psi + \dot{\psi}\mathrm{L}\cos\theta \\ \dot{z} - \dot{\theta}\mathrm{L}\cos\theta \end{Bmatrix}. \tag{3.3}$$

The $\bar{\mathbf{v}}_{\mathrm{E}}(\mathbf{P})$ components are linear forms of the first time derivatives $(\dot{\psi}, \dot{\theta}, \dot{\varphi}, \dot{x}, \dot{y}, \dot{z})$ of the configuration variables $(\psi, \theta, \varphi, x, y, z)$. The coefficients are a function of these variables (in that case, of ψ and θ). ◀

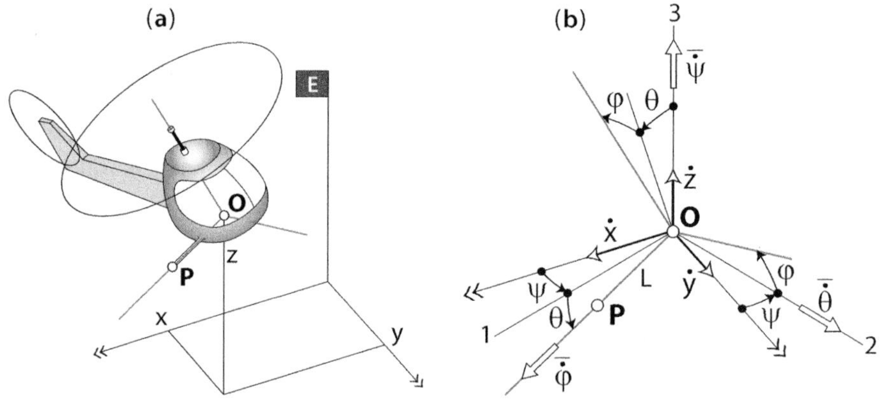

Fig. 3.2

---

[1] Note that a circular trajectory of a point may be associated with a translation or a rotation motion. In the former case, the curvature radius of all points in the rigid body is the same, whereas in the latter they have different values in principle.

The use of Eq. (3.1) calls for not only the expression of $\bar{\boldsymbol{\Omega}}_{\mathrm{R}}^{\mathrm{S}}$ but also the knowledge of the velocity of a point **Q** of the body. This information can be easily deduced in many cases.

► **Example 3.2** The velocity of the wheel center **C** can be determined from that of point $\mathbf{J}_s$, which is the wheel contact point with the ground (E). Under nonsliding conditions, $\bar{\mathbf{v}}_{\mathrm{E}}(\mathbf{J}_{\mathrm{s}}) = \bar{0}$. The angular velocity $\bar{\boldsymbol{\Omega}}_{\mathrm{E}}^{\mathrm{s}}$ can be expressed as a function of the time derivatives of the Euler angles (Fig. 3.3):

$$\bar{\mathbf{v}}_{\mathrm{E}}(\mathbf{C}) = \bar{\mathbf{v}}_{\mathrm{E}}(\mathbf{J}_{\mathrm{s}}) + \bar{\boldsymbol{\Omega}}_{\mathrm{E}}^{\mathrm{s}} \times \overline{\mathbf{J}_{\mathrm{s}}\mathbf{C}},$$

$$\{\bar{\mathbf{v}}_{\mathrm{E}}(\mathbf{C})\}_{\mathrm{B}} = \begin{Bmatrix} 0 \\ 0 \\ 0 \end{Bmatrix} + \begin{Bmatrix} \dot{\theta} \\ \dot{\varphi}\cos\theta \\ \dot{\psi} + \dot{\varphi}\sin\theta \end{Bmatrix} \times \begin{Bmatrix} 0 \\ -\mathrm{R}\sin\theta \\ \mathrm{R}\cos\theta \end{Bmatrix} = \begin{Bmatrix} \dot{\varphi}\mathrm{R} + \dot{\psi}\mathrm{R}\sin\theta \\ -\dot{\theta}\mathrm{R}\cos\theta \\ -\dot{\theta}\mathrm{R}\sin\theta \end{Bmatrix}. \tag{3.4}$$

◄

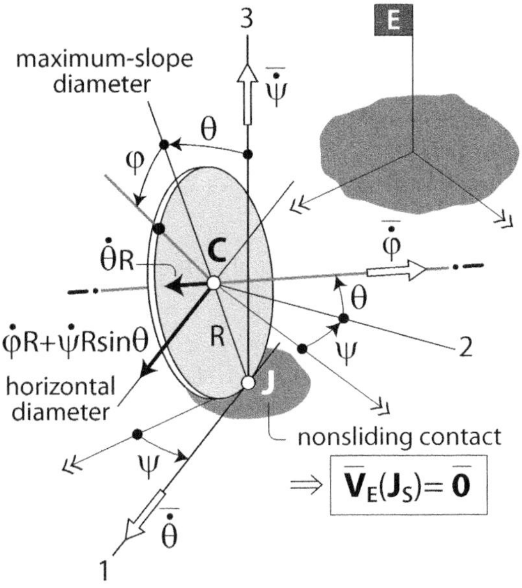

Fig. 3.3

The calculation of accelerations through Eq. (3.2) is not so usual: in addition to the expressions of $\bar{\boldsymbol{\Omega}}_{\mathrm{R}}^{\mathrm{S}}$ and $\bar{\boldsymbol{\alpha}}_{\mathrm{R}}^{\mathrm{S}}$, it requires the knowledge of the acceleration of a point $\mathbf{O}_{\mathrm{S}}$, and that information is seldom evident. Unlike Eq. (3.1), Eq. (3.2) is mainly used in simple cases where all the vectors involved can be described through arrows with an indication of value.

► **Example 3.3** The two blades rotate around a vertical axis under the action of a motor that introduces the angle $\psi(\mathrm{t})$. Their center **O** moves along a horizontal line according to a time law $\mathrm{s}(\mathrm{t})$ controlled by a hydraulic cylinder (Fig. 3.4).

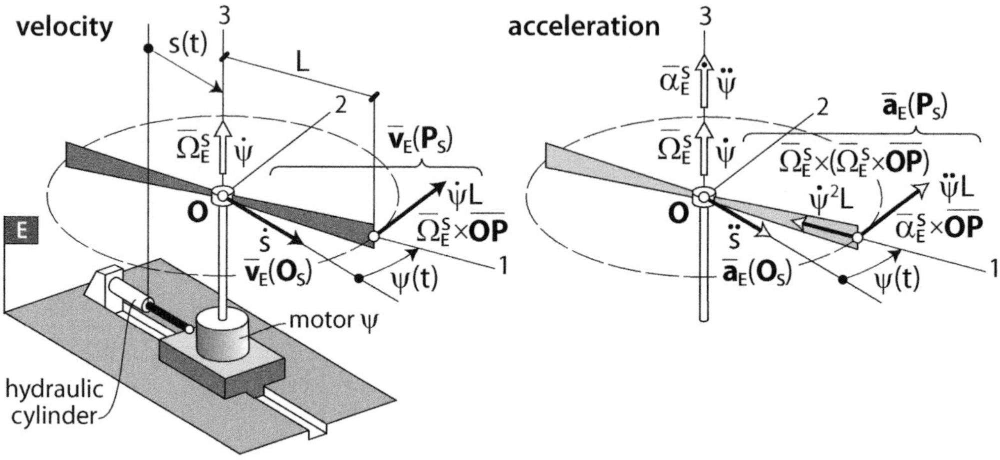

Fig. 3.4

The velocity and acceleration of a blade end point **P** with respect to the ground (E) can be calculated through the two equations of RBK. The same point **O** can be used in both equations, as its velocity $\dot{s}(t)$ and its acceleration $\ddot{s}(t)$ are known. The additive terms – containing vector products – can be easily determined graphically:

$$\bar{\mathbf{v}}_E(\mathbf{P}) = \bar{\mathbf{v}}_E(\mathbf{O}) + \bar{\boldsymbol{\Omega}}_E^s \times \overline{\mathbf{OP}} \Rightarrow \{\bar{\mathbf{v}}_E(\mathbf{P})\}_B = \left\{ \begin{matrix} \dot{s}\cos\psi \\ \dot{\psi}L - \dot{s}\sin\psi \\ 0 \end{matrix} \right\},$$

$$\bar{\mathbf{a}}_E(\mathbf{P}) = \bar{\mathbf{a}}_E(\mathbf{O}) + \bar{\boldsymbol{\alpha}}_E^s \times \overline{\mathbf{OP}} + \bar{\boldsymbol{\Omega}}_E^s \times \left(\bar{\boldsymbol{\Omega}}_E^s \times \overline{\mathbf{OP}}\right) \Rightarrow \{\bar{\mathbf{a}}_E(\mathbf{P})\}_B = \left\{ \begin{matrix} \ddot{s}\cos\psi - \dot{\psi}^2 L \\ \ddot{\psi}L - \ddot{s}\sin\psi \\ 0 \end{matrix} \right\}. \tag{3.5}$$

For specific configurations (corresponding to a single time instant), the advantage of using RBK to determine the acceleration – as an alternative to the time derivative of the velocity – can be even greater. This is the case, for example, for the specific configuration $\psi = 0$. ◀

Very often, the velocities are calculated through Eq. (3.1), and the corresponding accelerations are obtained as time derivatives of those velocity vectors. One has to make sure that the vector being derived represents the point velocity at any time instant (and not just at a particular time instant), since the time derivative process is based on the knowledge of the vector in two consecutive and very close time instants. Since Eq. (3.1) can be used to find particularized expressions (for a specific configuration) of velocities, it is easy to forget this precaution.

▶ **Example 3.4** A plane has a simple rotation movement with respect to the ground (E) (Fig. 3.5a) so that the center **O** of the propeller S describes a circular trajectory with center **C** (fixed to the ground) and radius R, and constant speed v_0. The propeller rotates relative to the plane with an angular velocity of constant value $\dot{\theta}_0$.

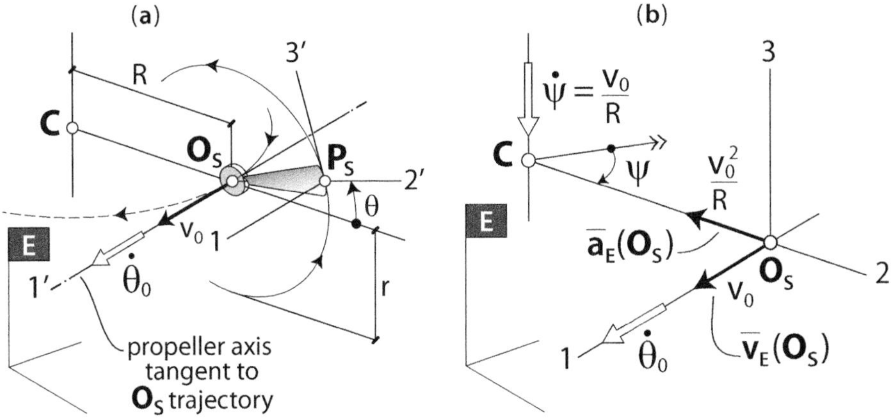

Fig. 3.5

The velocity and acceleration of the propeller end point **P** relative to the ground can be obtained through the time derivative of the position vector $\overline{\mathbf{CP}}(=\overline{\mathbf{CO}}+\overline{\mathbf{OP}})$, through composition of movements (as in Exercise 2.8), and through RBK.

In this example, $\bar{\mathbf{v}}_E(\mathbf{O})$ and $\bar{\mathbf{a}}_E(\mathbf{O})$ are well known (Fig. 3.5b). The intrinsic components of these vectors, projected on the vectors bases B and B′ fixed to the plane and the propeller, respectively, are:

$$\{\bar{\mathbf{v}}_E(\mathbf{O})\}_B = \{\bar{\mathbf{v}}_E(\mathbf{O})\}_{B'} = \begin{Bmatrix} v_0 \\ 0 \\ 0 \end{Bmatrix}, \quad \{\bar{\mathbf{a}}_E(\mathbf{O})\}_B = \begin{Bmatrix} 0 \\ -v_0^2/R \\ 0 \end{Bmatrix},$$

$$\{\bar{\mathbf{a}}_E(\mathbf{O})\}_{B'} = \frac{v_0^2}{R}\begin{Bmatrix} 0 \\ -\cos\theta \\ \sin\theta \end{Bmatrix}. \tag{3.6}$$

The angular velocity of the propeller relative to the ground $\left(\bar{\boldsymbol{\Omega}}_E^S\right)$ comes from the variation of angle ψ (which gives the position of **O** on its circular trajectory and plays the role of the first Euler angle) and that of angle θ describing the propeller rotation relative to the plane (second Euler angle): $\bar{\boldsymbol{\Omega}}_E^S = \bar{\dot{\boldsymbol{\psi}}}_0 + \bar{\dot{\boldsymbol{\theta}}}_0$. The first rotation is the angular velocity of the plane relative to the ground, and its value is $\dot{\psi}_0 = v_0/R$.

The angular acceleration $\bar{\boldsymbol{\alpha}}_E^S$ can be obtained through a geometric time derivative. The first rotation $\left(\bar{\dot{\boldsymbol{\psi}}}_0\right)$ is constant, whereas the second one $\left(\bar{\dot{\boldsymbol{\theta}}}_0\right)$ has a constant value but a variable direction (because of $\bar{\dot{\boldsymbol{\psi}}}_0$). Consequently, $\bar{\boldsymbol{\alpha}}_E^S$ is radial and points to the center **C**:

$$\{\bar{\boldsymbol{\alpha}}_E^S\}_B = \begin{Bmatrix} 0 \\ -\dot{\psi}_0\dot{\theta}_0 \\ 0 \end{Bmatrix} = \begin{Bmatrix} 0 \\ -(v_0/R)\dot{\theta}_0 \\ 0 \end{Bmatrix}, \quad \text{or} \quad \{\bar{\boldsymbol{\alpha}}_E^S\}_{B'} = \frac{v_0}{R}\dot{\theta}_0\begin{Bmatrix} 0 \\ -\cos\theta \\ \sin\theta \end{Bmatrix}. \tag{3.7}$$

Equations (3.1) and (3.2) yield:

$$\{\bar{\mathbf{v}}_E(\mathbf{P})\}_{B'} = \{\bar{\mathbf{v}}_E(\mathbf{O})\}_{B'} + \left\{\bar{\boldsymbol{\Omega}}_E^S \times \overline{\mathbf{OP}}\right\}_{B'} = \begin{Bmatrix} (v_0/R)(R + r\cos\theta) \\ 0 \\ r\dot{\theta}_0 \end{Bmatrix},$$

$$\{\bar{\mathbf{a}}_E(\mathbf{P})\}_{B'} = \{\bar{\mathbf{a}}_E(\mathbf{O})\}_{B'} + \left\{\bar{\boldsymbol{\alpha}}_E^S \times \overline{\mathbf{OP}}\right\}_{B'} + \left\{\bar{\boldsymbol{\Omega}}_E^S \times \left(\bar{\boldsymbol{\Omega}}_E^S \times \overline{\mathbf{OP}}\right)\right\}_{B'}$$

$$= \begin{Bmatrix} -2(v_0/R)r\dot{\theta}_0 \sin\theta \\ -(v_0/R)^2(R + r\cos\theta)\cos\theta - r\dot{\theta}_0^2 \\ (v_0/R)^2(R + r\cos\theta)\sin\theta \end{Bmatrix}. \tag{3.8}$$

It is easy to verify that the calculation of $\bar{\mathbf{a}}_E(\mathbf{P})$ is simpler if one proceeds to the time derivative of $\bar{\mathbf{v}}_E(\mathbf{P})$ (which is allowed since the velocity corresponds to a general configuration of the propeller):

$$\{\bar{\mathbf{a}}_E(\mathbf{P})\}_{B'} = \frac{d}{dt}\{\bar{\mathbf{v}}_E(\mathbf{P})\}_{B'} + \left\{\bar{\boldsymbol{\Omega}}_E^{B'} \times \bar{\mathbf{v}}_E(\mathbf{P})\right\}_{B'}$$

$$= \begin{Bmatrix} -(v_0/R)r\dot{\theta}\sin\theta \\ 0 \\ 0 \end{Bmatrix} + \begin{Bmatrix} \dot{\theta}_0 \\ -(v_0/R)\sin\theta \\ -(v_0/R)\cos\theta \end{Bmatrix} \times \begin{Bmatrix} (v_0/R)(R + r\cos\theta) \\ 0 \\ r\dot{\theta}_0 \end{Bmatrix}. \tag{3.9}$$

◀

Equation (3.1) can also be used to calculate the velocity of **geometric points**: points that do not belong to any material body. An example is the point that, at each time instant, corresponds to the contact of a wheel with the ground.

▶ **Example 3.5** A wheel describes a planar motion with a nonsliding contact on a horizontal floor (Fig. 3.6a). At each time instant, one can identify three points in the wheel–ground contact that evolve differently over time:

- point $\mathbf{J}_s$ of the wheel, which is instantaneously fixed to the ground as it is a nonsliding contact
- point $\mathbf{J}_E$ of the ground, which is permanently fixed to the ground
- point $\mathbf{J}_g$ (geometric contact point), which shows at each time instant the location of the wheel–ground contact; it moves relative to the ground and describes the wheel track on the horizontal floor

To better understand the movement of the geometric point $\mathbf{J}_g$, it is useful to imagine a rigid body with known kinematics to which it belongs: the square articulated to the wheel center $\mathbf{O}_S$, which keeps constant its orientation relative to the ground. As that square does not rotate, all its points have the same velocity:

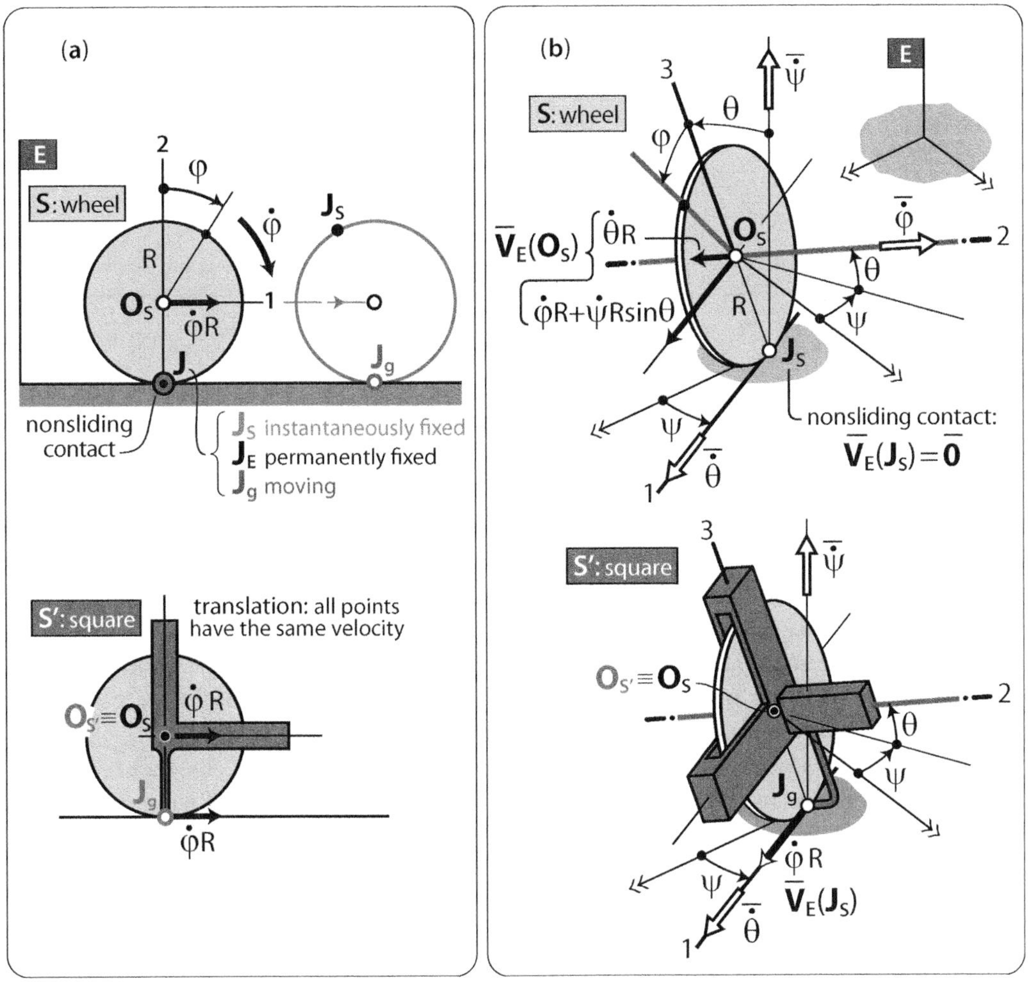

Fig. 3.6

$$\bar{v}_E(J_g) = \bar{v}_E(O_{s'}) = \bar{v}_E(O_s) \Rightarrow \{\bar{v}_E(J_g)\} = \begin{Bmatrix} \dot{\varphi}R \\ 0 \\ 0 \end{Bmatrix}. \tag{3.10}$$

In the general case of **perfect rolling** (rotation without sliding) of a wheel on a plane (Fig. 3.6b), the movement can be described through three Euler angles (as seen in Example 3.2). The wheel center velocity is:

$$\bar{v}_E(C) = \bar{\Omega}^s_E \times \overline{J_sC} \Rightarrow \{\bar{v}_E(C)\} = \begin{Bmatrix} \dot{\varphi}R + \dot{\psi}R\sin\theta \\ -\dot{\theta}R \\ 0 \end{Bmatrix}. \tag{3.11}$$

The J_g movement is the same as that of a point of a square articulated to the wheel center ($O_{s'} = O_s$) and with three axes directed according to the wheel axis, the horizontal diameter and that of maximum slope:

$$\{\overline{\mathbf{v}}_{\mathrm{E}}(\mathbf{J}_{\mathrm{g}})\} = \{\overline{\mathbf{v}}_{\mathrm{E}}(\mathbf{O}_{\mathrm{s}'})\} + \left\{\overline{\boldsymbol{\Omega}}_{\mathrm{E}}^{\mathrm{S}'} \times \overline{\mathbf{O}_{\mathrm{s}'}\mathbf{J}_{\mathrm{g}}}\right\} = \begin{Bmatrix} \dot{\varphi}\mathrm{R} + \dot{\psi}\mathrm{R}\sin\theta \\ -\dot{\theta}\mathrm{R} \\ 0 \end{Bmatrix} + \begin{Bmatrix} \dot{\theta} \\ \dot{\psi}\sin\theta \\ \dot{\psi}\cos\theta \end{Bmatrix} \times \begin{Bmatrix} 0 \\ 0 \\ -\mathrm{R} \end{Bmatrix}$$
$$= \begin{Bmatrix} \dot{\varphi}\mathrm{R} \\ 0 \\ 0 \end{Bmatrix}. \qquad (3.12)$$

The $\mathbf{J}_{\mathrm{g}}$ velocity along the wheel track depends only on $\dot{\varphi}$, and its value is $\dot{\varphi}\mathrm{R}$. ◀

3.3 Basic Constraint Conditions: Contact and No Sliding

The *impenetrability of material objects* has consequences when they come into contact (in one or more points): whether they are rigid or deformable, this property introduces a restriction in their relative movement. Another restriction comes from *surface roughness*: when two material objects are in contact, roughness may prevent the sliding between them.

Any restriction on the movement of rigid bodies constitutes a **constraint condition**. The conditions of contact and no sliding in a *single-point contact* between two rigid bodies are called **basic constraint conditions**.

The kinematic equations that express these limitations are called **constraint** or **restriction equations**. When dealing with rigid bodies, these equations are linear relationships among the variables used to describe the velocity of rigid bodies separately.

In a contact between the points $\mathbf{J}_1$ and $\mathbf{J}_2$ of the rigid bodies S1 and S2, respectively (Fig. 3.7), the constraints are expressed from the projections of the velocities of $\mathbf{J}_1$ and $\mathbf{J}_2$ on the normal direction and on the plane tangent to the contact:

- *contact*: $$\overline{\mathbf{v}}_{\mathrm{R2}}(\mathbf{J}_1)]_{\mathrm{n}} = 0 \left(\text{or } \overline{\mathbf{v}}_{\mathrm{R1}}(\mathbf{J}_2)]_{\mathrm{n}} = 0\right) \qquad (3.13)$$
- *no sliding*: $$\overline{\mathbf{v}}_{\mathrm{R2}}(\mathbf{J}_1)]_{\mathrm{tg}} = \overline{\mathbf{0}} \left(\text{or } \overline{\mathbf{v}}_{\mathrm{R1}}(\mathbf{J}_2)]_{\mathrm{tg}} = \overline{\mathbf{0}}\right) \qquad (3.14)$$

If it is a sliding contact, only the first equation holds. If it is a nonsliding contact, both equations apply simultaneously, and that is equivalent to the condition:

- *nonsliding contact*: $$\overline{\mathbf{v}}_{\mathrm{R2}}(\mathbf{J}_1) = \overline{\mathbf{0}} \left(\text{or } \overline{\mathbf{v}}_{\mathrm{R1}}(\mathbf{J}_2) = \overline{\mathbf{0}}\right) \qquad (3.15)$$

If this restriction occurs instantaneously in a succession of points of the rigid bodies, we say that there is **perfect rolling** between them.

These restriction equations can also be formulated in a general reference R. In that case, they are:

- *contact*: $$\overline{\mathbf{v}}_{\mathrm{R}}(\mathbf{J}_1)]_{\mathrm{n}} = \overline{\mathbf{v}}_{\mathrm{R}}(\mathbf{J}_2)]_{\mathrm{n}} \qquad (3.16)$$
- *no sliding*: $$\overline{\mathbf{v}}_{\mathrm{R}}(\mathbf{J}_1)]_{\mathrm{tg}} = \overline{\mathbf{v}}_{\mathrm{R}}(\mathbf{J}_2)]_{\mathrm{tg}} \qquad (3.17)$$
- *perfect rolling*: $$\overline{\mathbf{v}}_{\mathrm{R}}(\mathbf{J}_1) = \overline{\mathbf{v}}_{\mathrm{R}}(\mathbf{J}_2) \qquad (3.18)$$

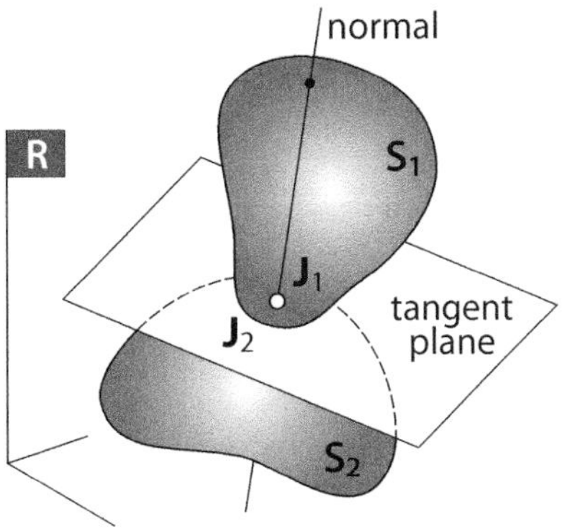

Fig. 3.7

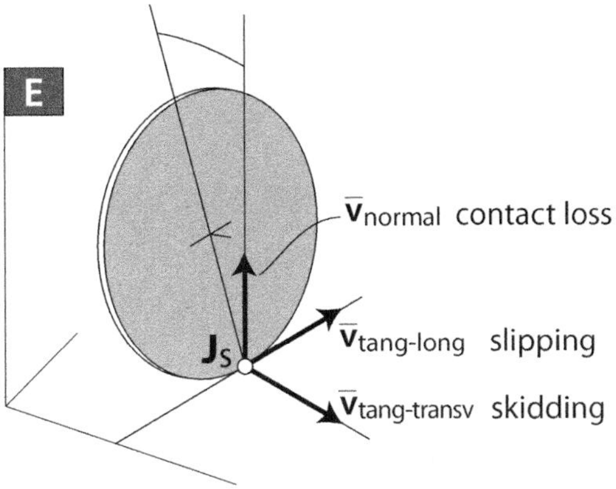

Fig. 3.8

For instance, when a wheel has perfect rolling motion on the ground E, the wheel point **J** in contact with the surface has zero velocity relative to it, and the instantaneous motion of the wheel is just a rotation around the contact point **J**: $\bar{\mathbf{v}}_{\mathrm{E}}(\mathbf{P}) = \bar{\mathbf{\Omega}}_{\mathrm{E}}^{\mathrm{s}} \times \bar{\mathbf{JP}}$.

This can be understood more easily by reasoning the impossibility of **J** having nonzero speed (Fig. 3.8) than by intuition. Intuition can lead to the erroneous idea that if **J** is at rest, the wheel rotation would cause a movement like the one shown in Fig. 3.9. This would be indeed the case if the contact point were always the same, but in perfect rolling this point is different at every time instant.

A spiked wheel provides an intuitive image of the process (Fig. 3.10). While the end of spike "a" (point $\mathbf{A}_{\mathrm{S}}$ of the wheel) is in contact with the ground, the wheel rotates around $\mathbf{A}_{\mathrm{S}}$ (which for a while is fixed to the ground). Later on, when point $\mathbf{B}_{\mathrm{S}}$ of spike "b" touches the ground, point $\mathbf{A}_{\mathrm{S}}$ leaves the ground, the wheel starts rotating around $\mathbf{B}_{\mathrm{S}}$, and so on. The limit case of a wheel with an infinite number of spikes provides the image of the perfect rolling motion of a wheel.

The general case of a rigid body with perfect rolling over another one can be illustrated in an analogous way (Fig. 3.11).

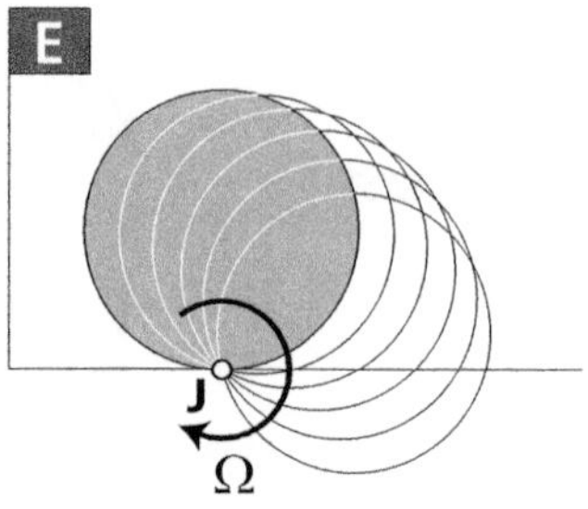

Fig. 3.9

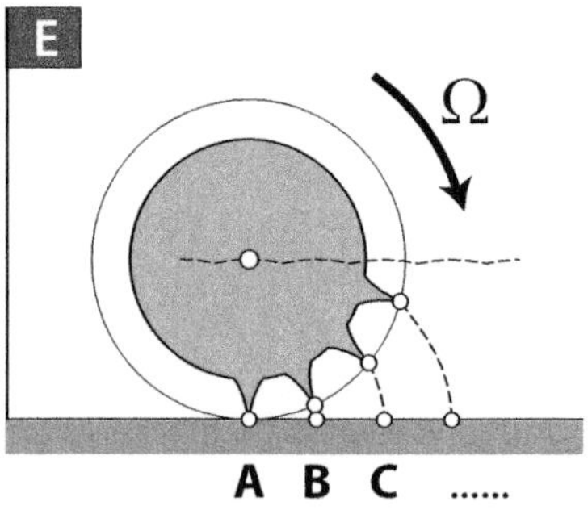

Fig. 3.10

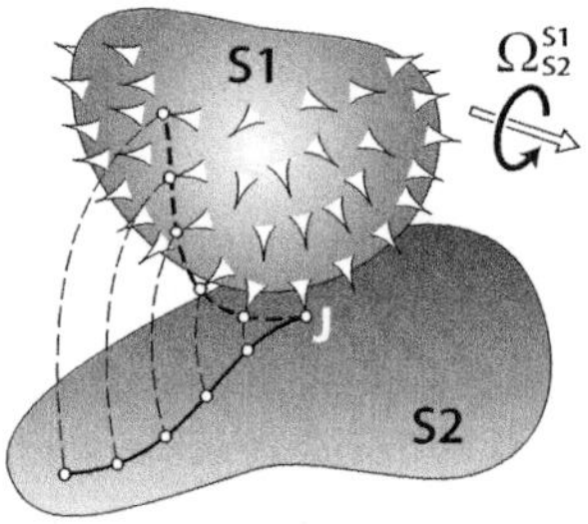

Fig. 3.11

In a rigid body with perfect rolling on a surface, although the instantaneous velocity of the contact point **J** relative to the surface is zero, the acceleration is not: just before contact, **J** is approaching the surface, and just after contact, it moves away from it. This change in speed implies an acceleration with a component normal to the surface (in the separating direction). This acceleration is formulated in Appendix 3A from the angular velocity of the body and the velocity of the geometric contact point.

3.4 Geometry of the Velocity Distribution: The Screw Axis

The term $\left(\bar{\mathbf{\Omega}}_{\mathrm{R}}^{\mathrm{S}} \times \overline{\mathbf{QP}}\right)$ appearing in the velocity distribution (Eq. (3.1)) is responsible for the differences among the velocities of points belonging to a same rigid body. As any vector product, it has the following properties:

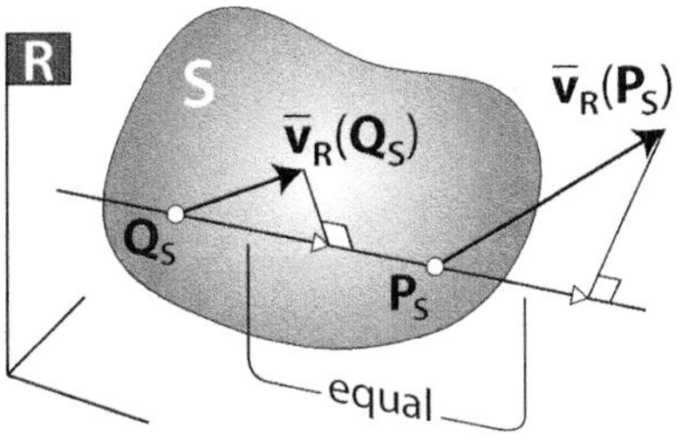

Fig. 3.12

(a) It is perpendicular to $\overline{\mathbf{QP}}$
(b) It is perpendicular to $\bar{\mathbf{\Omega}}_R^S$
(c) It is zero whenever $\bar{\mathbf{\Omega}}_R^S$ and $\overline{\mathbf{QP}}$ are parallel

These properties have consequences on the velocity distribution:

(a) The projections of the velocities of points **Q** and **P** on the line **QP** are equal (Fig. 3.12). This has to be so: if they were not equal, their difference would indicate the existence of an approaching or a separating speed between the two points
(b) The projections of the velocities of all points on the $\bar{\mathbf{\Omega}}_R^S$ direction is the same. When going from **P** to **Q** (Fig. 3.13), the additive term modifies only the component perpendicular to $\bar{\mathbf{\Omega}}_R^S$
(c) All points located on a straight line parallel to $\bar{\mathbf{\Omega}}_R^S$ have the same velocity (as for instance points **P**, **P**′, and **P**″ in Fig. 3.14)

Consider now straight lines of points parallel to $\bar{\mathbf{\Omega}}_R^S$. Due to properties (b) and (c), the corresponding velocity vectors can only differ in the component perpendicular to the line. This immediately raises a question: is there a line for which this component is zero (and therefore the velocity of its points has the direction of the line itself, which is that of $\bar{\mathbf{\Omega}}_R^S$)? The answer is affirmative: it is always possible to find a line of points with velocity parallel to $\bar{\mathbf{\Omega}}_R^S$.

Starting from point $\mathbf{Q}_s(=\mathbf{Q})$, looking for points $\mathbf{I}_s(=\mathbf{I})$ of the body having a velocity parallel to $\bar{\mathbf{\Omega}}_R^S$ is looking for points for which the additive term $\bar{\mathbf{\Omega}}_R^S \times \overline{\mathbf{QI}}$ is a vector opposite to $\bar{v}_R(\mathbf{Q})]_{\perp\Omega}$:

$$\bar{\mathbf{v}}_R(\mathbf{I})]_\Omega + \bar{\mathbf{v}}_R(\mathbf{I})]_{\perp\Omega} = \bar{\mathbf{v}}_R(\mathbf{Q})]_\Omega + \bar{\mathbf{v}}_R(\mathbf{Q})]_{\perp\Omega} + \bar{\mathbf{\Omega}}_R^S \times \overline{\mathbf{QI}}\Big]_{\perp\Omega},$$

$$\bar{\mathbf{v}}_R(\mathbf{I})]_{\perp\Omega} = 0 \Rightarrow \bar{\mathbf{v}}_R(\mathbf{Q})]_{\perp\Omega} = -\bar{\mathbf{\Omega}}_R^S \times \overline{\mathbf{QI}}. \tag{3.19}$$

In order that $\bar{\mathbf{\Omega}}_R^S \times \overline{\mathbf{QI}}$ has the suitable direction, **I** has to be located on the half plane defined by two straight lines going through **Q**: one parallel to $\bar{\mathbf{\Omega}}_R^S$ and the other one with the same direction as $\bar{\mathbf{\Omega}}_R^S \times \overline{\mathbf{QI}}$ (Fig. 3.15).

In order that $\bar{\mathbf{\Omega}}_R^S \times \overline{\mathbf{QI}}$ has the suitable module, **I** has to be at a distance ρ of the straight line parallel to $\bar{\mathbf{\Omega}}_R^S$ and going through **Q**:

$$\rho = \frac{\left|\bar{v}_R(\mathbf{Q})]_{\perp\Omega}\right|}{\left|\bar{\mathbf{\Omega}}_R^S\right|}. \tag{3.20}$$

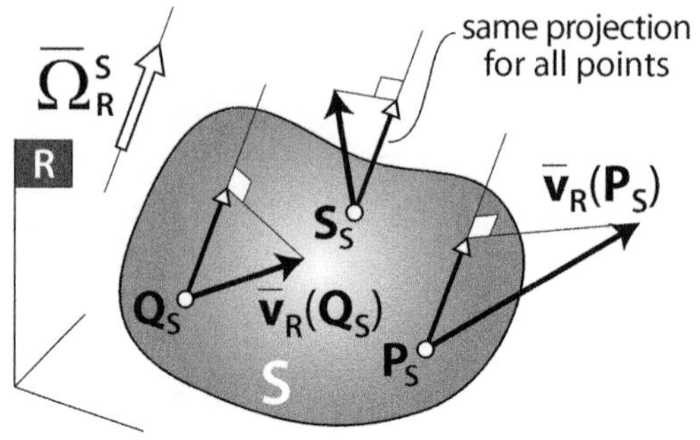

Fig. 3.13

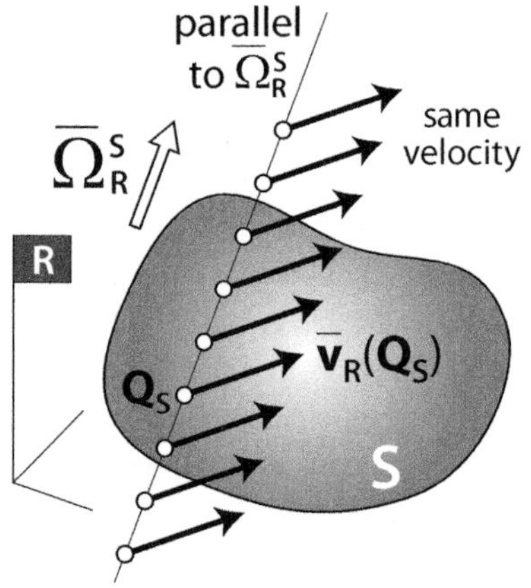

Fig. 3.14

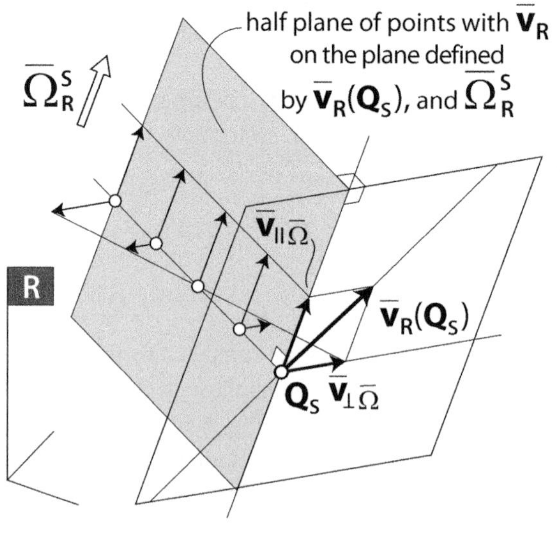

Fig. 3.15

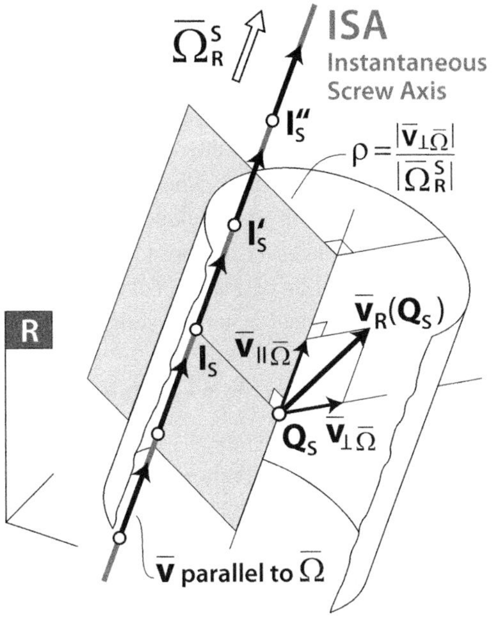

Fig. 3.16

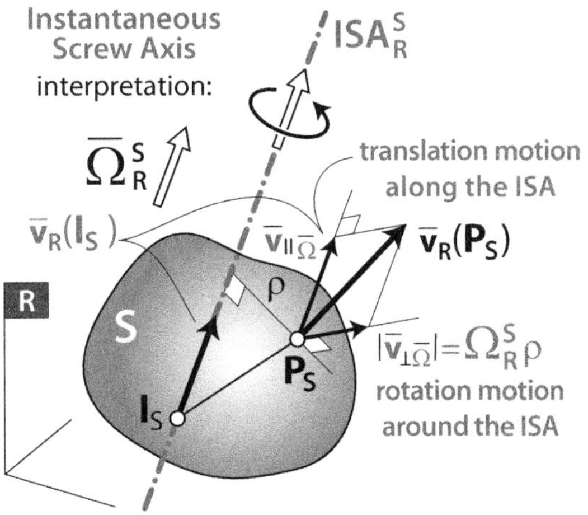

Fig. 3.17

This condition defines a cylinder with radius ρ whose axis of revolution is the line parallel to $\bar{\boldsymbol{\Omega}}_{\mathrm{R}}^{\mathrm{S}}$ and going through **Q** (Fig. 3.16). The intersection of that cylindrical surface and the aforementioned half plane determines the line of points with velocity parallel to $\bar{\boldsymbol{\Omega}}_{\mathrm{R}}^{\mathrm{S}}$.

When the velocity of any point **P** of the rigid body is calculated from the kinematic torsor at a point **I** of that line, $\left\{\bar{\mathrm{v}}_{\mathrm{R}}(\mathbf{I}), \bar{\boldsymbol{\Omega}}_{\mathrm{R}}^{\mathrm{S}}\right\}^{\mathrm{T}}$, that velocity has a simple description: it is the superposition of the velocity $\bar{\boldsymbol{\Omega}}_{\mathrm{R}}^{\mathrm{S}} \times \overline{\mathbf{I}\,\mathbf{P}}$ associated with the rotation around the line parallel to $\bar{\boldsymbol{\Omega}}_{\mathrm{R}}^{\mathrm{S}}$ going through **I**, and the velocity $\bar{\mathrm{v}}_{\mathrm{R}}(\mathbf{I})$ along that line (Fig. 3.17). For this reason, the line of points whose velocity is parallel to $\bar{\boldsymbol{\Omega}}_{\mathrm{R}}^{\mathrm{S}}$ is called **instantaneous screw**

axis (ISA). As far as the velocity distribution is concerned, the instantaneous body motion is a composition of a rotation around the ISA plus a sliding translation along the ISA.

One obvious consequence is that the ISA points have the minimum speed (not necessarily zero), since the sliding velocity along the ISA is the same for all points of the body and that associated with the rotation around the ISA is zero for the ISA points.

The simple description of the velocity distribution allowed by the ISA cannot be extended to the acceleration distribution, since the corresponding equation (Eq. [3.2]) contains more terms. Thus, it is an error to calculate the acceleration of a point **P** of the rigid body as a superposition of a tangential acceleration associated with the sliding along the ISA and a tangential and a normal component associated with the rotation around the ISA.

Determining the ISA is not always easy, but in some cases it is straightforward. For instance:

- When two points of the body have zero speed: as zero is the minimum possible value, those points have to be on the ISA. This is the case of a body moving on a surface fixed to the reference frame while keeping a multiple-point nonsliding contact with it
- When the direction of $\bar{\mathbf{\Omega}}_{\mathrm{R}}^{\mathrm{S}}$ is known and there is a point – or a straight line of points – whose velocity is parallel to $\bar{\mathbf{\Omega}}_{\mathrm{R}}^{\mathrm{S}}$: those points have just the sliding velocity along the ISA and therefore belong to the ISA. An example is a screw
- If the rigid body has two points **A** and **B** with the same velocity, the line **AB** is parallel to the ISA, and the **AB** direction is that of the $\bar{\mathbf{\Omega}}_{\mathrm{R}}^{\mathrm{S}}$ (Fig. 3.18a). Moreover, if the velocity of **A** and **B** is perpendicular to **AB**, that velocity comes exclusively

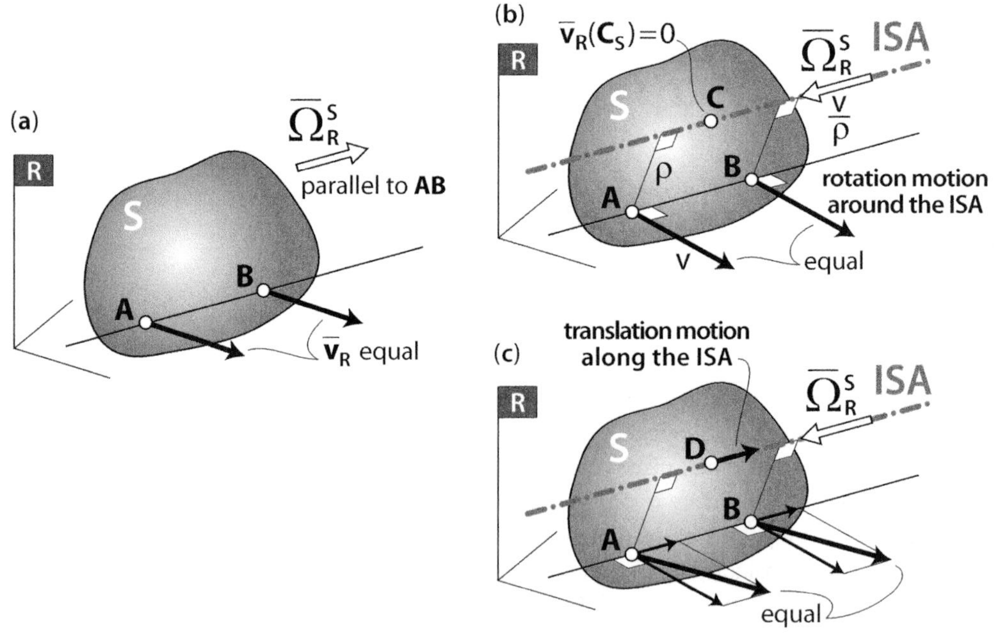

Fig. 3.18

from the "rotation about the ISA," and the points on the ISA have zero speed – instantaneously or permanently. It will be enough to locate a point **C** of the body with zero speed: the ISA at that time instant is parallel to the line **AB** going through **C** (Fig. 3.18b). If the velocity of **A** and **B** is not perpendicular to the line **AB**, its projection on **AB** is the velocity associated with the "sliding along the ISA." Finding a point **D** of the body with velocity parallel to **AB** determines the ISA, which is the line parallel to **AB** and going through **D** (Fig. 3.18c)

In principle, the screw axis changes along time. At each time instant, it coincides with a line of points of the reference frame R and with a line of points of the body frame. The temporal succession of these lines in the respective frames defines two ruled surfaces: the **fixed axode** – fixed to the reference frame R – and the **moving axode** – fixed to the body frame.

At each time instant, the two axodes share a contact generatrix, which is the ISA. The moving axode rolls on the fixed axode around that generatrix and slides along it. The movement of the moving axode on the fixed axode is identical to that of the rigid body relative to the reference frame R.

► **Example 3.6** The disk (Fig. 3.19a) rolls without sliding on the ground (which is the reference frame) driven by the rotating arm. Its angular velocity can be described as the superposition of two Euler rotations: a first Euler rotation $\overline{\dot\psi}$, associated with the movement of the arm around the vertical axis fixed to the ground, and a third Euler rotation $\overline{\dot\varphi}$ around the arm. The second Euler rotation $\overline{\dot\theta}$, which would correspond to the change of inclination of the disk relative to the plane, is zero in this case.

The nonsliding condition at the disk–ground contact imposes a dependency relationship between $\overline{\dot\psi}$ and $\overline{\dot\varphi}$. That relationship is formulated in a straightforward way through the ISA. The disk has two zero velocity points:

- point **J**, because of the perfect rolling condition between disk and ground
- point **A**, which belongs to the disk frame as its distance to any point on the disk is constant

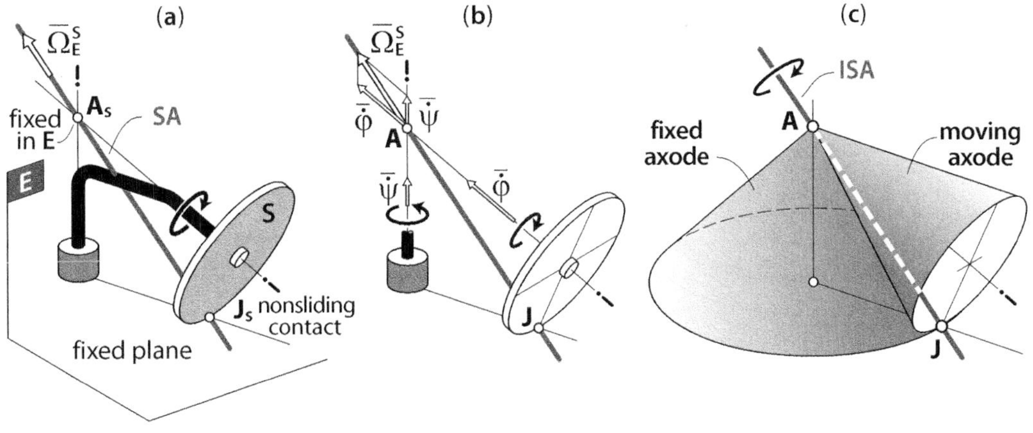

Fig. 3.19

The line **AJ** is the ISA. The relationship between $\overline{\dot{\psi}}$ and $\overline{\dot{\varphi}}$ is readily formulated by imposing that $\overline{\dot{\psi}}+\overline{\dot{\varphi}}$ has the ISA direction (Fig. 3.19b).

The fixed axode is a conical surface with vertical axis, while the moving axode is a conical surface whose axis is the disk axis. Both axodes share permanently the vertex **A** and have a nonsliding contact along the ISA (Fig. 3.19c). ◀

▶ **Example 3.7** The ball has nonsliding contacts at points **P** and **Q** with the ring track fixed to the ground E (Fig. 3.20a). As the ball evolves around the ring axis, it could be wrongly thought that $\bar{\mathbf{\Omega}}_{\mathrm{E}}^{\mathrm{ball}}$ is vertical. This is not so: as **P** and **Q** do not slide, their instantaneous speed relative to the ground is zero and consequently they determine the ISA. The ball angular velocity $\bar{\mathbf{\Omega}}_{\mathrm{E}}^{\mathrm{ball}}$ has the radial direction of **PQ** and goes always through the ring track center $\mathbf{O}_{\mathrm{E}}$ (Fig. 3.20b).

As the $\bar{\mathbf{v}}_{\mathrm{E}}(\mathbf{C})$ can be obtained as "rotation about the ISA," its modulus v is related to that of $\bar{\mathbf{\Omega}}_{\mathrm{E}}^{\mathrm{ball}}$ through the distance from **C** to the ISA (h). Thus:

$$\{\bar{\mathbf{v}}_{\mathrm{E}}(\mathbf{C})\}_{\mathrm{B}}=\begin{Bmatrix}0\\0\\\mathrm{v}\end{Bmatrix},\quad \left\{\bar{\mathbf{\Omega}}_{\mathrm{E}}^{\mathrm{ball}}\right\}_{\mathrm{B}}=\begin{Bmatrix}0\\-\mathrm{v}/\sqrt{\mathrm{r}^2-\mathrm{s}^2}\\0\end{Bmatrix}. \tag{3.21}$$

As point $\mathbf{O}_{\mathrm{E}}$ always belongs to the ISA, it is also a point $\mathbf{O}_{\mathrm{ball}}$ of the ball reference frame, which is permanently at rest $\left(\bar{\mathbf{v}}_{\mathrm{E}}(\mathbf{O}_{\mathrm{ball}})=\bar{\mathbf{0}}, \bar{\mathbf{a}}_{\mathrm{E}}(\mathbf{O}_{\mathrm{ball}})=\bar{\mathbf{0}}\right)$. This is a good clue to determine the axodes:

- Fixed axode: formed by horizontal radial lines going through $\mathbf{O}_{\mathrm{E}}$; it is a conical surface with vertex $\mathbf{O}_{\mathrm{E}}$, vertical axis and half aperture 90°
- Moving axode: it is also a conical surface with the same vertex $\mathbf{O}_{\mathrm{ball}}=\mathbf{O}_{\mathrm{E}}$; as the angle between the ISA and the $\mathbf{O}_{\mathrm{ball}}\mathbf{C}$ axis is constant, this axode is a

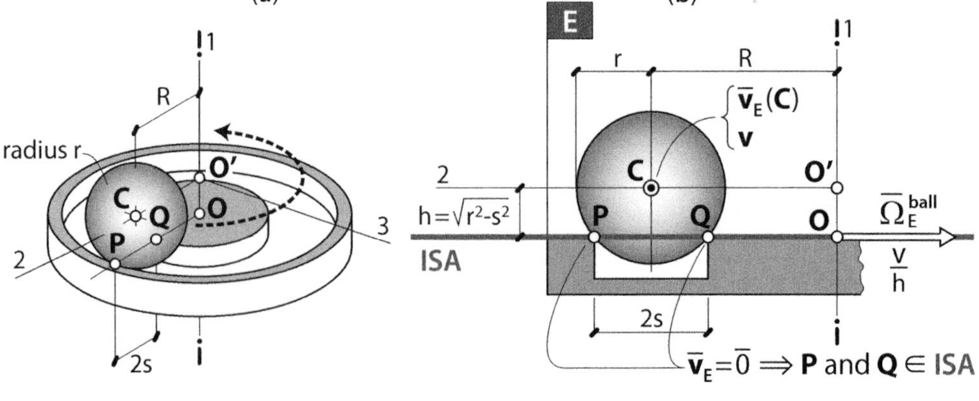

Fig. 3.20

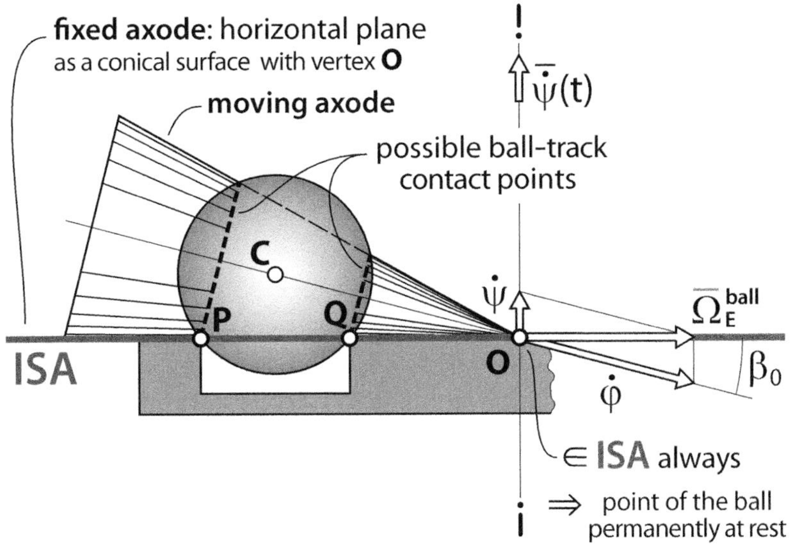

Fig. 3.21

revolution conical surface, with half aperture $\beta_0 = \text{a}\tan\left(R/\sqrt{r^2 - s^2}\right)$ (Fig. 3.21)

Describing the angular velocity of a ball in general through Euler rotations is not straightforward because of the ball spherical geometry: there is no singular axis in the ball, and the visualization of the third Euler rotation $\bar{\dot{\varphi}}$ is not simple. However, in this particular example, the description is evident once the axodes have been identified:

- First Euler rotation $\bar{\dot{\psi}}$: it is the rotation of the vertical plane containing **C**, **P**, and **Q** about the vertical axis going through $\mathbf{O}_E$; its value $\dot{\psi}$ is related to the **C** motion through $\dot{\psi} = v/R$
- Third Euler rotation $\bar{\dot{\varphi}}$: it is the rotation of the moving axode (and the ball!) about its axis **OC**. As the addition $\bar{\dot{\psi}} + \bar{\dot{\varphi}} = \bar{\mathbf{\Omega}}_E^{ball}$ has to be horizontal, its value $\dot{\varphi}$ is related to $\dot{\psi}$ through $\dot{\varphi} = \dot{\psi}/\sin\beta_0$

The second Euler angle corresponds to the inclination β_0 of the $\bar{\dot{\varphi}}$ axis relative to the horizontal plane, and it is constant $\left(\bar{\dot{\theta}} = \bar{0}\right)$.

Knowing the ISA and the axodes is also useful to discover other non evident aspects of kinematics. For instance:

- Point $\mathbf{O}'_{ball}$ of the ball that coincides instantaneously with $\mathbf{O}'_E$ does not have zero velocity as it is not located on the ISA
- Points **P** and **Q** of the ball have nonzero acceleration, as they belong to the moving axode and the only point on that surface that is permanently at rest is $\mathbf{O}_{ball}$
- The ball angular acceleration $\bar{\boldsymbol{\alpha}}_E^{ball}$ comes from the change in value and direction of $\bar{\mathbf{\Omega}}_E^{ball}$. The change in value generates a term parallel to $\bar{\mathbf{\Omega}}_E^{ball}$ with value $\dot{v}/\sqrt{r^2 - s^2}$. The change in direction is only associated with the horizontal component of $\bar{\dot{\varphi}}$ (with value $\dot{\varphi}\cos\beta_0$), which rotates with angular velocity $\bar{\dot{\psi}}$ relative to the ground

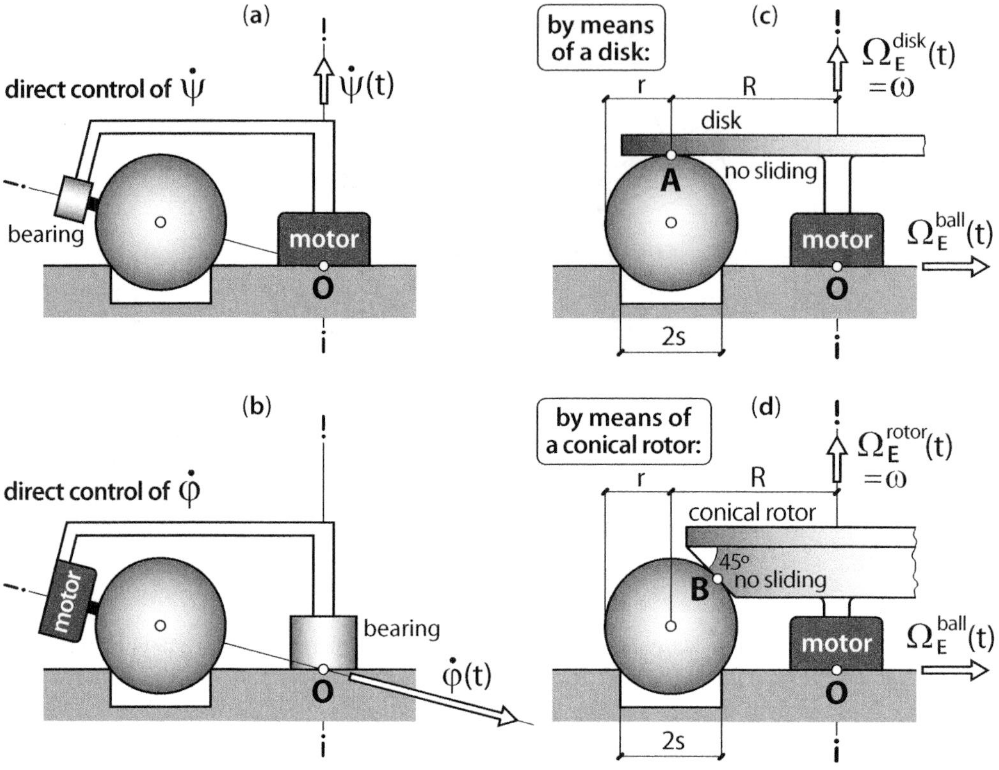

Fig. 3.22

Another interesting issue is that of the control of the ball motion. As it is a rotation, it can be achieved through a motor. The strategy is not unique:

- Direct control of $\dot{\psi}$: the motor stator is fixed to the ground, and the rotor coincides with the axis of the moving axode (Fig. 3.22a). The rotor speed ω coincides with $\dot{\psi}$
- Direct control of $\dot{\varphi}$: the motor stator is fixed to the axis of the moving axode, and the rotor coincides with the moving axode (Fig. 3.22b). The rotor speed ω coincides with $\dot{\varphi}$
- Control of an intermediate rotation through a new contact with the ball: the motor stator is fixed to the ground, and the rotor has a nonsliding single-point contact with the ball (Figs. 3.22c,d). The rotor speed ω coincides neither with $\dot{\psi}$ nor with $\dot{\varphi}$. Its value as a function of $\mathbf{\Omega}_{\mathrm{E}}^{\mathrm{ball}}$ can be obtained by imposing that $\bar{v}_{\mathrm{E}}(\mathbf{A}_{\mathrm{S}},\mathbf{B}_{\mathrm{S}}) = \bar{v}_{\mathrm{E}}(\mathbf{A}_{\mathrm{rotor}},\mathbf{B}_{\mathrm{rotor}})$ instantaneously:

$$v_{\mathrm{E}}(\mathbf{A}_{\mathrm{S}}) = \left(r + \sqrt{r^2 - s^2}\right)\mathbf{\Omega}_{\mathrm{E}}^{\mathrm{ball}} = v_{\mathrm{E}}(\mathbf{A}_{\mathrm{rotor}}) = R\omega,$$
$$v_{\mathrm{E}}(\mathbf{B}_{\mathrm{S}}) = \left(\frac{r}{\sqrt{2}} + \sqrt{r^2 - s^2}\right)\mathbf{\Omega}_{\mathrm{E}}^{\mathrm{ball}} = v_{\mathrm{E}}(\mathbf{B}_{\mathrm{rotor}}) = \left(R - \frac{r}{\sqrt{2}}\right)\omega.$$

◀

The axodes are useful to study the equivalence of mechanisms. Often, the mechanisms that restrict the movement of a rigid body determine the axodes univocally. In these cases, two mechanisms are kinematically equivalent if they lead to the same axodes.

► **Example 3.8** Figure 3.23 shows four different ways to articulate the rigid body S1 to the ground S2:

- S1 fixed inside a spherical surface that rotates without sliding on the circular groove with center **O**
- S1 fixed to a wheel axis and articulated to point **O** in the ground through a ball-and-socket joint
- S1 fixed to an axis going through **O** and fixed to two wheels with different radius
- S1 fixed inside a conical surface with vertex **O** that rotates without sliding on the ground

In all these cases, both axodes (fixed and moving) are determined by the constraints, and they are always the same. Therefore, the four mechanical designs are kinematically equivalent.

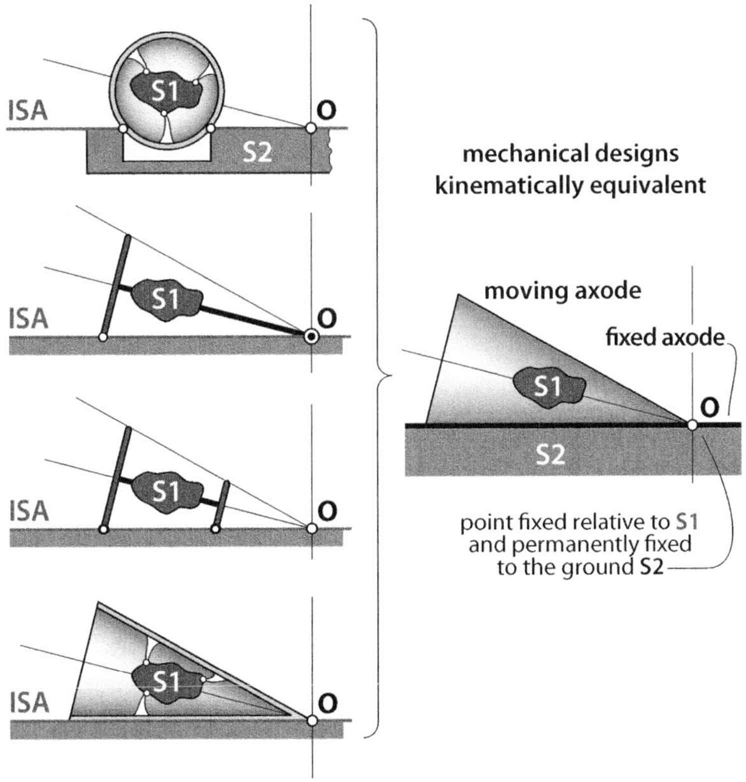

Fig. 3.23

◄

3.5 A Particular Case: Planar Motion, Instantaneous Center of Rotation

When all points in a rigid body have planar trajectories (necessarily contained in parallel planes), we say that the rigid body describes a **planar motion**. The direction of the body angular velocity $\bar{\boldsymbol{\Omega}}_{\mathrm{R}}^{\mathrm{S}}$ is constant and normal to those parallel planes (called **planes of motion**), and the sliding velocity along the ISA is zero.

When we apply Eqs. (3.1, 3.2) to the points of a section of the rigid body by a plane of motion, the velocity and acceleration distributions accept a particularly simple geometrical description because of the perpendicularity of $\bar{\boldsymbol{\Omega}}_{\mathrm{R}}^{\mathrm{S}}$ and $\bar{\boldsymbol{\alpha}}_{\mathrm{R}}^{\mathrm{S}}$ to the plane of motion (Figs. 3.24 and 3.25).

The interpretation of the velocity distribution based on the ISA is also simple because the constant direction of $\bar{\boldsymbol{\Omega}}_{\mathrm{R}}^{\mathrm{S}}$ implies that the axodes are cylindrical surfaces (Fig. 3.26). Since the sliding velocity along the ISA is zero, the moving axode rolls on the fixed axode without sliding along the contact generatrix.

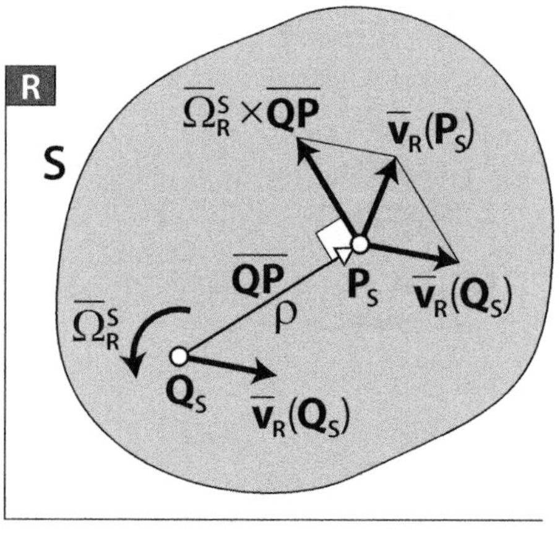

Fig. 3.24

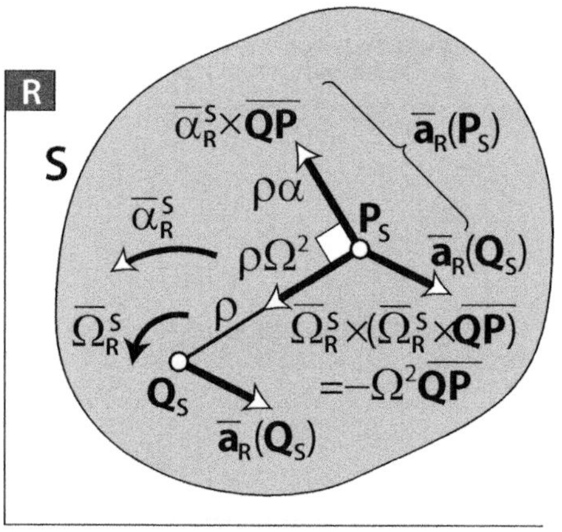

Fig. 3.25

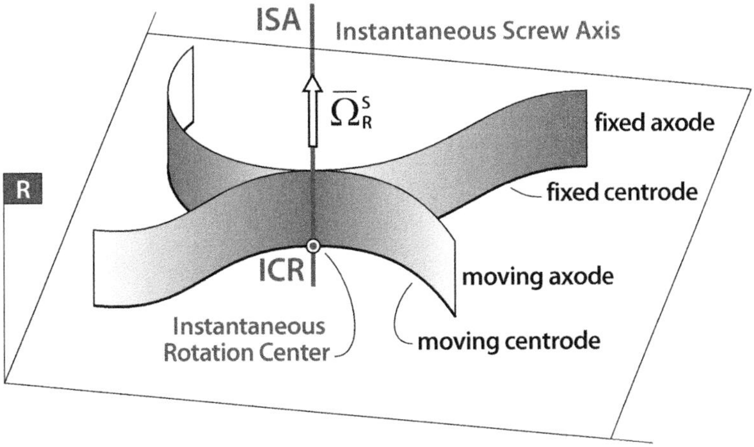

Fig. 3.26

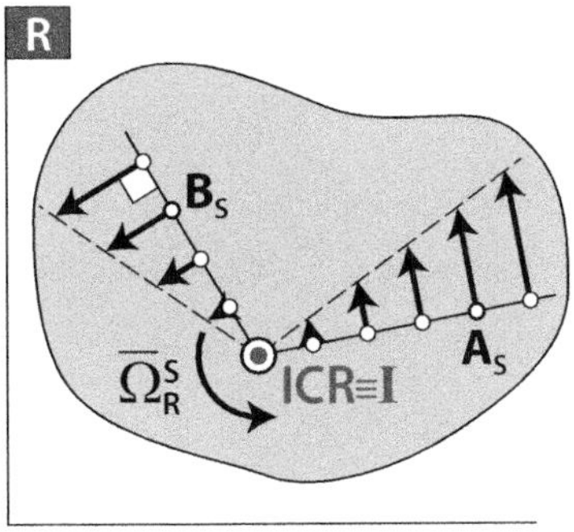

Fig. 3.27

Every plane of motion intersects the ISA and the axodes, and defines the following elements:

- **Instantaneous center of rotation** (**ICR**): it is the intersection of the plane of motion and the ISA, and has zero velocity. As far as velocities are concerned, the instantaneous movement of the body is a rotation around the **ICR**
- **Fixed centrode**: it is the intersection of the plane of motion and the fixed axode. It is the locus of the **ICR** in the reference frame
- **Moving centrode**: it is the intersection of the plane of motion and the moving axode. It is the locus of the **ICR** in the body reference frame

In each plane of motion, the moving centrode rolls without sliding on the fixed centrode. At each time instant, the point of tangency is the **ICR**.

The velocities of points $\mathbf{A}_S$, $\mathbf{B}_S$, etc. in a same plane of motion (Fig. 3.27) are associated with the rotation around the **ICR** (or **I**). They are perpendicular to vectors $\overline{\mathbf{IA}}$, $\overline{\mathbf{IB}}$, etc., and their modules are proportional to their distance to the **ICR** through the proportionality factor $\left|\bar{\mathbf{\Omega}}^S_R\right|$. This geometry facilitates the determination of the **ICR**.

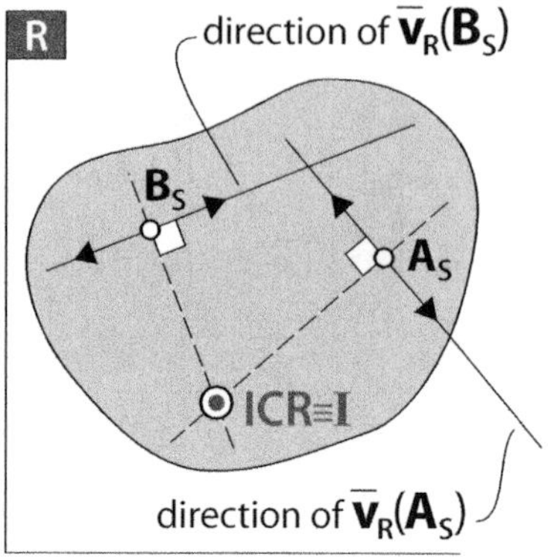

Fig. 3.28

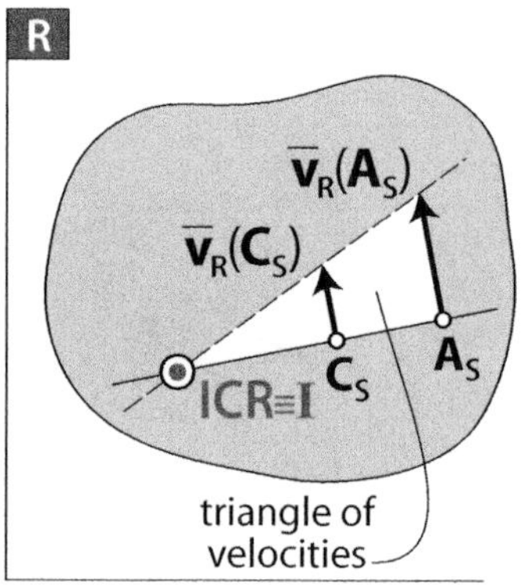

Fig. 3.29

Thus, if we know the direction of the velocities of two points $\mathbf{A}_S$ and $\mathbf{B}_S$ in the body (Fig. 3.28), the **ICR** is given by the intersection of the lines perpendicular to those velocities and passing through $\mathbf{A}_S$ and $\mathbf{B}_S$.

If we know the velocity of two points $\mathbf{A}_S$ and $\mathbf{C}_S$, and they are parallel (Fig. 3.29), those two points are aligned with the **ICR**. In the linear velocity distribution associated with the points located on the $\mathbf{A}_S\mathbf{C}_S$ line, the **ICR** is the zero velocity point.

► **Example 3.9** The end points **A** and **B** of the bar with length 2L slide along the orthogonal straight grooves a–a′ and b–b′, respectively, fixed to the ground (Fig. 3.30a). As the velocities $\bar{\mathbf{v}}_E(\mathbf{A})$ and $\bar{\mathbf{v}}_E(\mathbf{B})$ have the a–a′ and b–b′ direction, respectively, the intersection of the straight lines going through **A** and **B** and perpendicular to the respective grooves determines the **ICR** (Fig. 3.30b). Since the distance $\overline{\mathbf{OI}}$ remains constant and equal to 2L and **O** is fixed to the ground, the fixed centrode is the circumference with center **O** and radius 2L.

In the bar reference frame, the distance between the bar center **C** and the **ICR** is constant and equal to L. Consequently, the moving centrode is the circumference with center **C** and radius L.

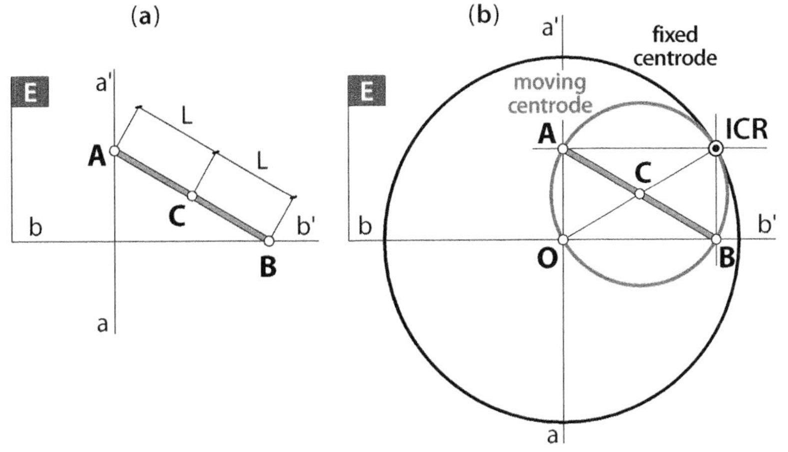

Fig. 3.30

If we removed the grooves and we fixed the bar to the moving centrode, it would be dragged by the latter (which rotates without sliding on the fixed centrode), and the resulting motion would be exactly the same.

The guiding mechanism based on the two orthogonal grooves is equivalent to a mechanism composed of a gear wheel with effective diameter 2L (one of whose diameters is the bar) rolling without sliding inside a ground-fixed crown with an effective diameter 4L. ◀

▶ **Example 3.10** The wheel P and the arm A rotate around the ground-fixed point **O** with angular velocities $\bar{\mathbf{\Omega}}_E^P$ and $\bar{\mathbf{\Omega}}_E^A$, respectively (Fig. 3.31a). The arm drags the wheel Q, which rolls without sliding relative to the wheel P.

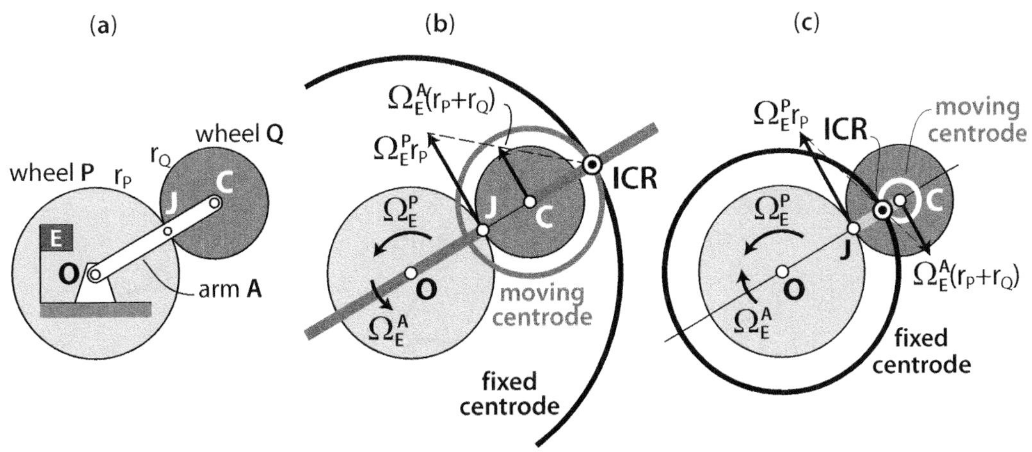

Fig. 3.31

The **ICR** of wheel Q can be determined from the velocities of its points **J** and **C**, which have to be equal to those of points **J** of wheel P, and **C** of the arm, respectively:

$$\bar{\mathbf{v}}_{\mathrm{E}}(\mathbf{J})=\bar{\boldsymbol{\Omega}}_{\mathrm{E}}^{\mathrm{P}}\times\overline{\mathbf{OJ}}\Rightarrow|\bar{\mathbf{v}}_{\mathrm{E}}(\mathbf{J})|=\mathrm{r_P}\boldsymbol{\Omega}_{\mathrm{E}}^{\mathrm{P}};\quad \bar{\mathbf{v}}_{\mathrm{E}}(\mathbf{C})=\bar{\boldsymbol{\Omega}}_{\mathrm{E}}^{\mathrm{A}}\times\overline{\mathbf{OC}}\Rightarrow|\bar{\mathbf{v}}_{\mathrm{E}}(\mathbf{C})|=(\mathrm{r_P}+\mathrm{r_Q})\boldsymbol{\Omega}_{\mathrm{E}}^{\mathrm{A}}.$$

Those velocities are parallel, and the **ICR** corresponds to the point on the line **JC** with zero velocity. Figure 3.31b corresponds to rotations $\bar{\boldsymbol{\Omega}}_{\mathrm{E}}^{\mathrm{P}}$ and $\bar{\boldsymbol{\Omega}}_{\mathrm{E}}^{\mathrm{A}}$ in the same direction. The locus of instantaneous centers is the straight line going through **J** and **C** (excepting the segment **JC**), and the specific location depends on the $\left(\Omega_{\mathrm{E}}^{\mathrm{P}}/\Omega_{\mathrm{E}}^{\mathrm{A}}\right)$ ratio.

Figure 3.31c corresponds to rotations $\bar{\boldsymbol{\Omega}}_{\mathrm{E}}^{\mathrm{P}}$ and $\bar{\boldsymbol{\Omega}}_{\mathrm{E}}^{\mathrm{A}}$ in opposite directions. The possible locations of instantaneous centers are now on the **JC** segment. ◀

► **Example 3.11** The wheel has a radius r and rolls without sliding on the ground (E), so $\mathbf{ICR}_{\mathrm{E}}^{\mathrm{wheel}}=\mathbf{J}$ (Fig. 3.32a). If its center **C** moves with constant speed $\mathrm{v_0}\ (=|\bar{\mathbf{v}}_{\mathrm{E}}(\mathbf{C})|)$, point **P** will have a speed $\mathrm{v_0}\sqrt{2}$ (as it is $\sqrt{2}$ times farther from **J** than **C**, Fig. 3.32b).

The acceleration $\bar{\mathbf{a}}_{\mathrm{E}}(\mathbf{P})$ cannot be obtained from that instantaneous velocity through a time derivative. Instead, RBK can be applied. As $\mathrm{v_0}$ is constant, $\bar{\mathbf{a}}_{\mathrm{E}}(\mathbf{C})=\bar{0}$ and $\bar{\boldsymbol{\alpha}}_{\mathrm{E}}^{\mathrm{wheel}}=\bar{0}$:

$$\bar{\mathbf{a}}_{\mathrm{E}}(\mathbf{P})=\bar{\boldsymbol{\Omega}}_{\mathrm{E}}^{\mathrm{wheel}}\times\left(\bar{\boldsymbol{\Omega}}_{\mathrm{E}}^{\mathrm{wheel}}\times\overline{\mathbf{CP}}\right)=\leftarrow\frac{\mathrm{v_0^2}}{\mathrm{r}}.$$

That acceleration is neither parallel nor perpendicular to $\bar{\mathbf{v}}_{\mathrm{E}}(\mathbf{P})$, so the normal component of $\bar{\mathbf{a}}_{\mathrm{E}}(\mathbf{P})$ is just a projection of that total acceleration: $\left|\mathrm{a}_{\mathrm{E}}^{\mathrm{n}}(\mathbf{P})\right|=\mathrm{v_0^2/r}\sqrt{2}$, in the $\overline{\mathbf{PJ}}$ direction (Fig. 3.32b). The radius of curvature is:

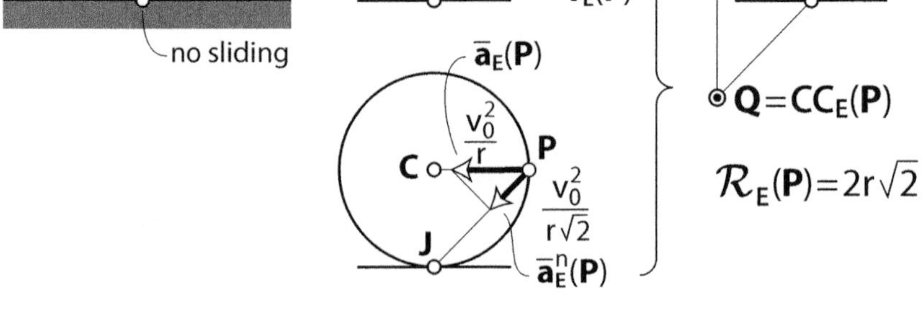

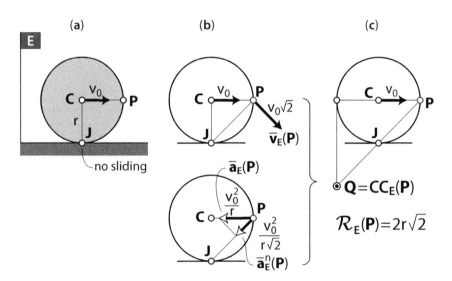

Fig. 3.32

$$\Re_E(\mathbf{P}) = \frac{v_E^2(\mathbf{P})}{|a_E^n(\mathbf{P})|} = \frac{2v_0^2}{v_0^2/r\sqrt{2}} = 2r\sqrt{2},$$

and the curvature center of its trajectory relative to the ground is **Q** (Fig. 3.32c). ◀

▶ **Example 3.12** The piston is articulated to the rotating bar **OP** and slides inside the cylinder, which in turn is articulated to the ground at point **Q** (Fig. 3.33a). The instantaneous center of rotation, relative to the ground, of the bar and the cylinder are straightforward ($\mathbf{ICR}_E^{bar} = \mathbf{O}$ and $\mathbf{ICR}_E^{cyl} = \mathbf{Q}$), but that of the piston is not.

There is just one point in the piston whose velocity relative to the ground has a well-known direction: as **P** belongs both to the piston an to the bar, its velocity is perpendicular to $\overline{\mathbf{OP}}$ as it is associated with the bar rotation about **O** (Fig. 3.33b). Thus, the $\mathbf{ICR}_E^{piston}$ has to be located on the straight line going through **O** and **P** (line 1).

The piston motion relative to the cylinder is very simple: a translation. If we take the cylinder as REL reference frame and the ground as AB reference frame:

- $\bar{v}_{REL}$, associated with the translation, is the same for all points in the piston
- $\bar{v}_{tr}$, associated with the rotation of the cylinder about **Q** (fixed to the ground), has the same direction for all points in the piston, and its value is proportional to the distance between the piston point and point **Q** (Fig. 3.33b)

If we consider point **P**$''$ of the piston (whose instantaneous location coincides with that of **Q**), $\bar{v}_{tr}(\mathbf{P}'') = \bar{0}$ and $\bar{v}_{AB}(\mathbf{P}'') = \bar{v}_{REL}(\mathbf{P}'')$. Thus, the $\mathbf{ICR}_E^{piston}$ has to be located on the straight line through **Q** perpendicular to the piston (line 2). The intersection between line 1 and line 2 is the $\mathbf{ICR}_E^{piston}$ (Fig. 3.33c).

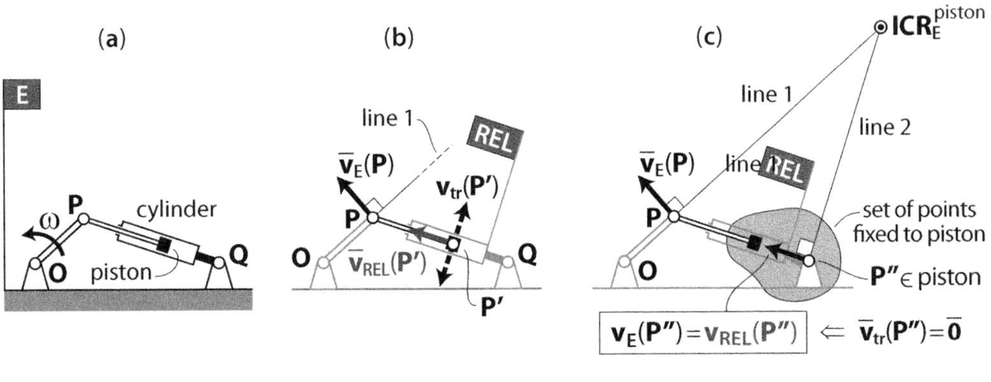

Fig. 3.33

◀

Appendix 3A Perfect Rolling: The Geometric Contact Point: Acceleration of Contact Points

The perfect rolling condition (nonsliding rotation) between two rigid bodies S1 and S2 can involve a single contact point or a line of contact points. All these points share the results presented below for single-point perfect rolling (Fig. 3A.1).

The geometric contact point $\mathbf{J}_g$ – showing the location of the contact at each time instant – has the same velocity in the body frames S1 and S2:

$$\bar{\mathbf{v}}_{S1}(\mathbf{J}_g) = \bar{\mathbf{v}}_{S2}(\mathbf{J}_g). \tag{3A.1}$$

This velocity is tangent to the respective tracks (trajectories of $\mathbf{J}_g$ on the rigid bodies surfaces).

The contact points $\mathbf{J}_{S1}$ and $\mathbf{J}_{S2}$, whose velocities fulfill the condition $\bar{\mathbf{v}}_{S2}(\mathbf{J}_{S1}) = \bar{\mathbf{v}}_{S1}(\mathbf{J}_{S2}) = \bar{\mathbf{0}}$, have nonzero accelerations in the S2 and S1 frames, respectively. Those accelerations can be calculated from the velocity of the geometric contact point:

$$\bar{\mathbf{a}}_{S2}(\mathbf{J}_{S1}) = -\bar{\boldsymbol{\Omega}}_{S2}^{S1} \times \bar{\mathbf{v}}_{S2}(\mathbf{J}_g) = -\bar{\mathbf{a}}_{S1}(\mathbf{J}_{S2}). \tag{3A.2}$$

They are symmetrical and perpendicular to $\bar{\mathbf{v}}_{S1,2}(\mathbf{J}_g)$ and to $\bar{\boldsymbol{\Omega}}_{S2}^{S1}$ and oriented according to the direction of mutual separation between S1 and S2: their projections on the direction normal to the contact tangent plane are separation accelerations.

♣ *Proof*

If we take S1 as *relative* frame and S2 as *absolute* frame, the velocity of the contact point $\mathbf{J}_g$ can be expressed as:

$$\bar{\mathbf{v}}_{S2}(\mathbf{J}_g) = \bar{\mathbf{v}}_{S1}(\mathbf{J}_g) + \bar{\mathbf{v}}_{tr}(\mathbf{J}_g). \tag{3A.3}$$

But $\bar{\mathbf{v}}_{tr}(\mathbf{J}_g) = \bar{\mathbf{v}}_{S2}(\mathbf{J}_{S1}) = \bar{\mathbf{0}}$, and Eq. (3A.1) holds.

To prove Eq. (3A.2), the starting point is the expression of $\bar{\mathbf{v}}_{S2}(\mathbf{P}_{S1})$ (where $\mathbf{P}_{S1}$ is any point of body S1) calculated at each instant from point $\mathbf{J}_{S1}$(which is the contact point at that precise time instant):

$$\bar{\mathbf{v}}_{S2}(\mathbf{P}_{S1}) = \bar{\boldsymbol{\Omega}}_{S2}^{S1} \times \overline{\mathbf{J}_{S1}\mathbf{P}_{S1}} \text{ at a precise time instant, and}$$

$$\bar{\mathbf{v}}_{S2}(\mathbf{P}_{S1}) = \bar{\boldsymbol{\Omega}}_{S2}^{S1} \times \overline{\mathbf{J}_g\mathbf{P}_{S1}} \text{ along time.} \tag{3A.4}$$

The time derivative of Eq. (3A.4) yields:

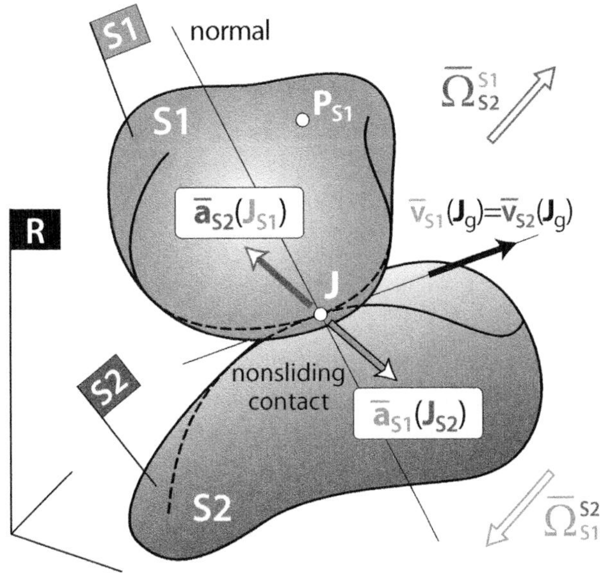

Fig. 3A.1

$$\bar{\mathbf{a}}_{S2}(\mathbf{P}_{S1}) = \bar{\boldsymbol{\alpha}}^{S1}_{S2} \times \overline{\mathbf{J}_g\mathbf{P}_{S1}} + \bar{\boldsymbol{\Omega}}^{S1}_{S2} \times \left.\frac{d\overline{\mathbf{J}_g\mathbf{P}_{S1}}}{dt}\right]_{S2}$$

$$= \bar{\boldsymbol{\alpha}}^{S1}_{S2} \times \overline{\mathbf{J}_g\mathbf{P}_{S1}} + \bar{\boldsymbol{\Omega}}^{S1}_{S2} \times \left(\bar{\mathbf{v}}_{S2}(\mathbf{P}_{S1}) - \bar{\mathbf{v}}_{S2}(\mathbf{J}_g)\right) \quad (3A.5)$$

If $\mathbf{P}_{S1}$ and $\mathbf{J}_{S1}$ coincide ($\mathbf{P}_{S1} \equiv \mathbf{J}_{S1}$), Eq. (A.I5) reduces to:

$$\bar{\mathbf{a}}_{S2}(\mathbf{J}_{S1}) = -\bar{\boldsymbol{\Omega}}^{S1}_{S2} \times \bar{\mathbf{v}}_{S2}(\mathbf{J}_g). \quad (3A.6)$$

Analogously:

$$\bar{\mathbf{a}}_{S1}(\mathbf{J}_{S2}) = -\bar{\boldsymbol{\Omega}}^{S2}_{S1} \times \bar{\mathbf{v}}_{S1}(\mathbf{J}_g) = +\bar{\boldsymbol{\Omega}}^{S1}_{S2} \times \bar{\mathbf{v}}_{S2}(\mathbf{J}_g) = -\bar{\mathbf{a}}_{S2}(\mathbf{J}_{S1}),$$

since $\bar{\boldsymbol{\Omega}}^{S2}_{S1} = -\bar{\boldsymbol{\Omega}}^{S1}_{S2}$ y $\bar{\mathbf{v}}_{S1}(\mathbf{J}_g) = \bar{\mathbf{v}}_{S2}(\mathbf{J}_g)$. ♣

The velocity $\bar{\mathbf{v}}_{S1,2}(\mathbf{J}_g)$ can be expressed as a function of $\bar{\boldsymbol{\Omega}}^{S1}_{S2}$ (or $\bar{\boldsymbol{\Omega}}^{S2}_{S1}$) and the curvature of S1 and S2 at the contact point (Euler–Savary equation for planar motion, and Euler–Savary generalized equation for 3D motion). This is interesting when studying the kinematics of mechanisms, but in a more basic context it is better to calculate the $\mathbf{J}_g$ velocity through the general methods of kinematics: time derivative of a position vector or RBK applied to a rigid body (or reference frame) containing $\mathbf{J}_g$.

Appendix 3B Maneuverability of Vehicles with Conventional Wheels

Figure 3B.1 shows the basic idealization of the nonsliding conventional wheel: rigid, with negligible thickness – and therefore with single-point contact with the ground – and perpendicular to the ground – considered to be flat and rigid. The center has one degree of freedom (DoF)[2] with value $R\dot{\varphi}$ in the direction of the horizontal diameter. It is controlled by the propulsion $\dot{\varphi}$ of the driving wheels. The planes containing the nonsteered wheels are fixed with respect to the chassis of the vehicle, while those containing the steered wheels rotate with angular velocity $\dot{\delta}$ relative to the chassis according to the steering control of the vehicle.

Less idealized descriptions (that include the flexibility of the wheel–ground contact) are necessary to study aspects such as the lateral sideslip of the wheels, and its consequences on the vehicle steering behavior (*understeer* and *oversteer*).

The next subsections are devoted to the kinematics of a wide variety of vehicles based on idealized conventional wheels.

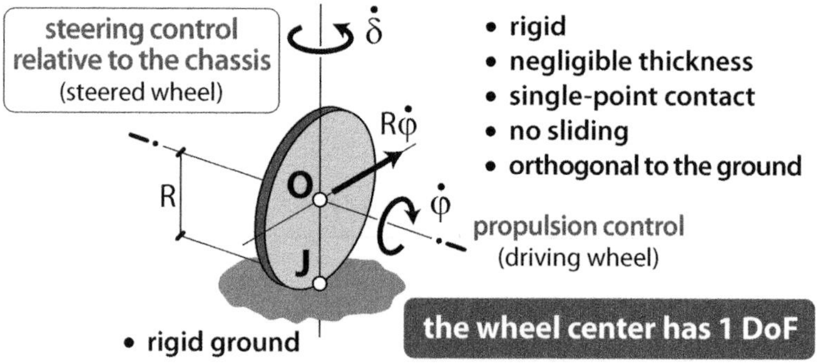

Fig. 3B.1

[2] A DoF is an independent velocity. This concept will be further discussed in Chapter 4.

3B.1 Vehicles with Two Steered Wheels and Two Nonsteered Wheels: 1-DoF Direction

It is the most usual type of vehicle – conventional cars, trucks, and buses. The two nonsteered wheels determine a chassis-fixed line that gives the possible locations of the chassis **ICR** (relative to the ground). The orientation of the two steered wheels with respect to the chassis – controlled by the DoF of the steering wheel – determines the precise position of this **ICR** on the straight line.

Figure 3B.2 shows that the two nonsteered wheels have to be coaxial; thus, all points of their common axis are possible **ICR** of the chassis with respect to the ground. If the two nonsteered wheels were fixed to the chassis (case b), the **ICR** would be determined by the intersection of their axes and would be fixed to the ground: the chassis would only be able to rotate around that point. If the axes were parallel but not coincident (case c), the **ICR** would be located at infinity, and the vehicle would only have one translation movement along its longitudinal axis.

The change of orientation of the two steered wheels with respect to the chassis (Fig. 3B.3) has to be different. If it were the same (Fig. 3B.3a), the intersection of each wheel axis with that of the nonsteered wheels would determine a point of zero speed, and the chassis would not be able to move without sliding. That difference in orientation constraints the design of the steering mechanism.

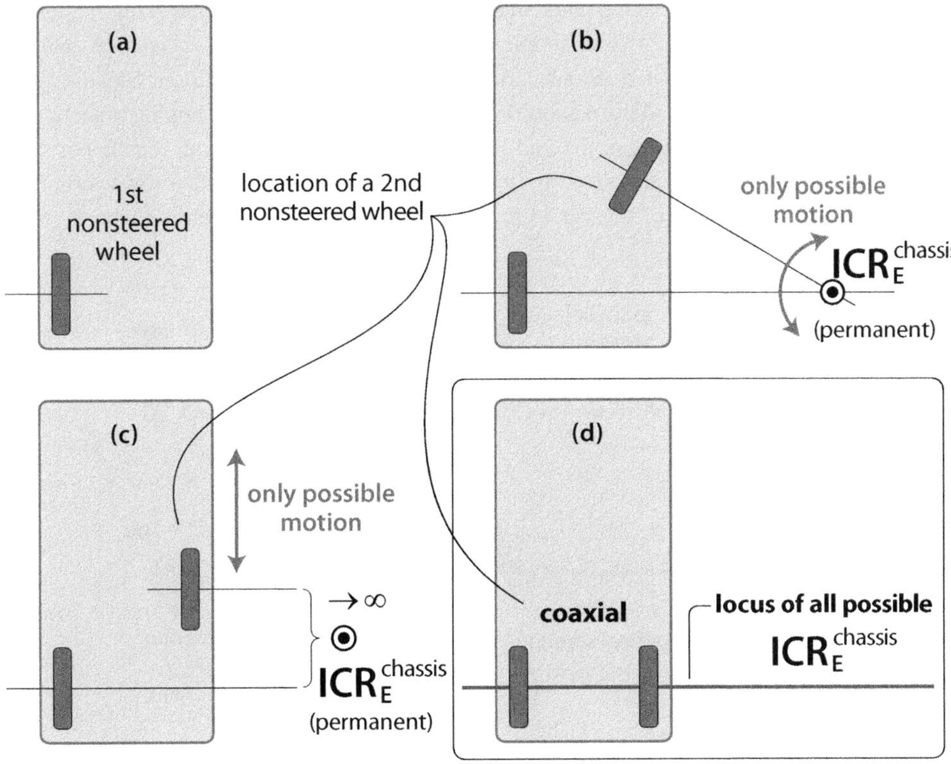

Fig. 3B.2

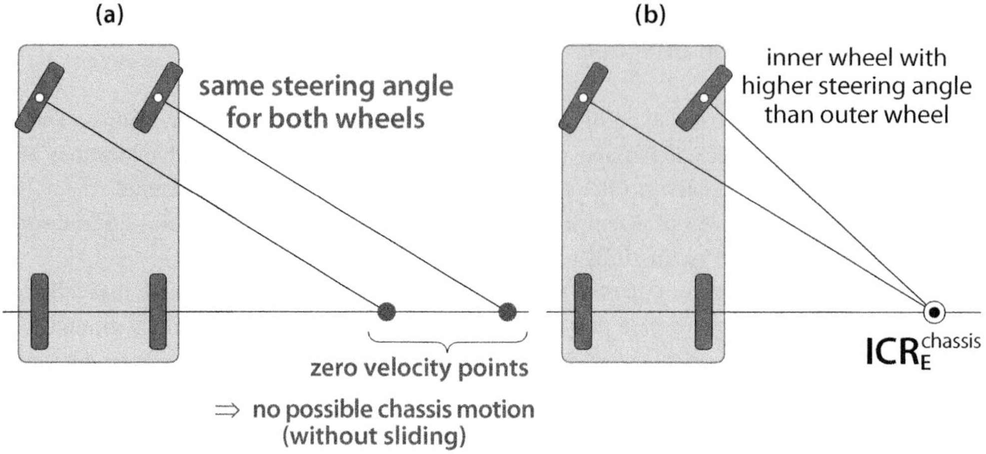

Fig. 3B.3

The Steering Trapeze

The rotation of the steered wheels must be such that their axes intersect the axis of the nonsteered wheels at a same point, which becomes the **ICR** of the chassis with respect to the ground. This is only achieved if the wheel at the inner side of a curve has a larger steer angle than that at the outer side (with a suitable value). The **steering trapeze** used in vehicles (Fig. 3B.4) is a simple mechanism that introduces acceptable changes of direction, although it does not strictly comply with that condition. Small errors are compensated by the deformation of the tires and do not imply slipping wheels.

In its simplest version, the end points of the cranks orientating the supports of the steered wheels are articulated by means of a bar that moves when the steering wheel turns. If the length of the bar were equal to the spacing between the wheels' rotation

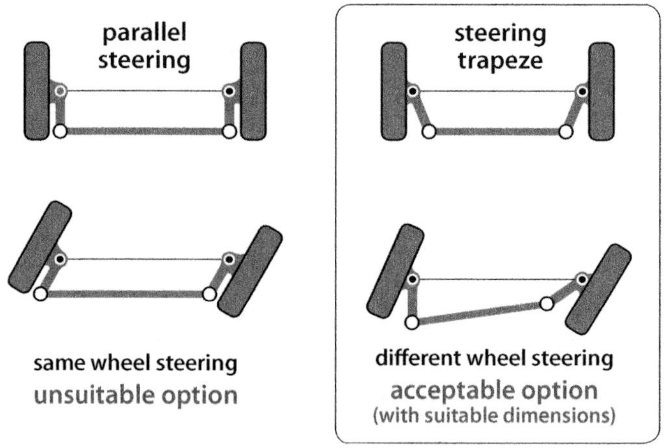

Fig. 3B.4

axes, the articulated quadrilateral would be a parallelogram in the central configuration, and the two wheels would have the same orientation change. It is possible to achieve a fairly acceptable approximate solution with a shorter bar (with a suitable length).

Front Steering and Rear Steering

Cars, trucks, and buses have front-steered wheels, while many maintenance vehicles (such as pallet trucks), public works, and agricultural vehicles have rear-steered wheels. In both cases the choice is based on kinematic reasons.

A **reason of convenience** to choose front steering is associated with the chassis velocity distribution in a turning vehicle: the velocity transverse component is higher at the vehicle end closest to the steered wheels than that at the end closest to the nonsteered wheels (Fig. 3B.5). If a vehicle has to move between other vehicles, it is convenient that the driver (oriented in the forward direction) sees ahead the vehicle zone with higher transverse motion to avoid collisions.

But there is a **reason of necessity** associated with the kinematic stability of the vehicle steering. One can analyze this stability in a simple way through the "bicycle" model (widely used in vehicle kinematics and dynamics).

If the front-steered wheel travels along a straight line (Fig. 3B.6), the angle ψ between the chassis longitudinal axis and that straight path decreases and tends asymptotically to zero: the chassis tends to be oriented in the direction of the forward movement of the steered wheel. It is a **directionally stable steering**. But if it is a rear-steered wheel, the angle θ increases and the chassis tends to deviate from the direction of vehicle motion: it is a **directionally unstable steering**.

Those stable/unstable tendencies also appear when the steered wheel follows a curved path. Figures 3B.7 and 3.B.8 show two cases where a front-steered wheel travels a circumference of radius R greater than the distance L between the front and rear axles. In Fig. 3B.7, the initial distance from the center **P** of the steered wheel to the **ICR** of the chassis with respect to the ground is less than the radius of the circular path of **P** with respect to the ground, while in Fig. 3B.7 it is greater.

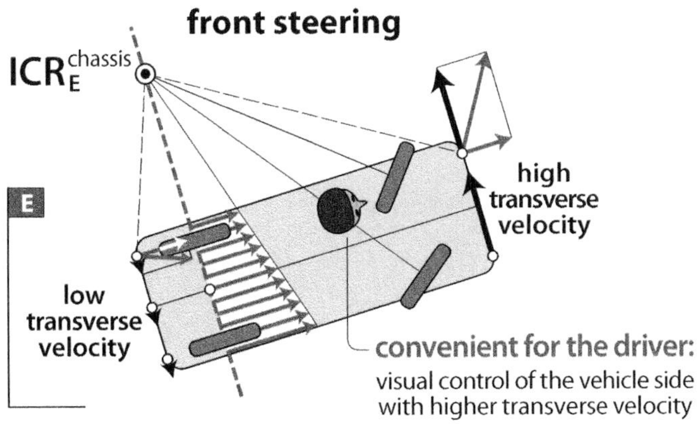

Fig. 3B.5

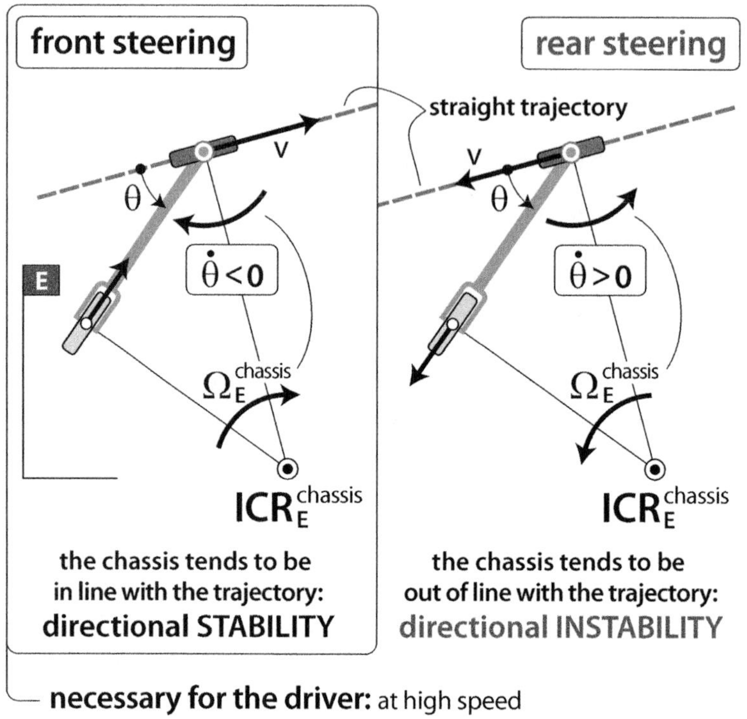

Fig. 3B.6

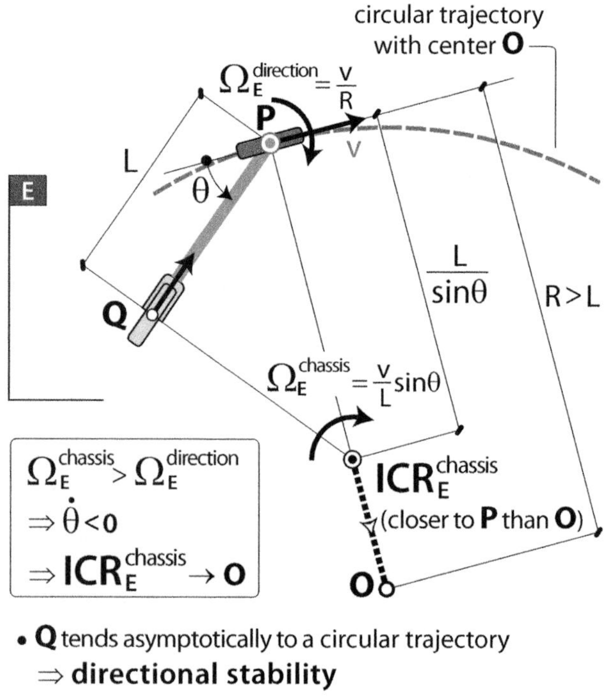

Fig. 3B.7

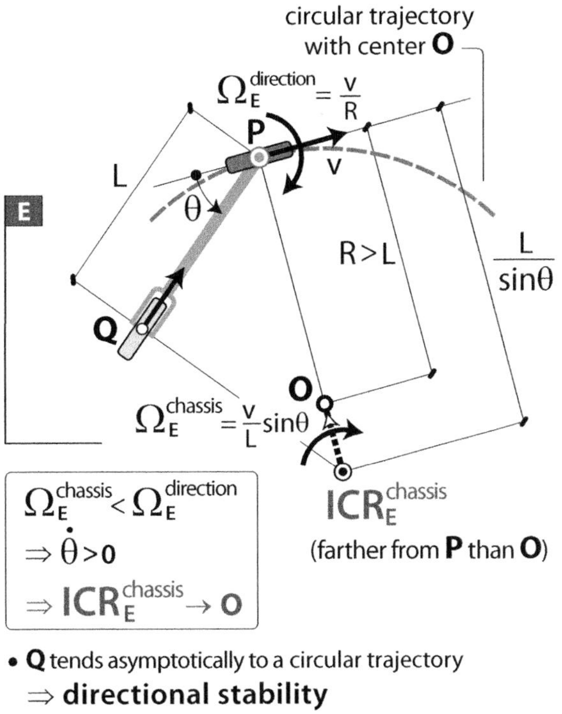
circular trajectory
with center O
$\Omega_E^{direction} = \frac{v}{R}$
P
v
L
θ
R>L
$\frac{L}{\sin\theta}$
E
Q
O
$\Omega_E^{chassis} = \frac{v}{L}\sin\theta$
$ICR_E^{chassis}$
(farther from P than O)
$\Omega_E^{chassis} < \Omega_E^{direction}$
$\Rightarrow \dot{\theta} > 0$
$\Rightarrow ICR_E^{chassis} \rightarrow O$
• Q tends asymptotically to a circular trajectory
$\Rightarrow$ directional stability

Fig. 3B.8

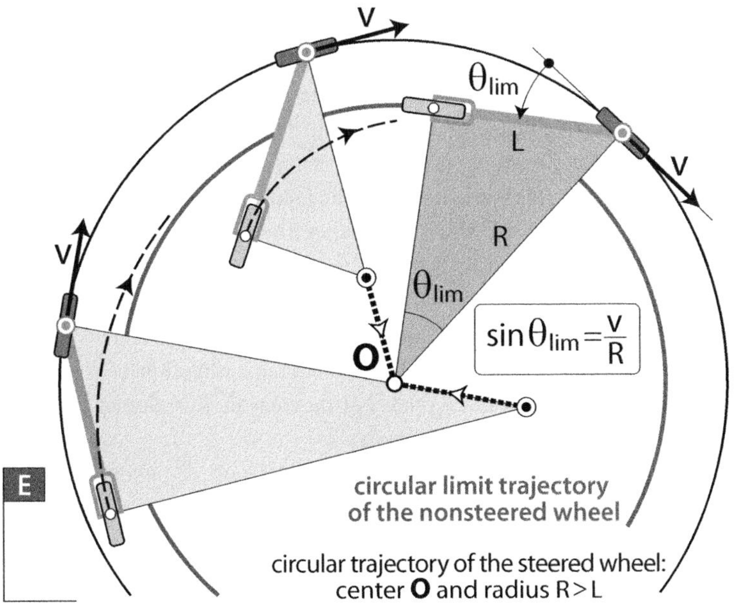
v
θ_{lim}
L
v
R
v
θ_{lim}
$\sin\theta_{lim} = \frac{v}{R}$
O
E
circular limit trajectory
of the nonsteered wheel
circular trajectory of the steered wheel:
center O and radius R>L

Fig. 3B.9

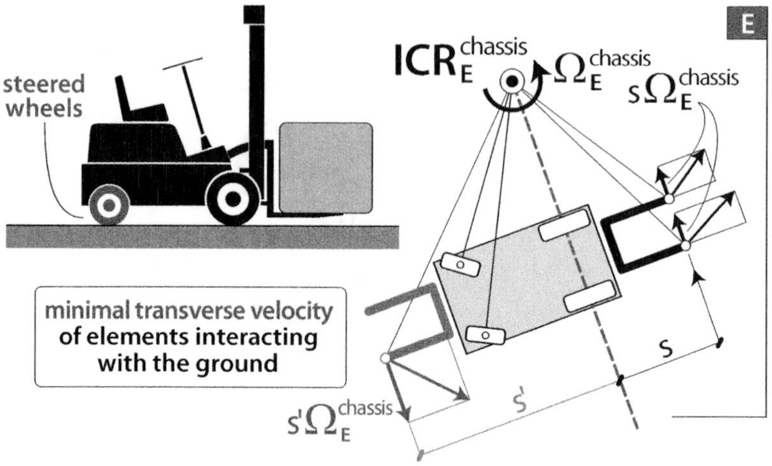

Fig. 3B.10

In both cases, the steering is directionally stable and, as shown in Fig. 3B.9:

- The initial **ICR** of the chassis relative to the ground evolves and tends asymptotically toward the center **O** of the circular trajectory of the steered wheel center
- The trajectory of the nonsteered wheel tends to a circumference with center **O** and radius smaller than that of the circumference traveled by the steered wheel

A vehicle with rear steering should be driven forward only at low speed: the correction of the directional deviations of the chassis is borne entirely by the driver reflexes. At high speeds, the reaction time of the driver would be insufficient, and the correction would not be possible. This is why all vehicles with a **maximum speed higher than a certain value** (roughly above 20 km/h) **must have front steering.**[3]

The backward motion of a vehicle with front steering is equivalent to the forward motion of a vehicle with rear steering. The inherent instability is evident in the deviation of the vehicle from the intended direction: it tends to cross to the direction of travel. But that backward motion is usually done at low speed, and the driver reflexes may stabilize it.

The choice of rear steering in pallet trucks, excavator machines, and cereal-harvesting machines is associated with the existence of an element (coupled to the chassis of the vehicle) that has to interact with fixed objects on the ground – the clip of the pallet truck, the excavator shovel, the collector saw with the spikes. This interaction requires that the speed component transverse to the advance direction be kept to a minimum when the vehicle follows a curved path. For this reason, this element is located at the vehicle end closest to the nonsteered wheels.

The pallet trucks (Fig. 3B.10) load and unload pallets along a warehouse aisle, and this implies a very tight cornering maneuver. As the aisles of warehouses have to be narrow and the pallets must be stacked as closely together as possible, it is necessary that the pallet have little transverse displacement during this maneuver. If the clamp

[3] This problem does not disappear in automated guided vehicles: the sensors providing the information of configuration and movement and the actuators also have nonzero time delays.

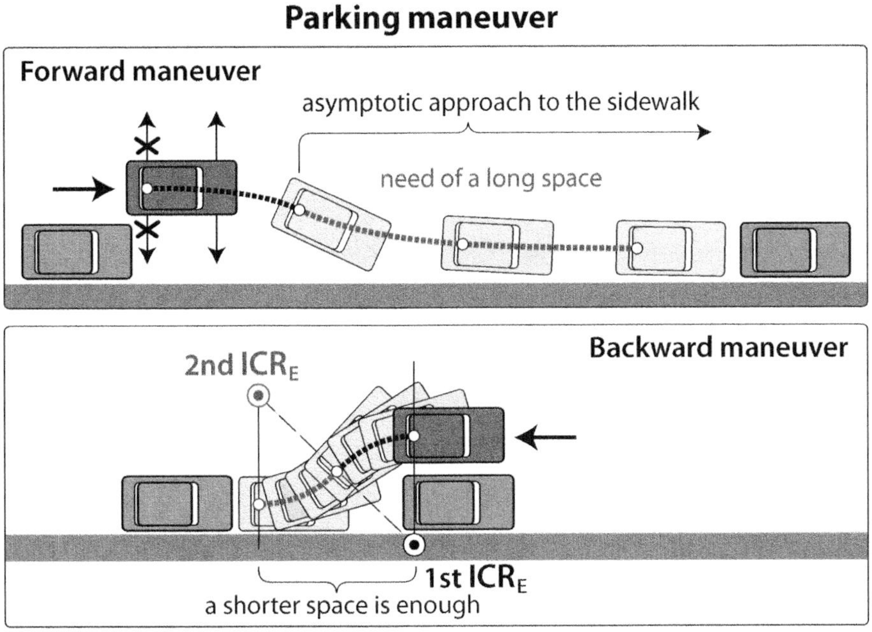

Fig. 3B.11

were at the end close to the steered wheels, the warehouse aisle would have to be wider to allow the maneuver.

Although the pallet trucks are slow and thus can have rear steering, they usually go backward when they follow straight paths along the corridors. The reason is evident: in that driving direction, the motion is directionally stable, and they can move at maximum speed with no concern about stability problems.

When it comes to cars, the rear steering is adequate in one maneuver: that of parking laterally next to a sidewalk (Fig. 3B.11). Since the car cannot move sideways, you need an approaching maneuver.

If you try to park forward, the wheels that come closer to the sidewalk initially are the steered wheels. If you do not want to climb on the sidewalk, you have to steer them in the opposite direction, and that slows the approaching speed of the rear wheels to the sidewalk. You would end up parking asymptotically, and that would require an unusual length of free sidewalk.

The right decision is to park backward. In a first phase, nonsteered wheels will come closer to the sidewalk; in a second phase, the front and rear wheels will touch the sidewalk simultaneously.

3B.2 Vehicles with Four Steered Wheels: 1-DoF Direction

Four-wheel steering (which consists in changing the orientation of the four wheels relative to the chassis simultaneously) is used in some slow-moving industrial vehicles and in some cars. In the first case, the front and rear wheels turn in opposite directions (Fig. 3B.12) with

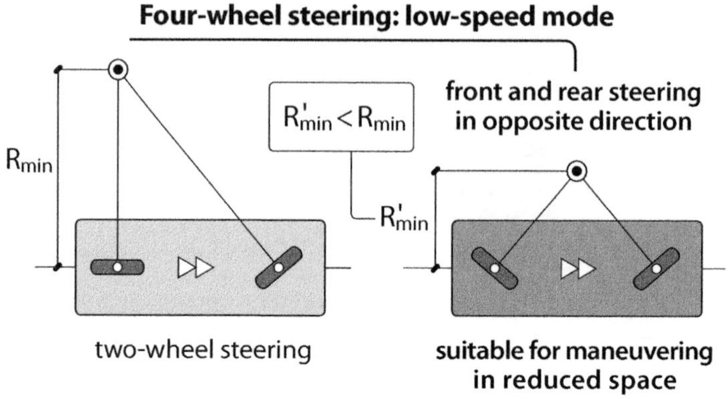

Fig. 3B.12

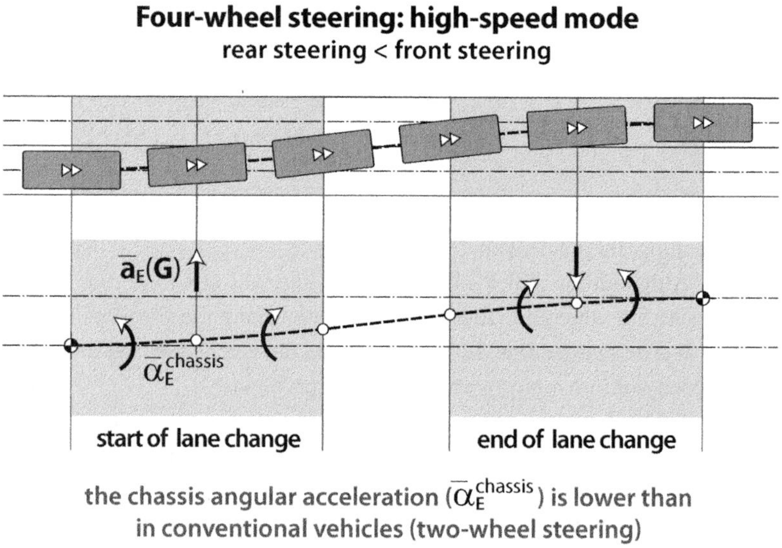

Fig. 3B.13

an obvious objective: increase the maneuverability in reduced spaces (since the minimum turning radius decreases). In addition, the more centered position of the **ICR** (compared to that of conventional vehicles) decreases the area swept by the turning vehicle.

In cars with four-wheel steering, front and rear wheels turn in opposite directions at low speeds: it is assumed that at low speeds the vehicle may be maneuvering in a reduced space.

At high speeds, however, all wheels turn to the same side. Now, the objective is very different: reducing the risk of skidding in overtaking (which is the most dangerous maneuver at high speeds). On a motorway, the radius of the curves is large enough to avoid the risk of skidding, but in the overtaking maneuver the driver chooses the radius

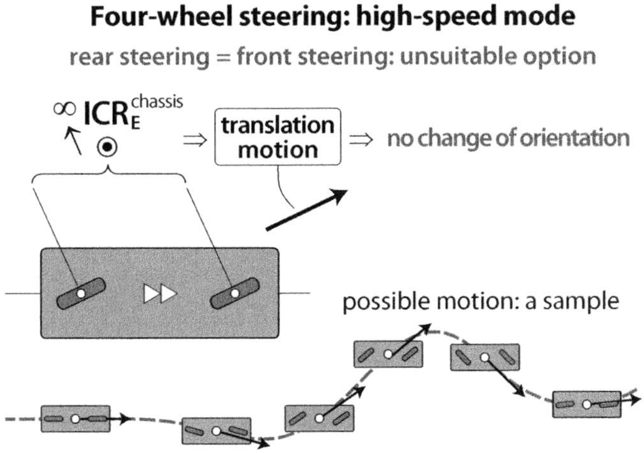

Fig. 3B.14

of the curve. If you choose too small a radius, you may start skidding, and this may lead to an accident very easily.

In the overtaking maneuver, the vehicle turns twice in one direction and twice in the opposite direction every time you change lanes (Fig. 3B.13). Every turn implies not only an angular acceleration of the chassis but also a linear acceleration of the center of mass of the vehicle.

As we will see in chapter 3 of *Rigid Body Dynamics* [Cambridge University Press, forthcoming], both the angular accelerations of the chassis and the transverse accelerations of the center of mass require forces transverse to the wheels (associated with the pneumatic–ground contact). If these forces exceed a certain value, skidding begins.

The acceleration of the center of mass is inevitable. But if the changes of orientation of the chassis are reduced, the value of the angular accelerations decreases, and so does that of the forces required by those accelerations. This reduction is what is sought by rotating the four wheels in the same direction.

One option would be to eliminate completely the orientation changes of the chassis. In a straight section, it is possible to move to the neighboring lane without changing the vehicle orientation: all you need is a same rotation in all four wheels (Fig. 3B.14). This option is not suitable for curved trajectories, as the car should always remain with its longitudinal axis approximately parallel to that of the road.

If the steering system allowed the wheels to rotate full turns, the car would be able to go through a roundabout, but it would be orthogonal to the lane at one-fourth of the roundabout, and at one-half the driver would be looking at the road left behind!

The solution proposed is to rotate the rear wheels less than the front wheels (rear angle about a half of the front angle, Fig. 3B.15). Thus, as the rear part of the car moves toward the neighboring lane, its orientation change is lower than that of a conventional car (with just front steering). Moreover, as the transverse displacement of the front side is greater than that of the rear side, the road is approximately ahead.

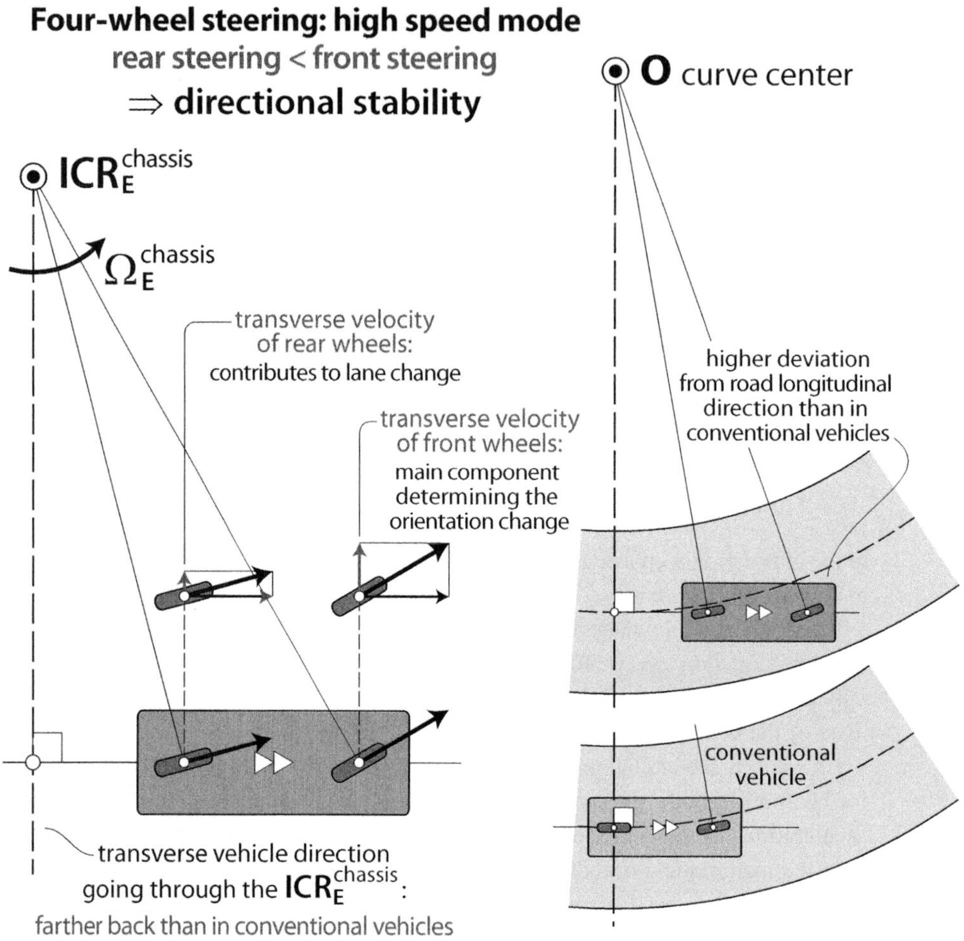

Fig. 3B.15

The **ICR** of the chassis relative to the ground remains at a finite distance, although displaced backward with respect to the position of the rear wheels. When following a circular path, the **ICR** ends up being in the center of the curve **O**. Thus, the car is advanced with respect to its transverse direction going through **O** (which is where the rear wheels of a conventional car would be). In the real case of large radius curves, this is irrelevant.

Another good point is that, as the front wheels rotate more than the rear ones, the forward motion is directionally stable.

Despite the good points of this option, it has been scarcely introduced in the real world of automobiles, but it is still a pole of attraction in automotive fairs.

If the rear wheels rotated more than the front wheels, the rear side of the car would move quicker toward the neighboring lane than the front side. Then, the chassis would turn in the opposite direction of that suggested by the rotation of the front wheels (Fig. 3B.16). The **ICR** of the chassis relative to the ground would be at a finite distance,

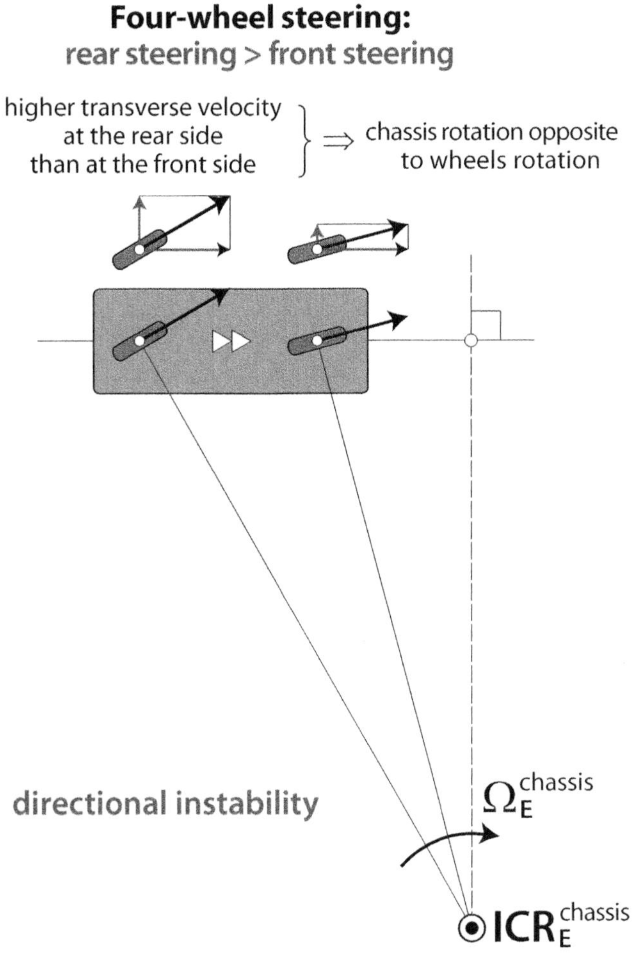

Fig. 3B.16

but on the opposite side to what the front wheels would seem to suggest. On the other hand, the forward motion would be directionally unstable. For all these reasons, this option is not adequate.

3B.3 Vehicles with Articulated Chassis: 1-DoF Direction

Vehicles with articulated chassis – formed by two subchassis articulated in a point (Fig. 3B.17) – have a maneuverability far superior to the conventional ones, although they do not have steered wheels.

Those vehicles do not have one **ICR** but two: one for each subchassis, located on the corresponding axle. It is straightforward to verify that these two **ICR** are aligned with the articulation center. Besides this condition, there is no other restriction regarding the positioning of the **ICR** on the axles.

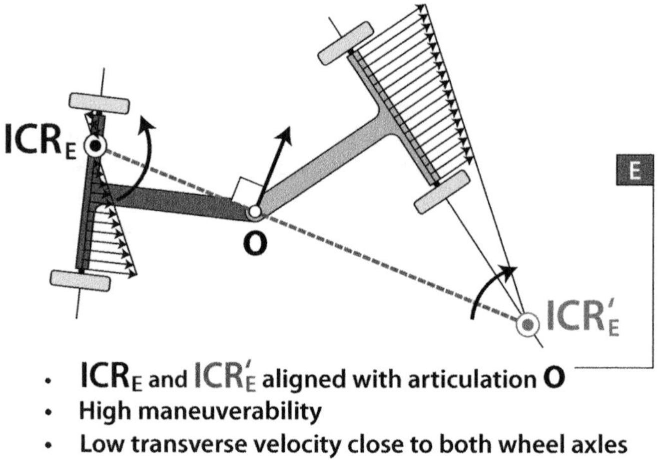

Fig. 3B.17

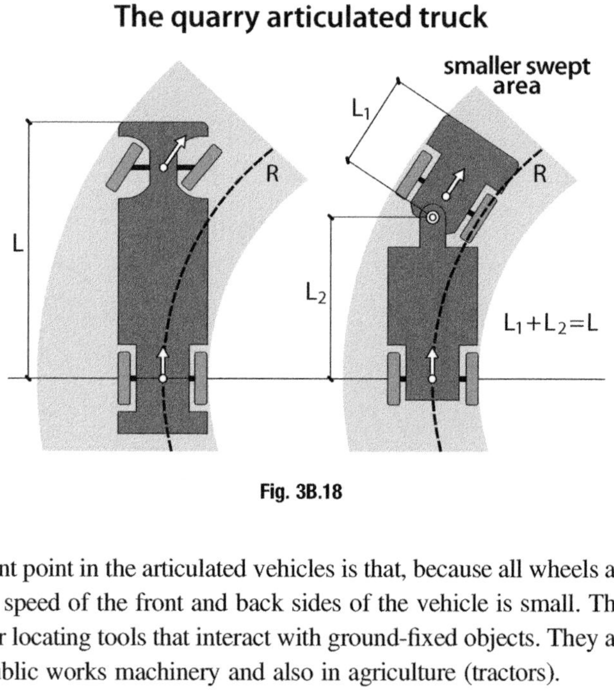

Fig. 3B.18

An important point in the articulated vehicles is that, because all wheels are nonsteered, the transverse speed of the front and back sides of the vehicle is small. Thus, both sides are suitable for locating tools that interact with ground-fixed objects. They are common in the field of public works machinery and also in agriculture (tractors).

The quarry trucks are very often articulated vehicles (although they do not have tools interacting with the ground). These vehicles need a greater maneuverability than conventional trucks in order to sweep a smaller track width at tight turns. In the quarries, the tracks have many hairpin bends and the slope of the terrain transverse to the track is usually pronounced, and so a reduced track width is convenient (Fig. 3B.18).

The **directional driving** of articulated vehicles is more complicated than that of conventional vehicles (with just front-steered wheels), where the steering angle

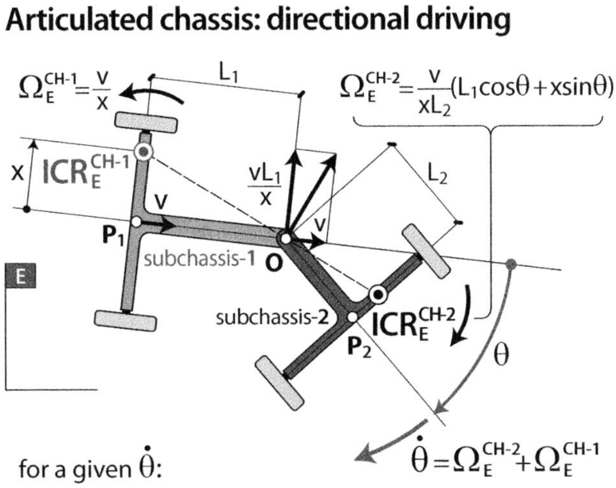
Articulated chassis: directional driving
$\Omega_E^{CH\text{-}1} = \frac{v}{x}$
$\Omega_E^{CH\text{-}2} = \frac{v}{xL_2}(L_1\cos\theta + x\sin\theta)$
L_1
L_2
x
$ICR_E^{CH\text{-}1}$
$\frac{vL_1}{x}$
v
P_1
subchassis-1
O
E
subchassis-2
$ICR_E^{CH\text{-}2}$
P_2
θ
for a given $\dot{\theta}$:
$\dot{\theta} = \Omega_E^{CH\text{-}2} + \Omega_E^{CH\text{-}1}$

$$x = \frac{L_2 + L_1\cos\theta}{\lambda - \sin\theta} \quad ; \quad \text{with } \lambda \equiv \frac{\dot{\theta}L_2}{v} = \frac{v_{CH\text{-}1}(\mathbf{P}_2)}{v_E(\mathbf{P}_1)}$$

the $ICR_E^{CH\text{-}1}$ position depends on θ and $\dot{\theta}L_2/v$

Fig. 3B.19

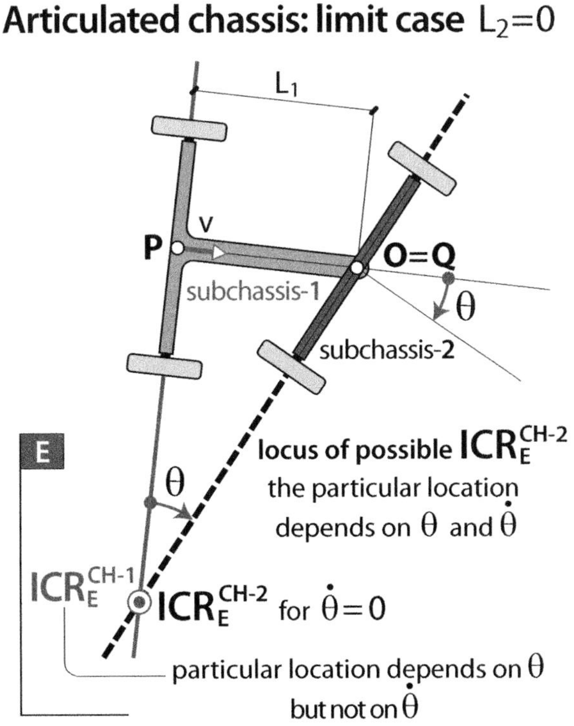
Articulated chassis: limit case $L_2=0$
L_1
v
P
O=Q
subchassis-1
θ
subchassis-2
E
locus of possible $ICR_E^{CH\text{-}2}$
the particular location
depends on θ and $\dot{\theta}$
θ
$ICR_E^{CH\text{-}1}$
$ICR_E^{CH\text{-}2}$ for $\dot{\theta}=0$
particular location depends on θ
but not on $\dot{\theta}$

Fig. 3B.20

determines unambiguously the position of the **ICR** of the chassis with respect to the ground. In an articulated vehicle, the position of the **ICR** of any of the two subchassis is a function not only of the angle θ between the two subchassis but also of its time derivative $\dot{\theta}$. For the vehicle described in Fig. 3B.19, the position x of the **ICR** of subchassis 1 is:

$$x = v\frac{L_2 + L_1 \cos\theta}{\lambda - \sin\theta}, \quad \text{with } \lambda \equiv \frac{L_2\dot{\theta}}{v} = \frac{v_{CH-1}(\mathbf{P}_2)}{v_E(\mathbf{P}_1)}. \tag{3B.1}$$

In the limit case $L_2 = 0$ shown in Fig. 3B.20, the position of the **ICR** of subchassis 1 – which can in fact be considered the chassis of the vehicle – is independent of $\dot{\theta}$ and v. This is analogous to what happens in vehicles with front-steered wheels. The **ICR** that does depend on $\dot{\theta}$ and v is that of subchassis 2 – or **steering wheel bar** – which can be located anywhere on the wheels' axle.

For $\dot{\theta} = 0$, the **ICR** of subchassis 2 coincides with that of subchassis 1.

3B.4 Platforms with 2-DoF Direction

In all previous vehicles, the location of the chassis **ICR** – or of any subchassis – is restricted to a line fixed to the chassis. The driver chooses the particular **ICR** location on that line through the driving controls. For this reason, they are called vehicles with a 1-DoF direction.

In vehicles with conventional wheels, the maximum maneuverability is obtained when the **ICR** of the chassis can be any point on the ground plane. This requires that the wheels' axles intersect at that point (Fig. 3B.21). In the case at hand, we need a control device with 2-DoF – or two coordinates, such as a *joystick* – instead of a steering wheel (which allows the **ICR** to be placed only on a line fixed to the chassis).

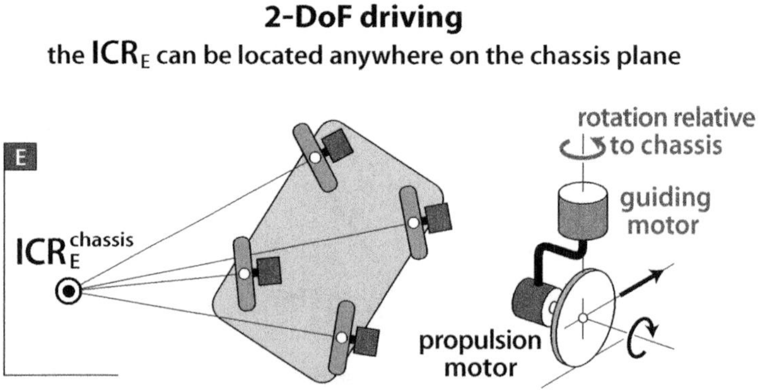

Fig. 3B.21

As there is no simple mechanism to introduce the appropriate orientation to each wheel from the 2-DoF control signal, the usual solution is a guiding motor per wheel. As the wheels steering angle must have a range of 360°, the propulsion of all wheels with a single engine is not possible. This is why each wheel is usually provided with a propulsion engine. Because of the high cost – eight engines for a four-wheel platform instead of a single engine as in the case of automobiles – this option is limited to specialized platforms.

They are usually called 3-DoF platforms, but they are not strictly 3-DoF. For example, they cannot start a displacement in any direction without orienting the wheels previously. The omnidirectional wheels (presented in Appendix 3C) do allow the realization of strictly 3-DoF vehicles.

Appendix 3C Maneuverability of Vehicles with Omnidirectional Wheels

Figure 3C.1 shows the basic idealization of an omnidirectional wheel. It has a single-point nonsliding contact with the ground – assumed to be flat and rigid – and its center has 2DoF: one corresponds to the motorized motion $R\dot{\varphi}$ along a chassis-fixed direction – like a nonsteered wheel – and the other one corresponds to a free motion in a direction defining a constant angle α with that of the motorized motion (in some cases it is not strictly constant, but it is unique for each position of the wheel).

A possible realization consists of a wheel coated with rollers with free rotation around their axes (which are fixed to the wheel). Those axes are perpendicular to the wheel radius that contains the roller center, and they define an angle α with the wheel axis, which is the direction of the motorized rotation. There is no slipping at the roller–ground contact.

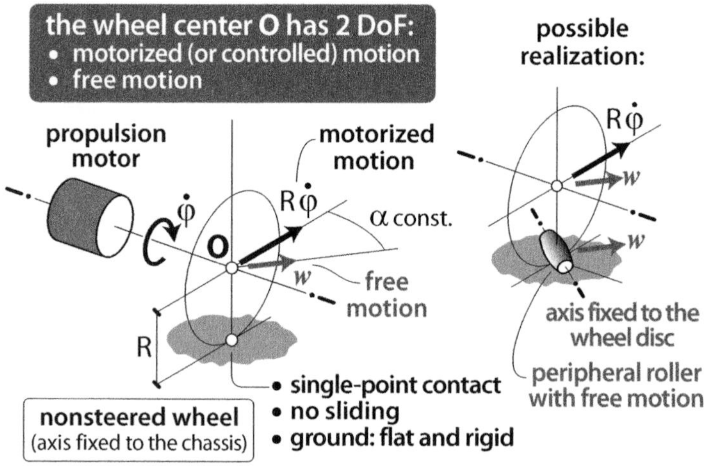

Fig. 3C.1

3C.1 Omnidirectional Wheels with Rollers at 45°

Figure 3C.2 shows the basic design of those wheels and their characteristics. The length and diameter of the rollers ensure that there is always roller–ground contact for all angular positions of the wheel, with a certain overlap between consecutive rollers.

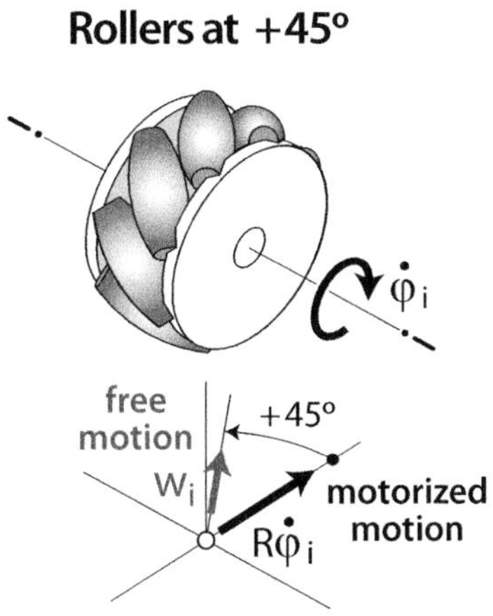

Fig. 3C.2

The angle α – which is nominally 45° – as well as the position of the roller–ground contact point relative to the chassis – which is nominally the projection of the wheel center on the ground – fluctuates with the angular position of the wheel. To simplify the kinematic study for control, this fluctuation is neglected (which is equivalent to assuming that there are infinite rolls of infinitesimal width at the wheel periphery). The positional inaccuracies associated with the finite dimensions of the rollers are not a concern in vehicles controlled by a driver.

These wheels were developed in the 1970s and used in pallet trucks to allow lateral motion and rotation around their center (so that their maneuverability is much higher than that of conventional ones). Its use has been practically limited to vehicles with four wheels and rectangular chassis (Fig. 3C.3).

The 3-DoF of the chassis can be associated to the longitudinal and transverse speeds of its center and to the speed of change of orientation. They are controlled by the motorized velocities of the wheels. Although three motorized wheels would be sufficient – the rotation of the fourth could be left free – the usual option is to motorize all four. In that case, their motorized velocities are not independent and have to fulfill a linear relationship.

There are many possible choices for the position and orientation of the wheels relative to the chassis, but a mandatory condition is that there be no free (or uncontrolled) motion. This requires that the directions perpendicular to the free velocities of the rollers' centers do not intersect all at the same point (which would be the **ICR** of a free or uncontrolled motion of the chassis, Fig. 3C.4).

The control of the 3-DoF of the chassis is based on the determination of the corresponding motorized velocities. Since the kinematics is algebraically linear, it is possible to calculate the velocity of the center of each wheel from the velocity associated with each DoF of the chassis. That velocity is a linear combination of its

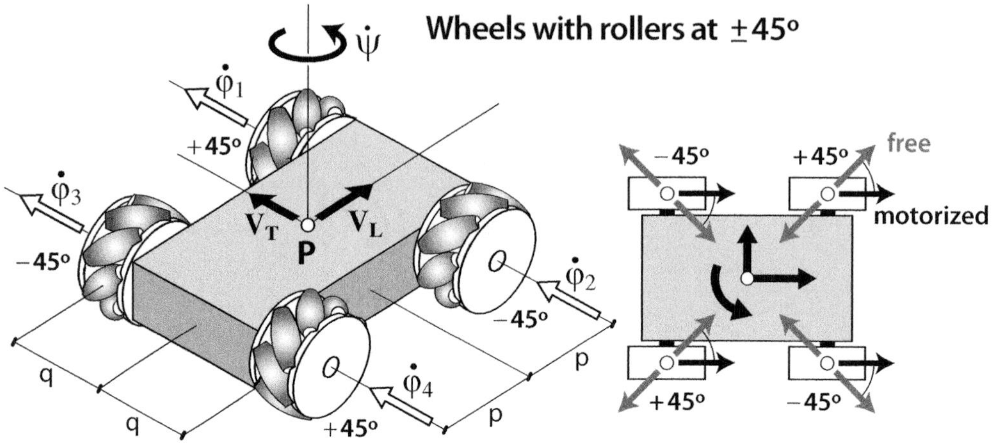

Fig. 3C.3

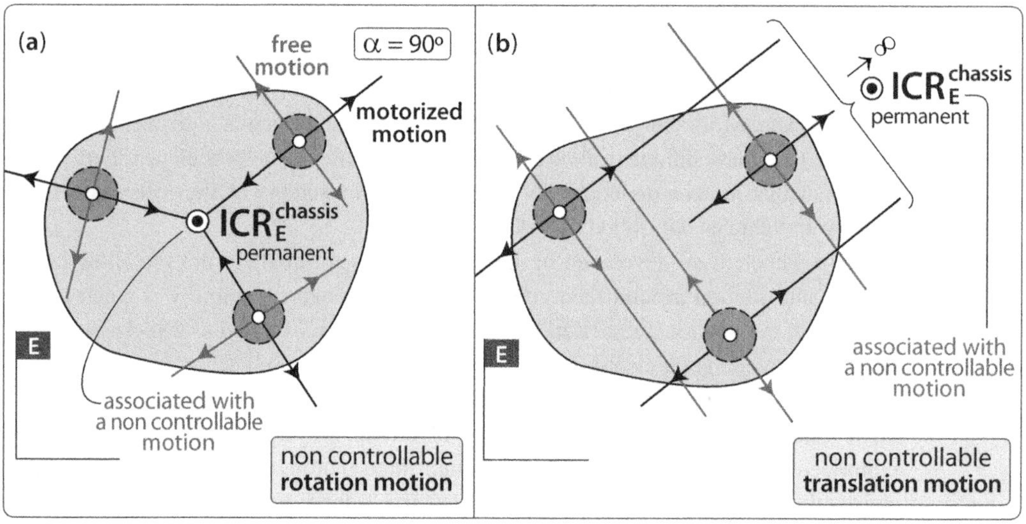

Fig. 3C.4

motorized and free velocities (Figs. 3C.5–3C.8). The motorized velocity for each wheel is the sum of the three motorized velocities required for each independent velocity of the chassis.

The control of the longitudinal, transverse, and rotation velocities through the motorized velocities of the wheels – without involving any positional coordinate of the wheels relative to the chassis – makes these vehicles authentically 3-DoF vehicles: starting from rest, they can initiate the movement in any direction without having to position previously any driving element (contrary to the case of the driving-steered wheels of vehicles with conventional wheels and 2-DoF driving).

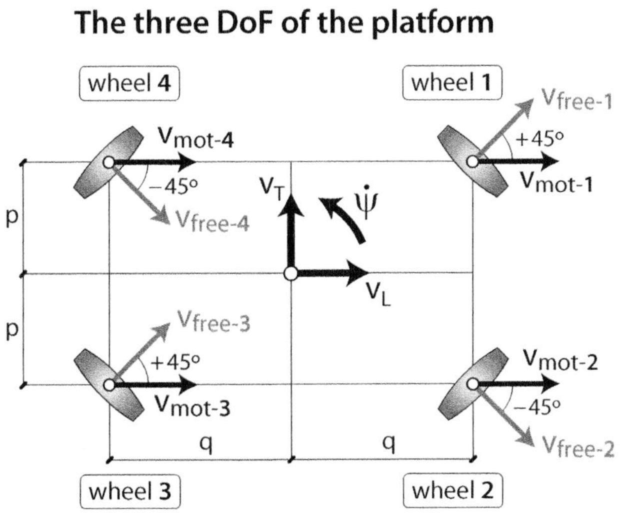

Fig. 3C.5

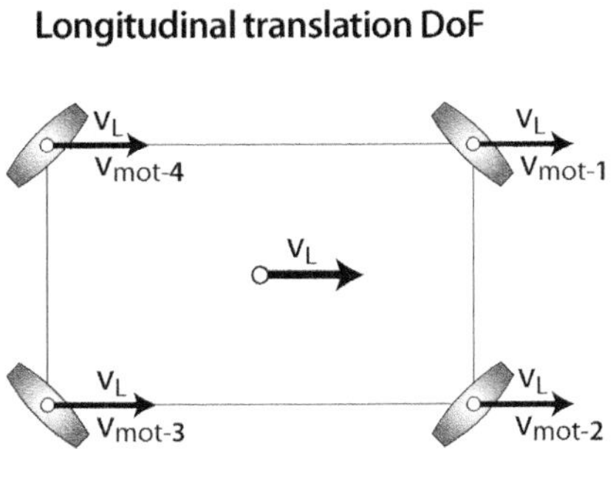

Fig. 3C.6

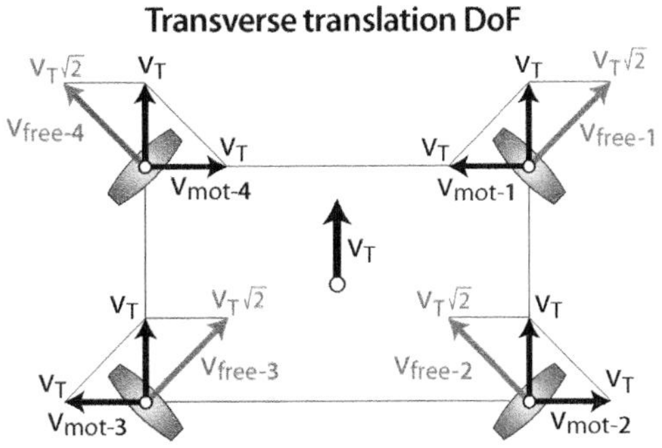

Fig. 3C.7

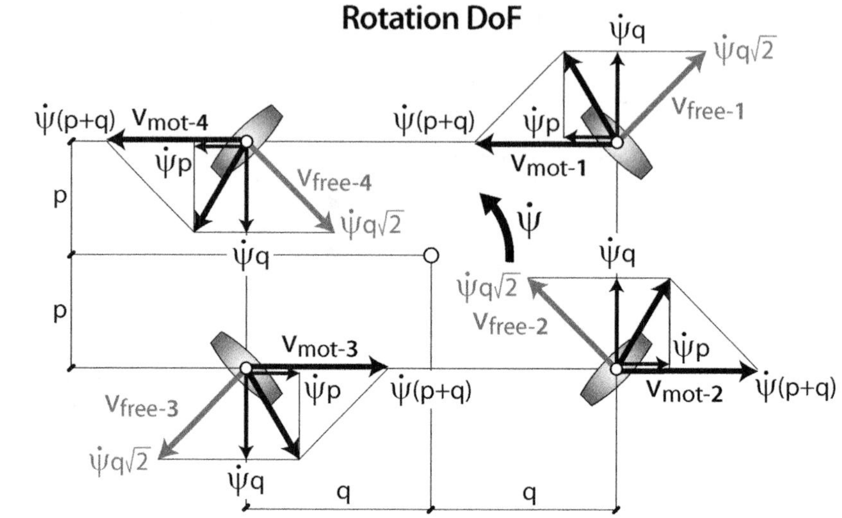

Fig. 3C.8

3C.2 Omnidirectional Wheels with Rollers at 90°

If the rollers' axes are at 90° with respect to the wheel axis (Fig. 3C.9), the free motion has the direction of the wheel axis – it is like a free skid on a conventional nonskidding wheel (that is, where the permitted sliding is strictly transverse to the wheel).

With 90° rollers, the periphery of the wheel cannot be completely covered by one single disk of rollers, and this is a disadvantage as compared to the 45° roller wheels. We need two coaxial disks of rollers that alternate the roller–ground contact (with a certain overlapping between the consecutive ground contact of two rollers, one of each

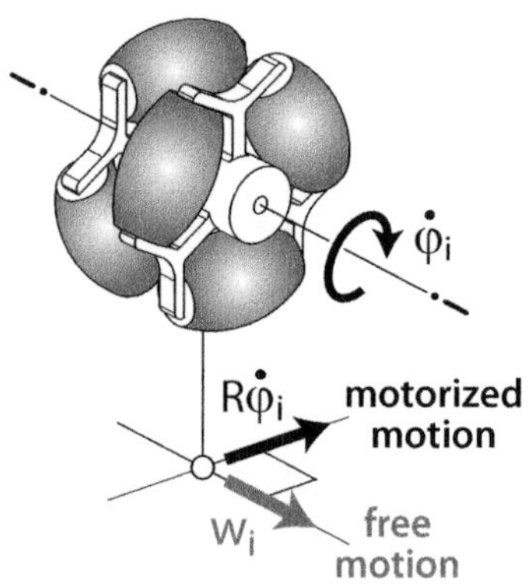

Fig. 3C.9

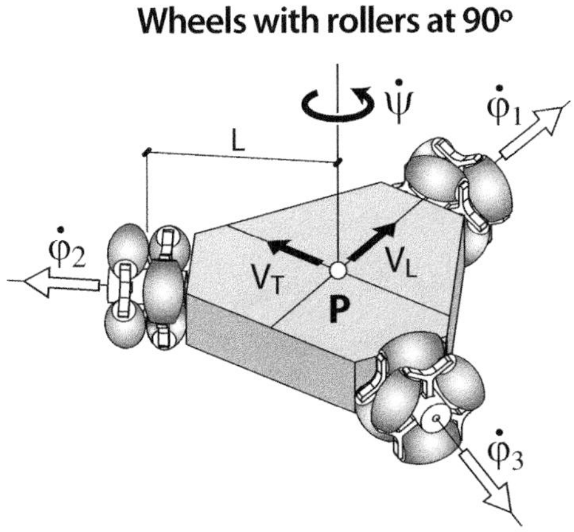

Fig. 3C.10

disk). An advantage with respect to the 45° roller wheels is that the ground–roller contact point is located exactly on the projection of the center of the roller disk on the ground and that the angle α is strictly constant.

The overlapping during consecutive roller–ground contacts is the main drawback of these wheels: if the chassis rotates, the centers of the two roller disks have different motorized velocities in principle, and therefore their angular speed $\dot{\varphi}$ would be different. However, that angular speed is the same for both of them. Consequently, slippage in one or both roller–ground contacts is unavoidable.

The constructive simplicity of these wheels together with the simplicity of their kinematic analysis has made them popular in the field of automation. They are frequently found in robotic platforms (such as the one in Fig. 3C.10) and in mobile robots.

3C.3 Omnidirectional Wheels with Spherical Rollers

If the number of 90° rollers per disk is minimized (only one roller), it becomes spherical. We still need to use two spherical roller disks – with truncated poles – because one single roller cannot maintain contact with the ground along a full turn of the motorized motion: the proximity of the ground–roller contact point to one of the poles of the roller would require high speeds of free rotation to achieve a finite free speed. The truncation is close to 45° but leaves a certain overlapping in the contact of the two rollers with the ground.

This design has constructive advantages that come both from the geometric simplicity of the roller and from the reduced number of bearings required per wheel (in that case, the bearings can be larger).[4]

[4] Regardless of the number of rollers, the normal force exerted by the ground on the wheel is entirely supported by the bearings of the roller in contact with the ground.

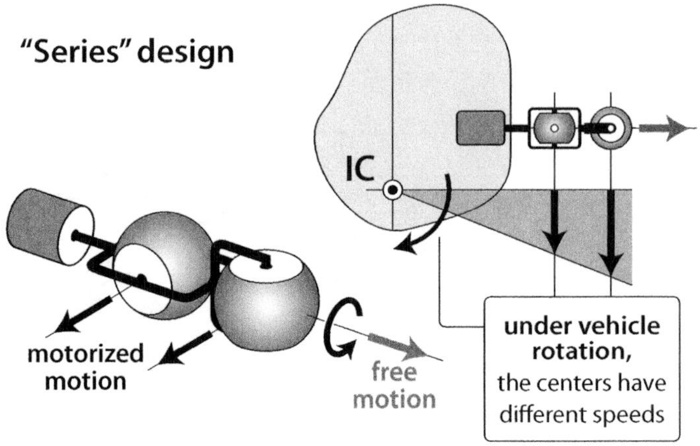

Fig. 3C.11

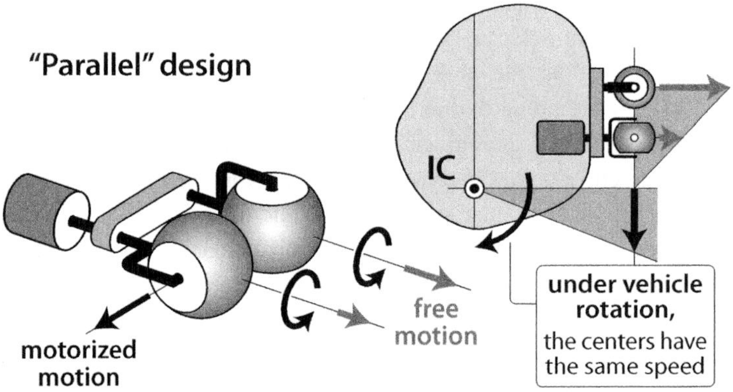

Fig. 3C.12

The design based on two coaxial disks of rollers leads to the spherical roller wheel shown in Fig. 3C.11. The problem of the aforementioned overlapping is evident. If the vehicle turns, the motorized speed of the center of the two rollers is different (and it would be necessary to control it according to what roller is in contact with the ground at each moment). But during the overlapping, the motorized rotation speed of the two rollers is the same. If there is movement, then sliding on one or both rollers is unavoidable (and that affects negatively the positional accuracy of the vehicle).

This problem is solved in the parallel design in Fig. 3C.12 (which is equivalent to using two coplanar disks instead of coaxial rollers). The rollers' centers are aligned with the direction of the motorized motion – the line of centers is perpendicular to the axis of the motorized rotation. In this design, the required motorized speed is always the same for the two rollers, even if the vehicle turns. The free velocities, however, are different when the vehicle turns, but they appear spontaneously on each roller according to the movement of the chassis.

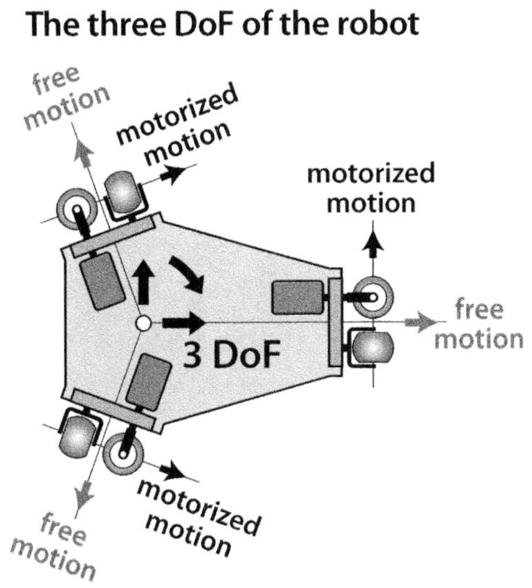

Fig. 3C.13

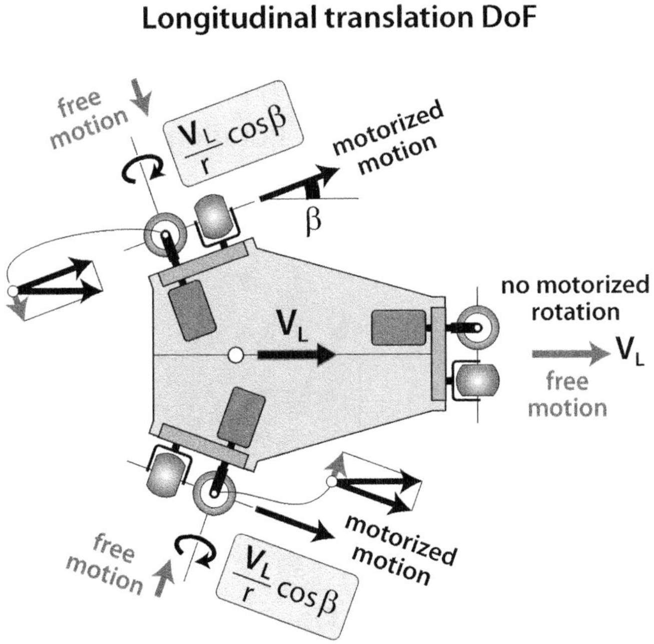

Fig. 3C.14

Figure 3C.13 shows a mobile robot with omnidirectional wheels with "series" spherical rollers. Figures 3C.14–3C.16 illustrate the determination of the motorized speeds required to achieve the movement of the robot according to each of the 3-DoF (longitudinal and transverse translational movements, and rotation movement).

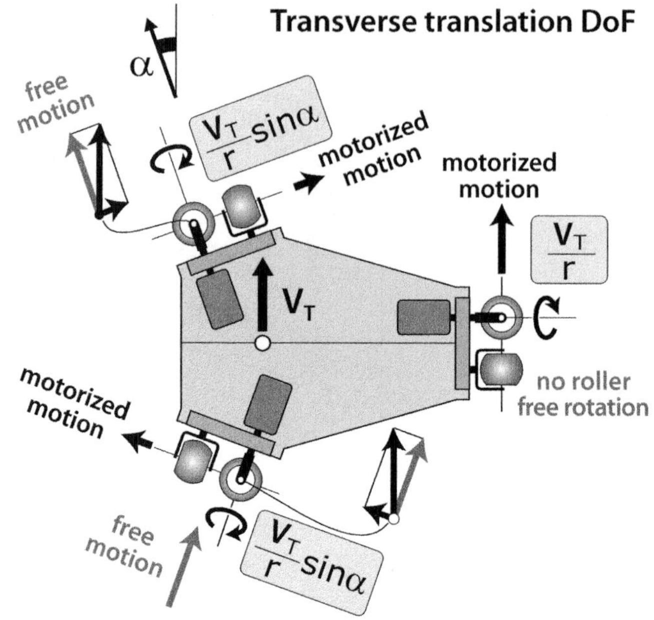

Fig. 3C.15

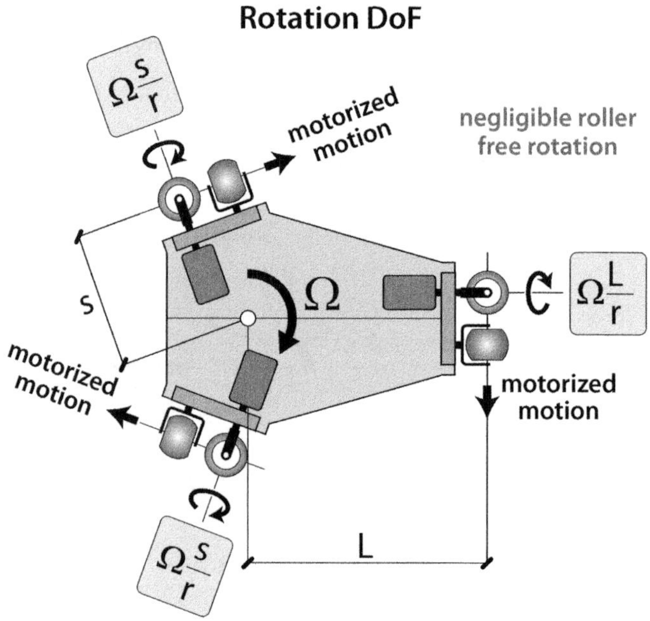

Fig. 3C.16

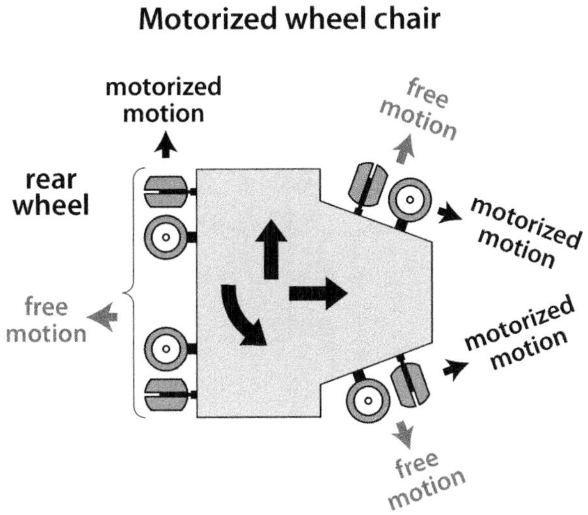

Fig. 3C.17

Figure 3C.17 shows the schematic design of a motorized chair based on that type of wheel. The rear wheel consists of four rollers to increase the stability in curve. This chair is designed to facilitate the labor insertion of disabled people. Work activities often require the person to move to lateral positions (for example, along a table). With a conventional motorized chair, this calls for a maneuver that requires a wide space and several starts and stops (and the corresponding battery consumption).

In addition, this chair requires smaller spaces in turning motions than conventional chairs. The minimum sweeping area in a conventional wheelchair has the **ICR** located at the midpoint between the wheels' centers (which is almost behind the user). As the feet are slightly ahead of the user body, their tips have to describe a trajectory that is the periphery of the swept circumference. The radius of this circumference usually makes it impossible to turn 180° inside an elevator: if you enter the elevator forward, you have to leave it backward (which is a risk when the landing communicates with stairs). On the other hand, it makes it impossible to use elevators with perpendicular entry and exit doors. The motorized chair with omnidirectional wheels allows turning around the central point. That minimizes the swept area and facilitates the use of the usual elevators.

The command needed to drive this type of chair is not trivial. Using the 3-DoF simultaneously (as in mobile robots) is a problem for the user: How do you choose properly the two translation speeds and the rotation speed at every time instant? Last but not least: the command must have three inputs.

These drawbacks can be overcome by taking into account three motion modalities that cover the user's needs (Fig. 3C.18):

- Longitudinal motion with change of orientation (longitudinal translation; it corresponds to the motion of conventional wheel chairs)

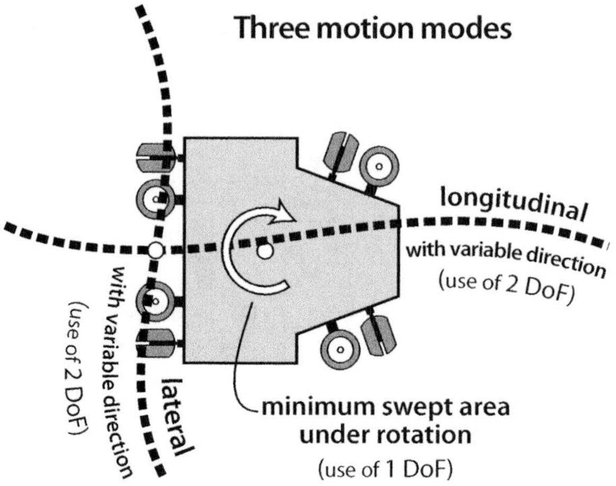

Fig. 3C.18

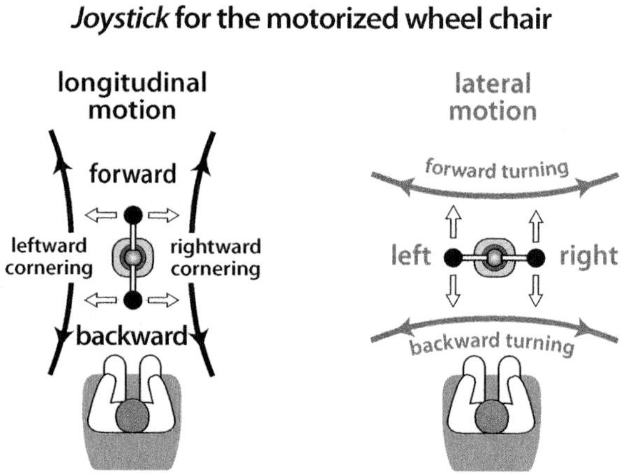

Fig. 3C.19

- Lateral motion with change of orientation (transverse translation; this is new!). The chair can move around a table while facing it constantly
- Rotation around the point guaranteeing a minimum sweeping area

The longitudinal and lateral motions with change of orientation are 2-DoF motions and require a double-entry command such as a joystick. For the minimum-sweeping rotation, a single input is enough. Driving is easier if one movement of the joystick controls the speed and the other one controls the change of orientation. In the two motion modes (longitudinal and lateral), the joystick movements controlling the speed and the orientation are exchanged (Fig. 3C.19).

Quiz Questions

For the sake of brevity, the following assumptions are made throughout the whole collection of questions unless stated otherwise:

- Threads, ropes, strings, and cables are inextensible
- The wheels of the vehicles do not slide on the ground

The ground reference frame is always denoted as E. All data are declared in the figures, but only part of them are declared in the text.

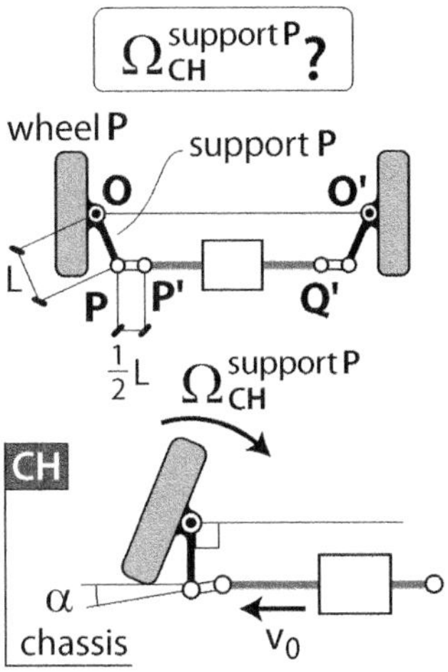

3.1 In the steering mechanism of a vehicle, points **O** and **O′** are fixed to the chassis (CH). The crank **OP** is fixed to the support of wheel P. The bar **P′Q′** moves transversely under the control of the steering wheel. What is the angular velocity of the support P relative to the chassis (CH) when the crank **OP** is perpendicular to the line **OO′** and the speed of bar **P′Q′** relative to the ground is v_0?

A $(1/2)(v_0/L)$
B v_0/L
C $2v_0/L$
D $(v_0/L)\cos\alpha$
E $(v_0/L)(1 - \sin\alpha)$

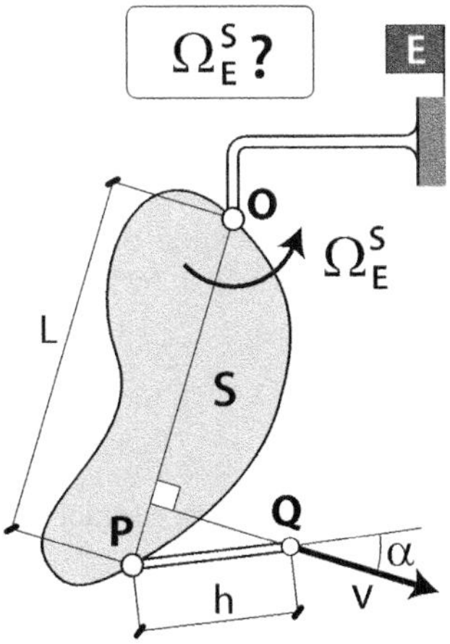

3.2 When the crank **PQ** moves, the rigid body S rotates around **O** (fixed to the ground). What is the value of the S angular velocity relative to the ground for the given configuration?

A v_0/L
B $(v_0 \cos\alpha)/L$
C $(v_0 \cos^2\alpha)/L$
D $v_0/(L - h\sin\alpha)$
E $(v_0 \cos\alpha)/(L - h\sin\alpha)$

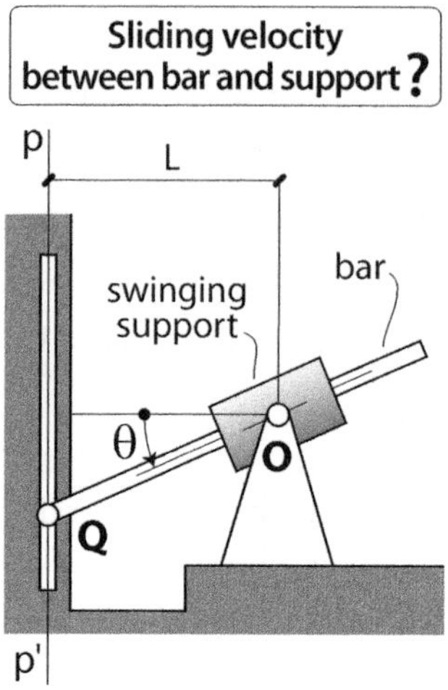

3.3 The bar slides inside the support that rotates around point **O** (fixed to the ground), and its end point **Q** moves along the straight line p–p′. What is the sliding velocity between bar and support?

A $L\dot{\theta}\sin\theta$
B $L\dot{\theta}\tan\theta$
C $L\dot{\theta}(\tan\theta/\cos\theta)$
D $L\dot{\theta}$
E $L\dot{\theta}/\cos^2\theta$

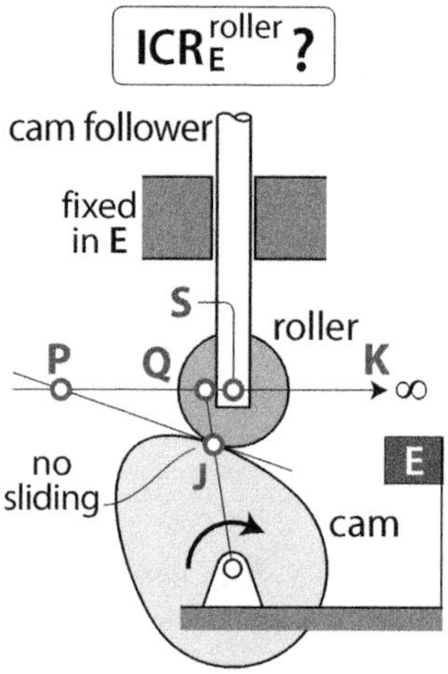

3.4 A rotating cam pushes a cam follower through a roller that has a nonsliding contact with the rotating cam. Which point is the $\mathbf{ICR}_E^{roller}$?

A **J**
B **P**
C **Q**
D **S**
E **K**

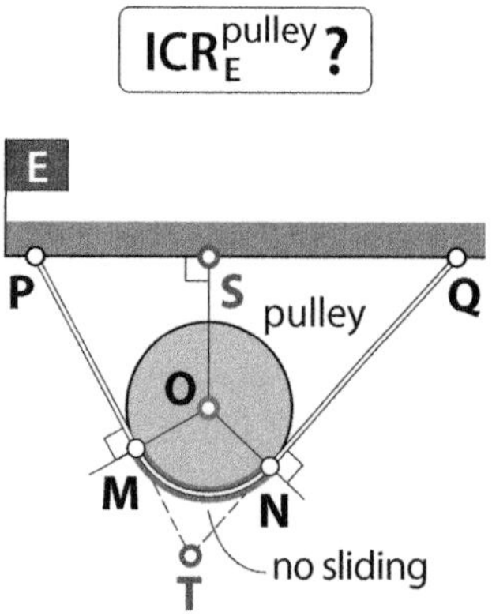

3.5 The pulley does not slide on the cable. Which point is the $\mathbf{ICR}_E^{pulley}$?

A **O**
B **S**
C **T**
D A point on the **MN** arch in contact with the pulley
E All points on the **MN** arch

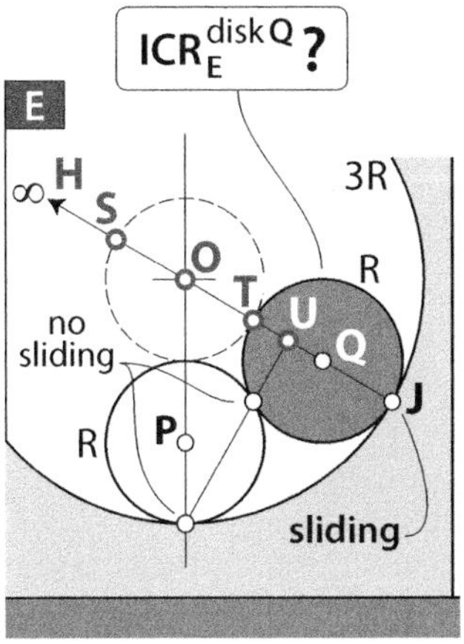

3.6 The disks have a nonsliding mutual contact. Disk P does not slide on the circular support, while disk Q does slide on that support. Which point is the $\mathbf{ICR}_E^{\text{disk Q}}$?

A **H**
B **S**
C **O**
D **K**
E **U**

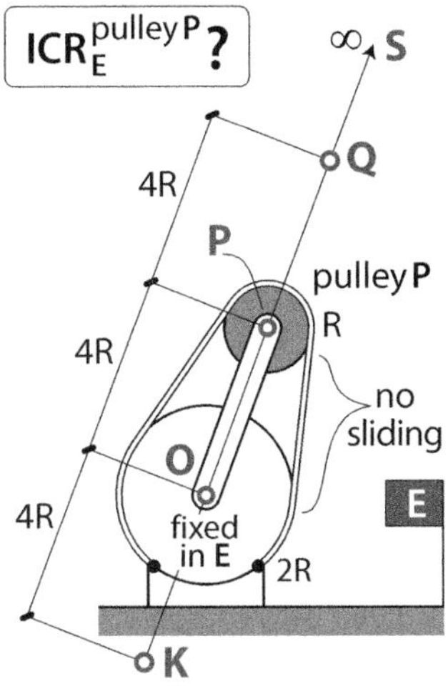

3.7 If there is no sliding between belt and pulleys, which point is the $\mathbf{ICR}_E^{\text{pulley P}}$?

A **O**
B **P**
C **Q**
D **S**
E **K**

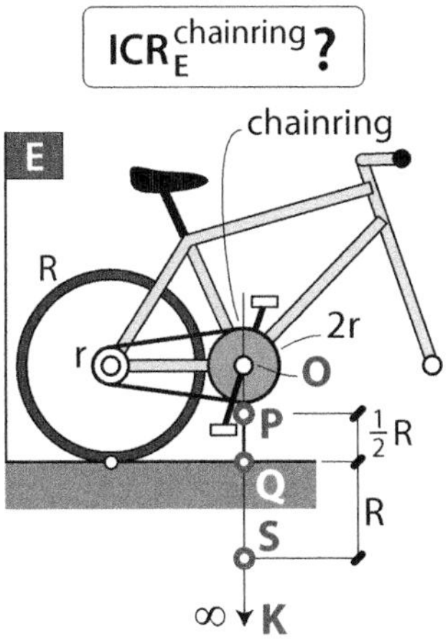

3.8 The bicycle moves without sliding on the ground. If the radius of the chainring is twice that of the sprocket, which point is the $\mathbf{ICR}_E^{\text{chainring}}$?

A **O**
B **P**
C **Q**
D **S**
E **K**

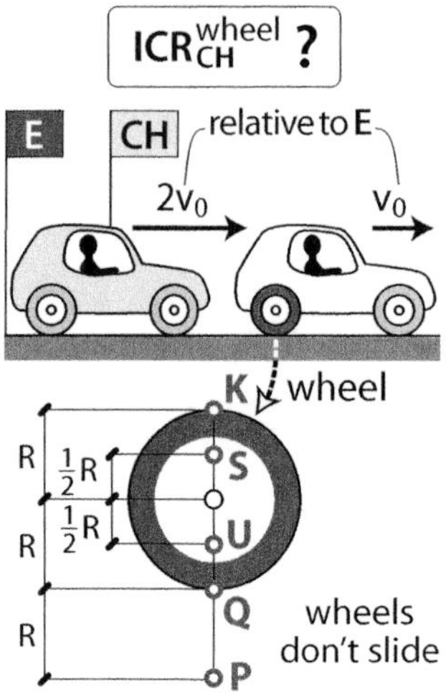

3.9 The two vehicles move with speeds $2v_0$ and v_0 relative to the ground. Which point is the $\mathbf{ICR}^{wheel}_{CH}$?

A **P**
B **Q**
C **U**
D **S**
E **K**

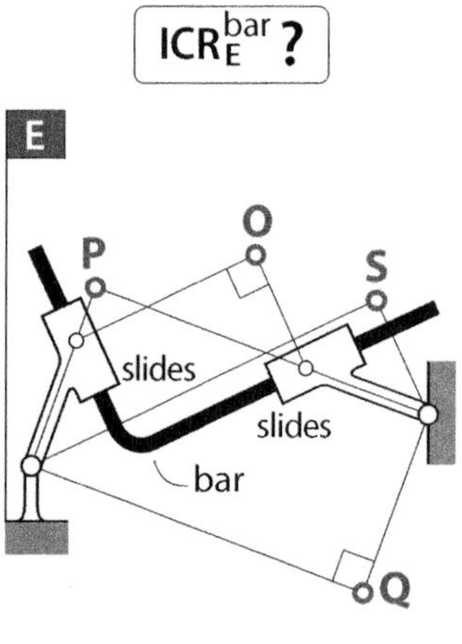

3.10 The bar slides inside the rotating supports articulated to the ground. Which point is the $\mathbf{ICR}^{bar}_{E}$?

A **O**
B **P**
C **Q**
D **S**
E It is not defined because the bar cannot move

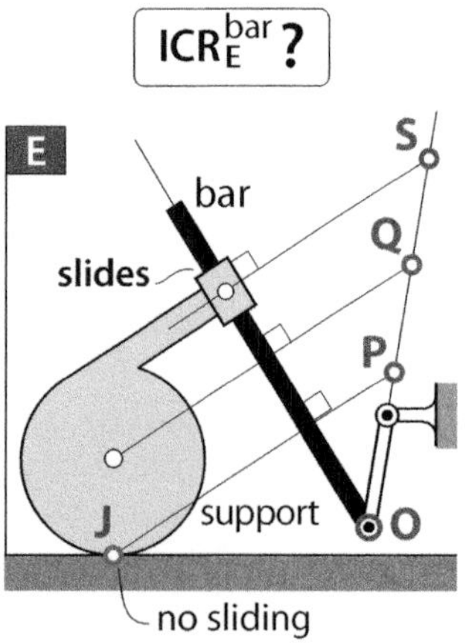

3.11 The bar slides inside the support that has a perfect rolling motion on the ground. The bar end point **O** is articulated to a crank, which in turn is articulated to the ground. Which point is the $\mathbf{ICR}^{bar}_{E}$?

A **O**
B **P**
C **Q**
D **S**
E **J**

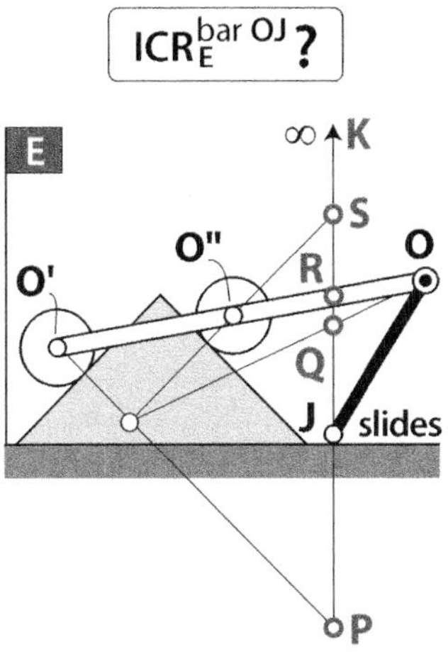

3.12 The bar **OO′** is articulated at points **O′** and **O″** to two identical rollers that rotate freely on two inclined planes (fixed to the ground). The bar **OJ** is articulated at point **O** to the bar **O′O**, and its end point **J** slides on the ground. Which point is the $\mathbf{ICR}_E^{bar\ \mathbf{OJ}}$?

A **H**
B **Q**
C **S**
D **K**
E **P**

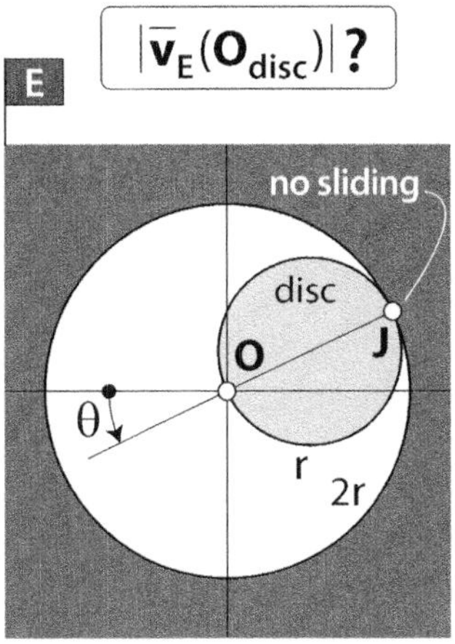

3.13 The disk has a perfect rolling motion on the circular contour. What is the speed of point **O** of the disk relative to the ground?

A 0
B $r\dot{\theta}$
C $r\dot{\theta}/2$
D $2r\dot{\theta}$
E $4r\dot{\theta}$

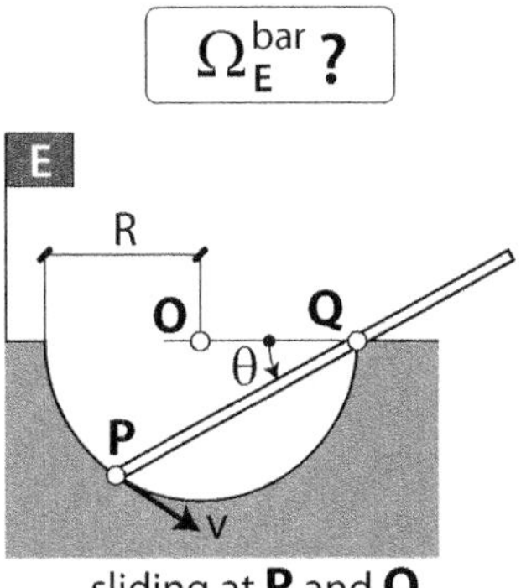

3.14 The bar has a planar motion, and its points **P** and **Q** slide on the semicircular support fixed to the ground. If the **P** sliding velocity is v, what is the value of the angular velocity of the bar relative to the ground?

A $\sqrt{2}v/R$
B v/R
C $2v/R$
D $v/(\sqrt{2}R)$
E $v/(2R)$

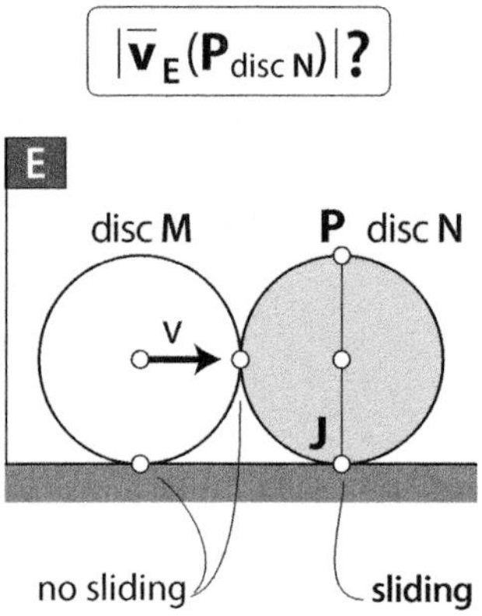

3.15 If the speed (relative to the ground) of the center of disk M is v, what is the speed of point **P** (highest point of disk N)?

A 0
B v
C 2v
D v/2
E $v\sqrt{2}$

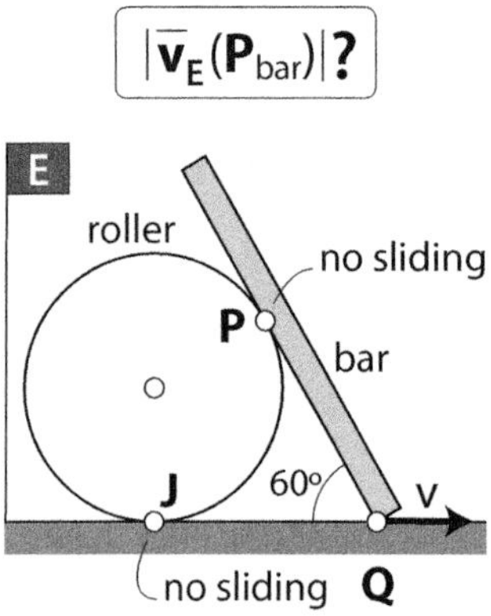

3.16 If the bar end point **Q** slides on the ground with speed v, what is the speed of point **P** of the bar (relative to the ground)?

A 0
B v
C v/2
D $v/\sqrt{3}$
E $v(\sqrt{3}/2)$

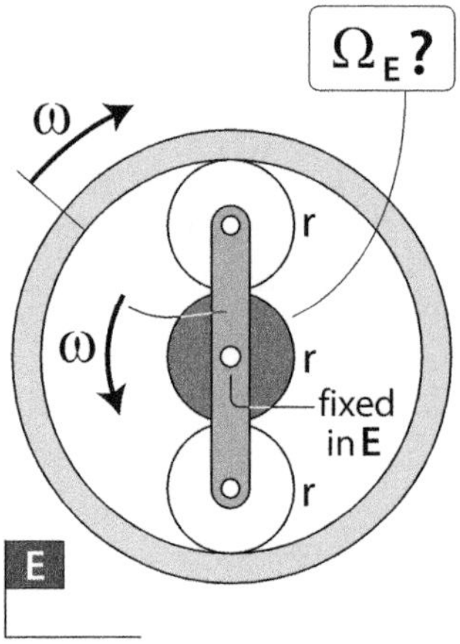

3.17 The wheels in the mechanism do not slide at their contact points. The outer wheel rotates with angular velocity ω relative to the ground in the clockwise direction, while the arm rotates with the same angular velocity ω but in opposite direction. What is the angular velocity of the central wheel relative to the ground?

A ω, clockwise
B (2/3)ω, clockwise
C 7ω, counterclockwise
D 0
E 5ω, counterclockwise

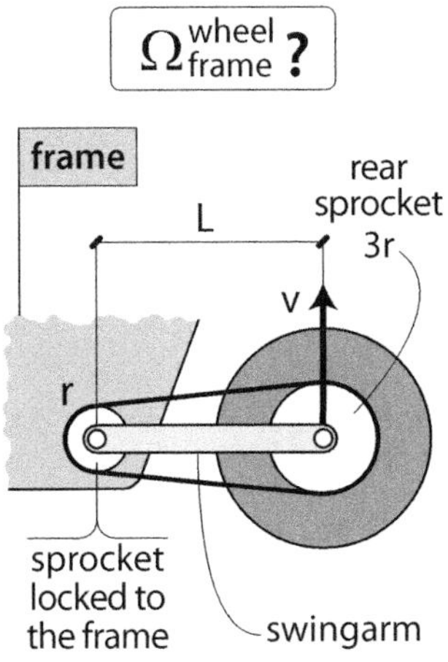

3.18 In the transmission of the rear wheel of a motorcycle, the rear sprocket is blocked relative to the frame and the wheel center has a vertical velocity v relative to the frame. What is the angular velocity of the wheel relative to the frame $\left(\bar{\boldsymbol{\Omega}}^{\text{wheel}}_{\text{frame}}\right)$?

A 0
B (1/3)(v/L), clockwise
C (1/3)(v/L), counterclockwise
D (2/3)(v/L), clockwise
E (2/3)(v/L), counterclockwise

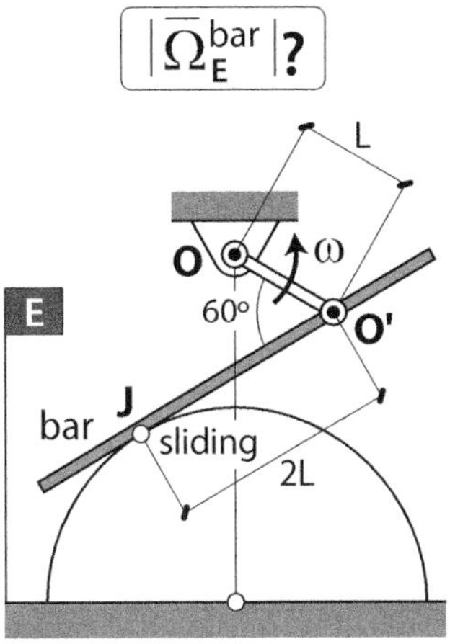

3.19 The crank **OO′** rotates with angular velocity ω about its ground-fixed end point **O**. Its end point **O′** is articulated to the bar, which slides at **J** on the semicircular support fixed to the ground. What is the module of $\left|\bar{\boldsymbol{\Omega}}^{\text{bar}}_{\text{E}}\right|$ for the shown configuration?

A 0
B ω
C ω/2
D ω/3
E ω/4

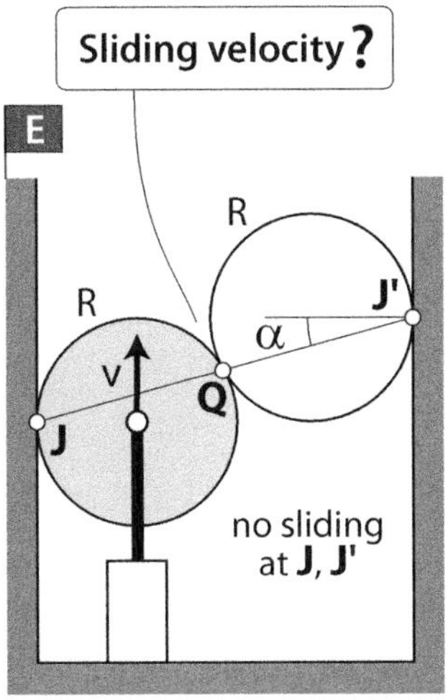

3.20 The wheels rotate without sliding at their contacts with the vertical walls under the action of a hydraulic cylinder, which controls the velocity v of the center of the lower wheel relative to the ground. What is the sliding velocity at **Q**?

A 2v cos α
B v cos 2α
C 4v cos α
D 2v cos 2α
E 0

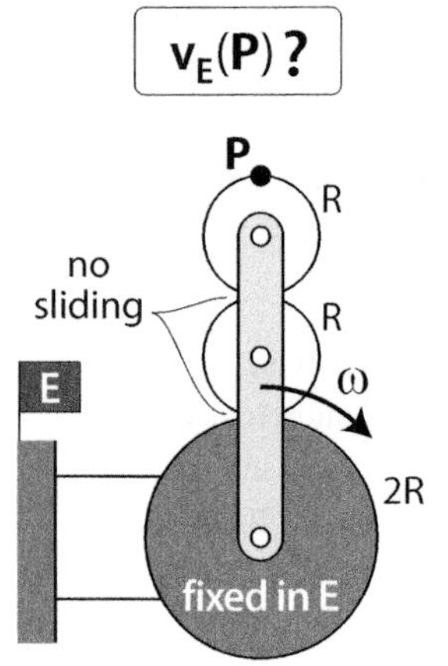

3.21 The small wheels are articulated to the arm that rotates with angular velocity ω relative to the ground. If there is no sliding at the wheels contact points, what is the velocity of point **P** relative to the ground?

A 0
B $2\omega R \rightarrow$
C $4\omega R \rightarrow$
D $5\omega R \rightarrow$
E $6\omega R \rightarrow 0$

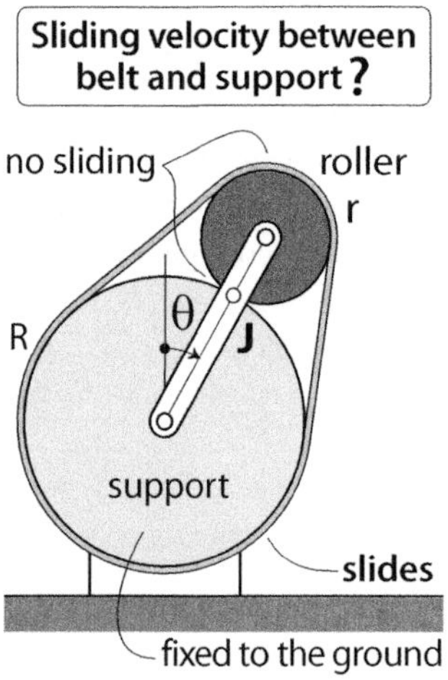

3.22 The roller rotates without sliding on the circular support fixed to the ground. The arm is articulated to the center of the roller and the support. The roller motion is constrained by a belt, which does not slide on the roller but does slide on the support. What is the sliding velocity between belt and support?

A $\dot{\theta}r$
B $\dot{\theta}R$
C $2\dot{\theta}r$
D $2\dot{\theta}R$
E $\dot{\theta}(R+r)$

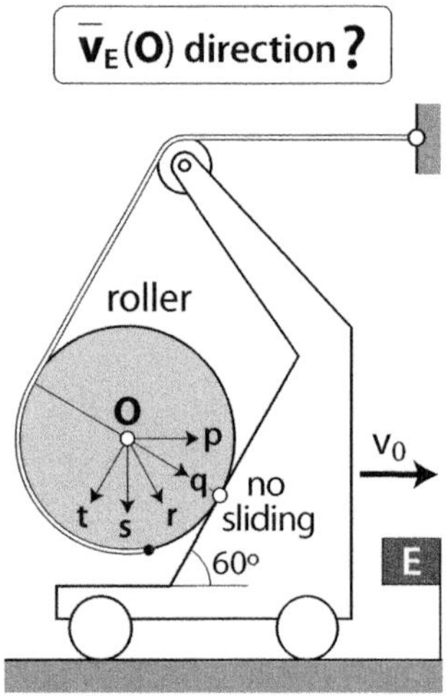

3.23 The carriage moves with speed v_0 relative to the ground (E). The roller has a nonsliding contact with the carriage. The cable has an end point fixed to the ground and another one fixed to the roller. The section between the ground and the pulley is horizontal, and that between the pulley and the roller is parallel to the inclined wall of the carriage. What is the direction of $\bar{v}_E(\mathbf{O})$?

A **p**
B **q**
C **r**
D **s**
E **t**

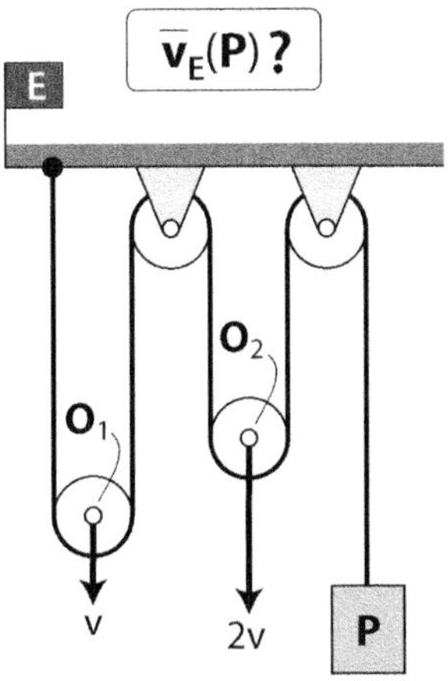

3.24 The centers $\mathbf{O_1}$ and $\mathbf{O_2}$ of the pulleys move downward with velocities v and 2v relative to the ground, respectively. If the cable is inextensible and does not slide on the pulleys, what is the speed of block **P** relative to the ground?

A (3/2)v
B 3v
C 4v
D 5v
E 6v

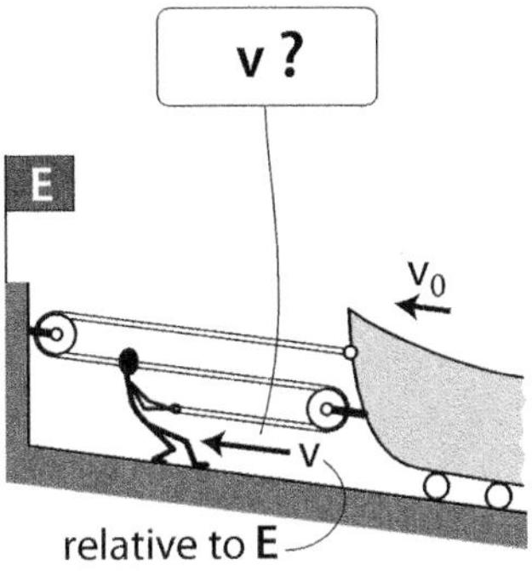

3.25 If the cable does not slide on the pulleys, what velocity (relative to the ground) has to be imposed on the cable end point so that the boat moves up with speed v_0 relative to the ground?

A v_0
B $2v_0$
C $3v_0$
D $(3/2)v_0$
E $(1/2)v_0$

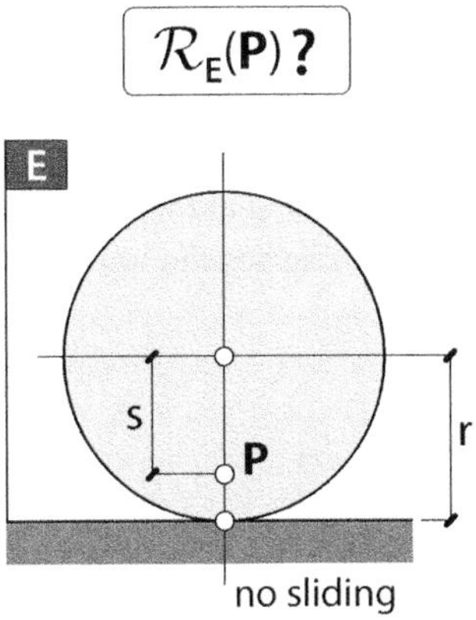

3.26 What is the curvature radius of the trajectory of point **P** relative to the ground?

A r − s
B s
C ∞
D $s^2/(r-s)$
E $(r-s)^2/s$

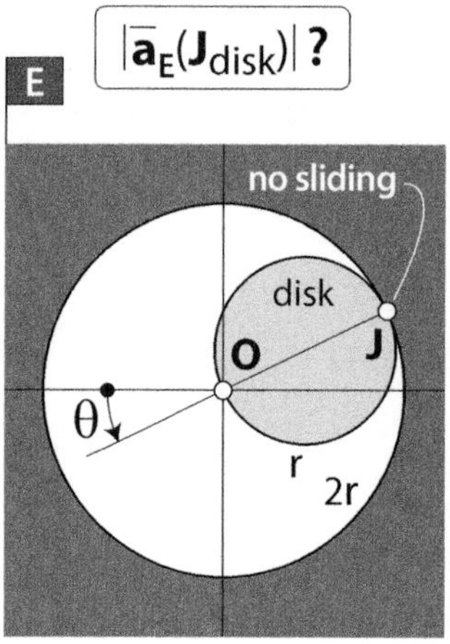

3.27 What is the module of the acceleration of point **J** of the disk, relative to the ground, for the given configuration?

A 0
B ∞
C $r\dot{\theta}^2$
D $2r\dot{\theta}^2$
E $\sqrt{2}r\dot{\theta}^2$

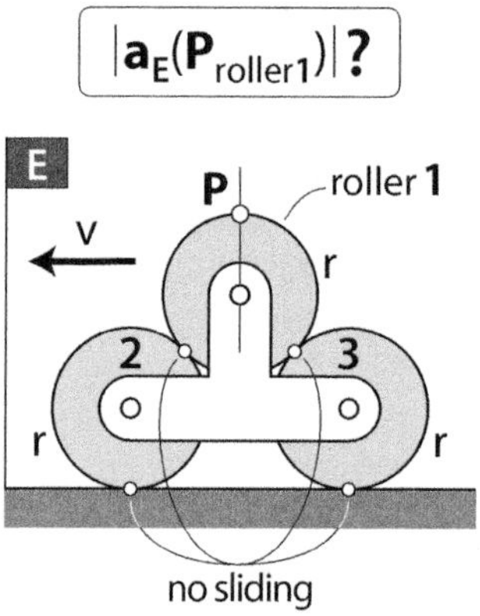

3.28 The three rollers are identical, and their centers move with the same speed v relative to the ground. What is the module of the acceleration of point **P** of roller 1 relative to the ground?

A 0
B v^2/r
C $2v^2/r$
D $4v^2/r$
E $9v^2/r$

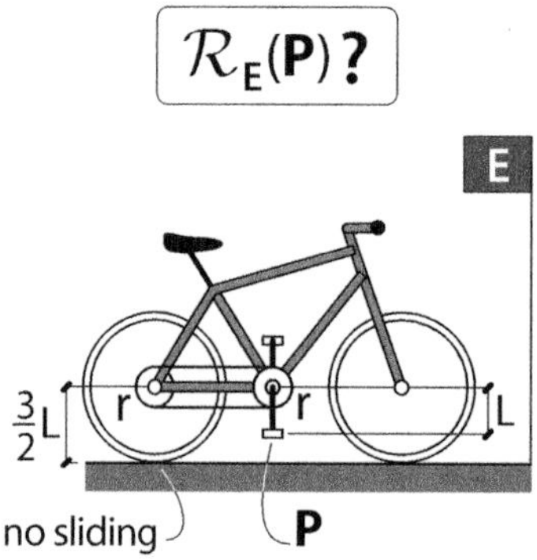

3.29 The sprocket and the chainring of the bicycle have the same radius. What is the curvature radius of the trajectory relative to the ground of point **P** of the pedal when it reaches the lowest position?

A L
B (1/2)L
C (1/4)L
D (2/3)L
E (1/3)L

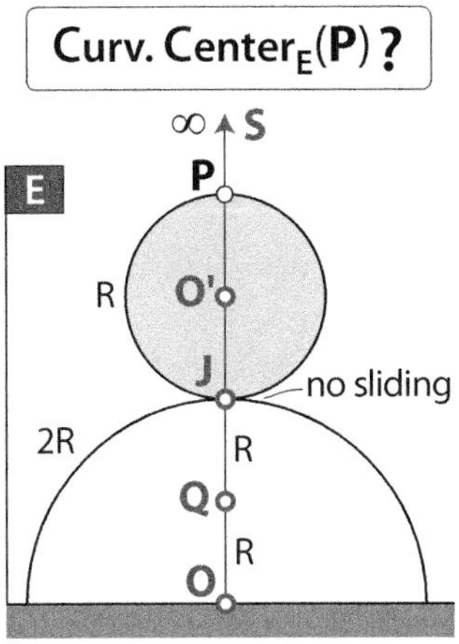

3.30 What point is the curvature center of the trajectory relative to the ground of point **P** of the wheel when it reaches the highest position?

A r
B 2r
C 3r
D 4r
E 6r

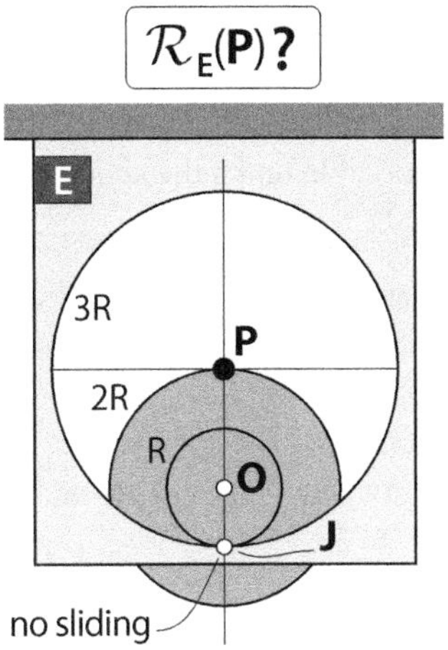

3.31 What is the curvature radius of the trajectory relative to the ground of point **P** of the roller when it reaches the highest position?

A R
B 2R
C 3R
D 4R
E 6R

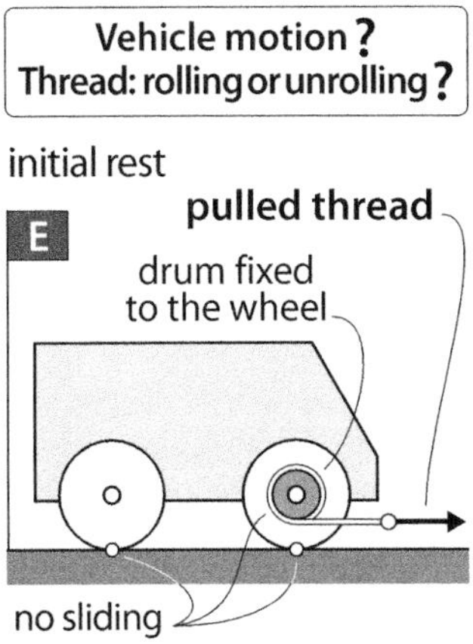

3.32 The thread does not slide on the drum (fixed to the front wheel). If we pull it forward, how is the vehicle motion relative to the ground? Does the thread roll or unroll on the drum?

A The vehicle moves forward, and the thread rolls up on the drum
B The vehicle moves forward, and the thread unrolls
C The vehicle moves forward, and the thread neither rolls up nor unrolls
D The vehicle does not move, and the thread unrolls
E The vehicle slides on the ground necessarily, and the thread unrolls

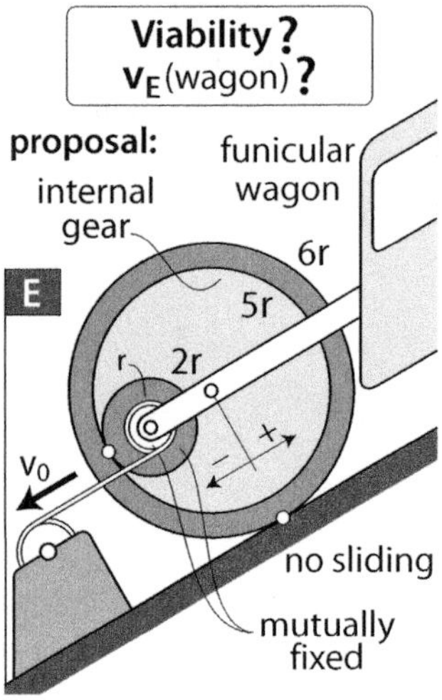

3.33 Someone proposes to move upward the cable railway by pulling downward the thread rolled up on the gear. Is it a viable project? What would be the velocity of the cable railway relative to the ground?

A It is not viable; the cable railway would move downward freely

B It is viable; the cable railway moves up with $(12/7)v_0$

C It is viable; the cable railway moves up with $(7/12)v_0$

D It is viable; the cable railway moves up with $(12/5)v_0$

E It is viable; the cable railway moves up with $(5/12)v_0$

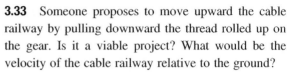

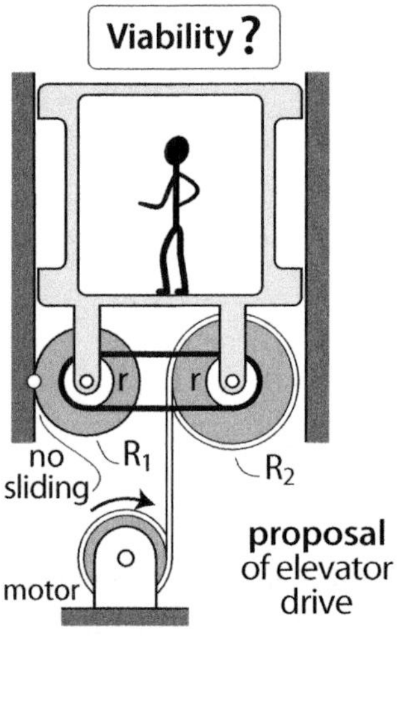

3.34 Someone proposes to move upward the elevator by rolling up the cable under the action of a motor. Is it a viable project?

A It is not viable

B It is viable if $R_1 = R_2$

C It is viable if $R_1 > R_2$

D It is viable if $R_1 < R_2$

E It is viable for any values of R_1 and R_2

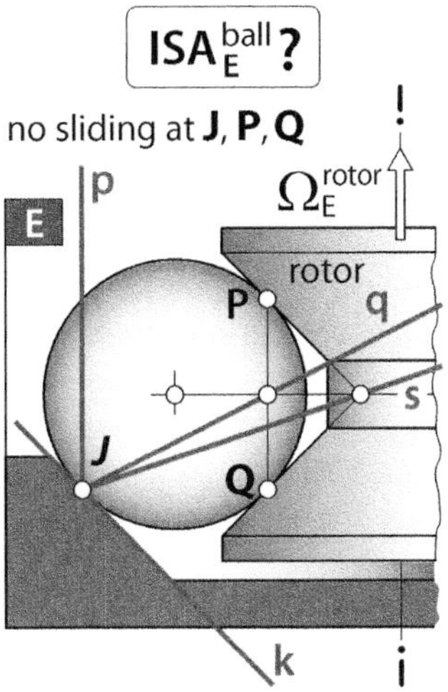

3.35 The ball has nonsliding contacts with the ground (E) and the rotor. Which line is the ISA_E^{ball}?

A Line **p**
B Line **q**
C Line **s**
D Line **t**
E It is not defined because the ball cannot move under those nonsliding conditions

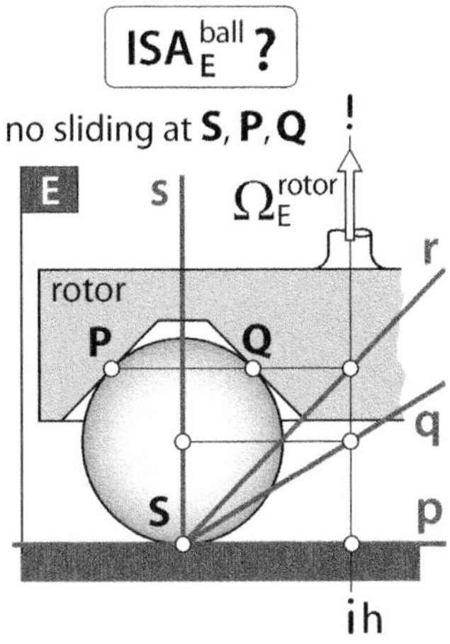

3.36 The ball has nonsliding contacts with the ground (E) and the rotor. Which line is the ISA_E^{ball}?

A Line **p**
B Line **q**
C Line **r**
D Line **s**
E It is not defined because the ball cannot move under those nonsliding conditions

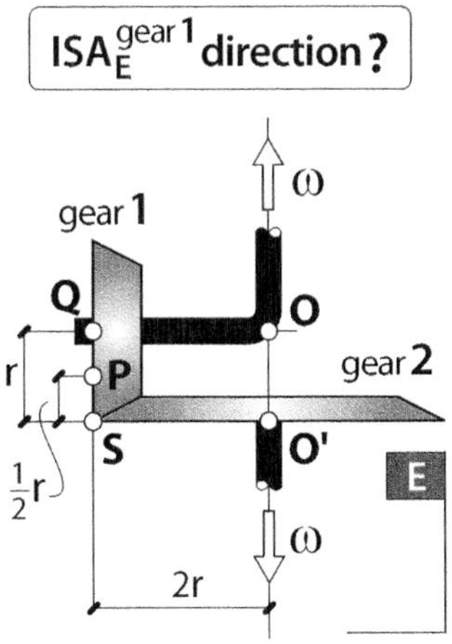

3.37 In the bevel gearing, the two coaxial axes rotate with the same angular velocity ω relative to the ground (E) but opposite directions. Which line is the $ISA_E^{gear\ 1}$?

A Line **QO**
B Line **SO′**
C Line **SO**
D Line **PO**
E Line **OO′**

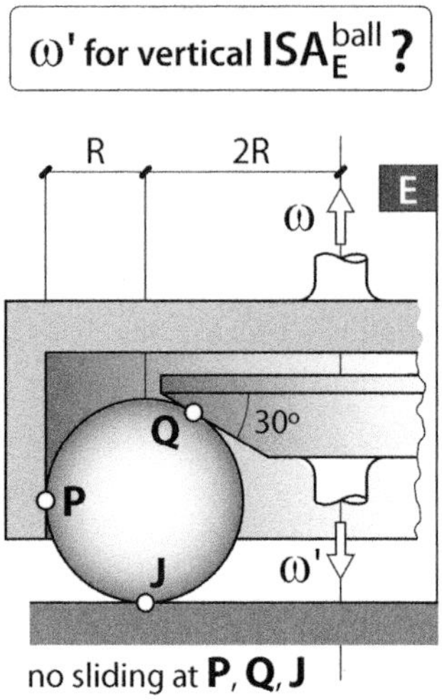

3.38 The ball has nonsliding contacts with the ground (E) and the two rotors. For which ω' value would the ISA_E^{ball} be vertical?

A 3ω
B ω
C -3ω
D The ISA_E^{ball} is vertical for any value of ω'
E $-\omega$

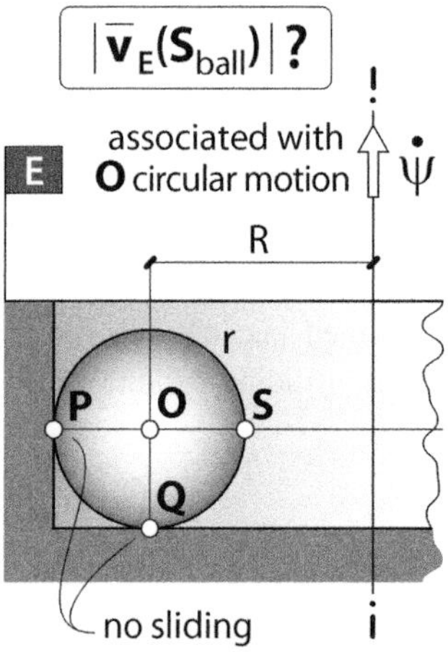

3.39 The ball has two nonsliding contacts with the ground (E). If $\dot{\psi}$ is the angular velocity associated with the circular motion of its center **O** relative to the ground, what is the value of $|\bar{\mathbf{v}}_E(\mathbf{S}_{ball})|$?

A $R\dot{\psi}$
B $2R\dot{\psi}$
C $\sqrt{2}R\dot{\psi}$
D $(R-r)\dot{\psi}$
E $\sqrt{2}(R-r)(R/r)\dot{\psi}$

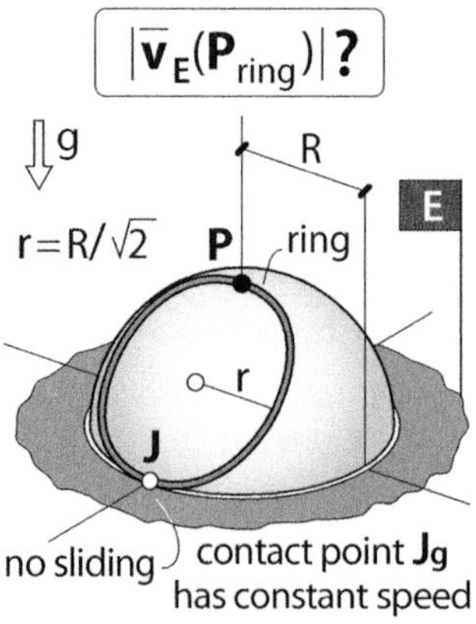

3.40 The ring moves on the semispherical support (fixed to the ground) without sliding at **J**. If the geometrical contact point has a constant speed v_0 relative to the ground (E), what is the value of $|\bar{\mathbf{v}}_E(\mathbf{P}_{ring})|$ for that particular configuration?

A 0
B v_0
C $v_0/2$
D $v_0/\sqrt{2}$
E $2v_0$

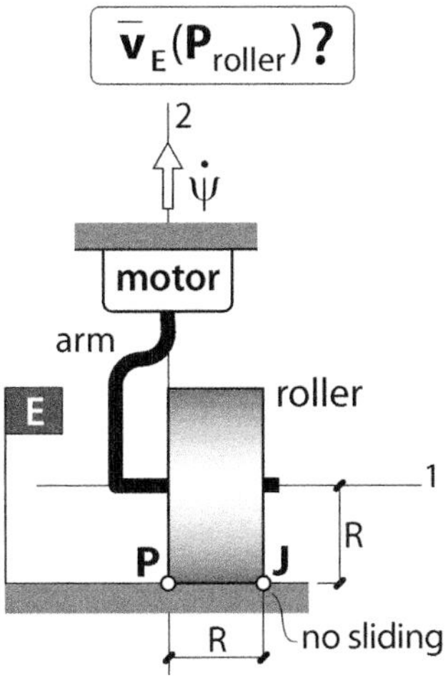

3.41 The roller moves under the action of a motor, which controls the arm rotation $\dot{\psi}$. If there is no sliding at point **J**, what is the velocity of point **P** of the roller relative to the ground?

A 0
B $\dot{\psi}R \odot$
C $\dot{\psi}R \otimes$
D $2\dot{\psi}R \otimes$
E $2\dot{\psi}R \odot$

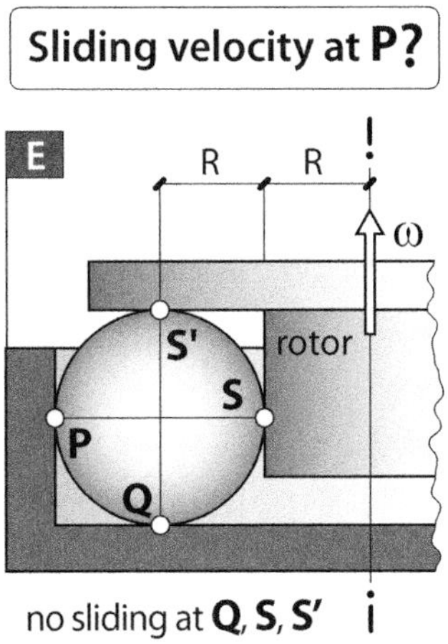

3.42 The ball has a nonsliding contact with the ground at **Q**, and two nonsliding ones at **S** and **S′** with the rotor. What is the value of the sliding velocity at **P**?

A 0
B ωR
C $(1/2)\omega R$
D $2\omega R$
E $\sqrt{2}\omega R$

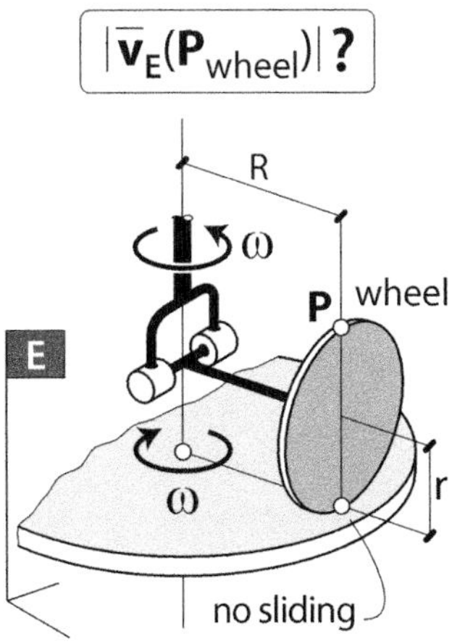

3.43 The wheel axis and the circular platform rotate in opposite directions with angular velocity ω. If the wheel does not slide on the platform, what is the speed of point **P** of the wheel, relative to the ground, when it reaches the highest position?

A $\omega(R + r)$
B $\omega(R - r)$
C $2\omega R$
D $3\omega R$
E $4\omega R$

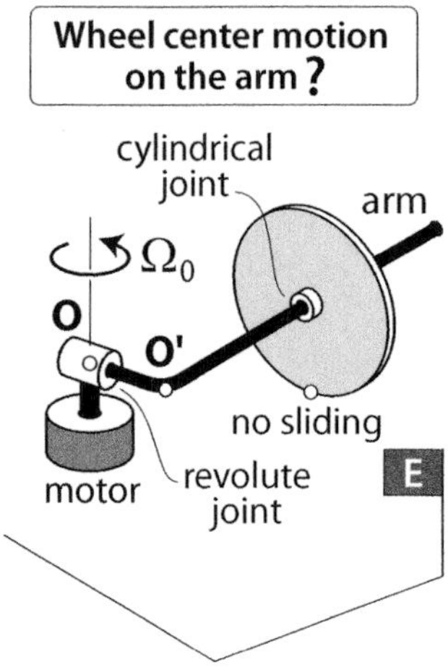

3.44 The wheel moves without sliding on the ground driven by the arm, which rotates with constant angular velocity Ω_0 relative to the ground. What is the motion of the wheel center relative to the arm?

A It moves away from **O**′ with constant speed
B It moves away from **O**′ with increasing speed
C It moves toward **O**′ with constant speed
D It moves toward **O**′ with decreasing speed
E It does not move relative to the arm, and sliding on the ground is inevitable

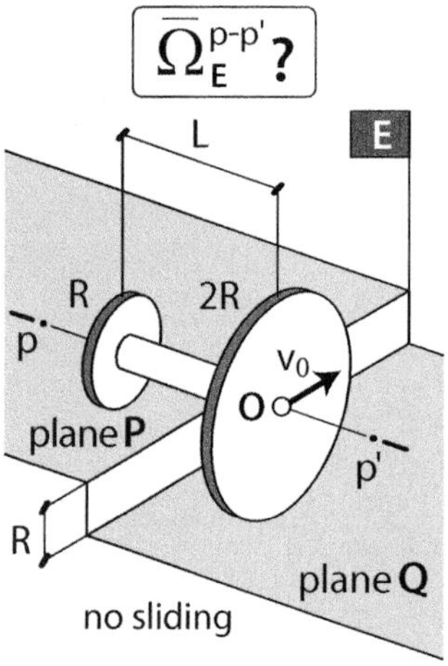

3.45 The two disks are mutually fixed and move without sliding on the ground. What is the angular velocity (relative to the ground) of their common axis p–p′?

A It may have any value
B It is zero as the axis has a rectilinear translation motion
C It is v_0/L
D It is $v_0/2L$
E It is $v_0/(L^2 + R^2)$

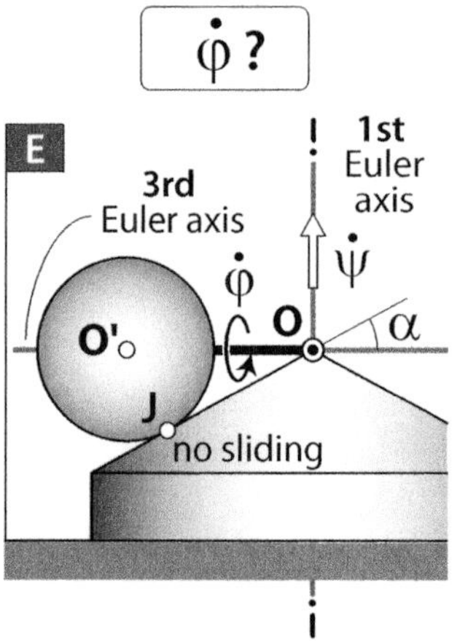

3.46 The ball is articulated to the ground through a ball-and-socket joint at **O**. Its angular velocity is described by two Euler rotations: a first rotation $\dot{\psi}$ around the vertical axis, and a third rotation $\dot{\varphi}$ about its axis **OO**′. What is the value of $\dot{\varphi}$?

A 0
B $\dot{\psi}$
C $\dot{\psi}\tan\alpha$
D $\dot{\psi}/\tan\alpha$
E $\dot{\psi}/\sin\alpha$

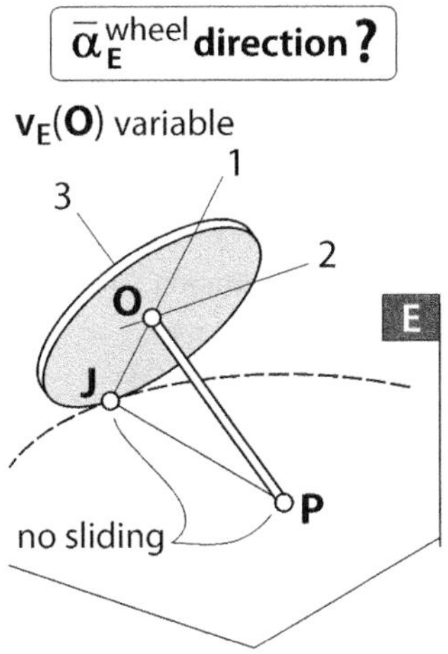

3.47 The wheel moves on the ground without sliding at its two contact points **J** and **P**. If the speed of its center **O** is not constant, what is the direction of its angular acceleration relative to the ground?

A It is vertical
B It is horizontal with a positive second component
C It is horizontal with a negative second component
D It is not constrained
E It is always in the 1–3 plane

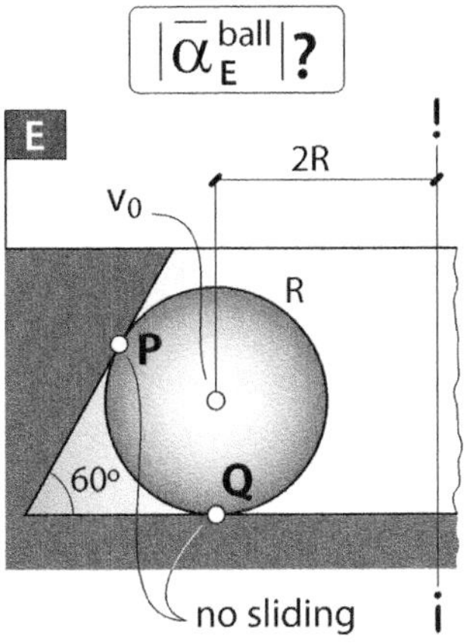

3.48 The ball has two nonsliding contacts with the ground-fixed axisymmetric support. Its center moves with constant speed v_0 relative to the ground. What is the value of its angular acceleration relative to the ground?

A 0
B $(v_0/R)^2$
C $(1/2)(v_0/R)^2$
D $\sqrt{3}(v_0/R)^2$
E $(\sqrt{3}/2)(v_0/R)^2$

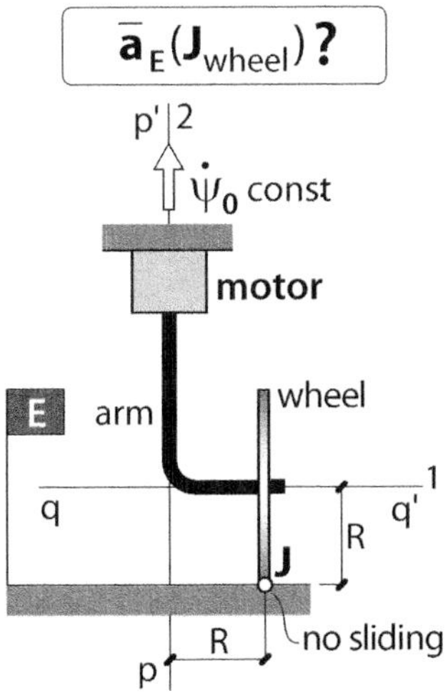

3.49 The wheel moves without sliding on the ground under the action of the motor, which controls the arm constant rotation $\dot{\psi}_0$. What are the components (a_1, a_2) of acceleration of point **J** of the wheel relative to the ground?

A $(\dot{\psi}_0^2 R, 0)$
B $(0, \dot{\psi}_0^2 R)$
C $(-\dot{\psi}_0^2 R, 0)$
D $(\dot{\psi}_0^2 R, \dot{\psi}_0^2 R)$
E $(-\dot{\psi}_0^2 R, \dot{\psi}_0^2 R)$

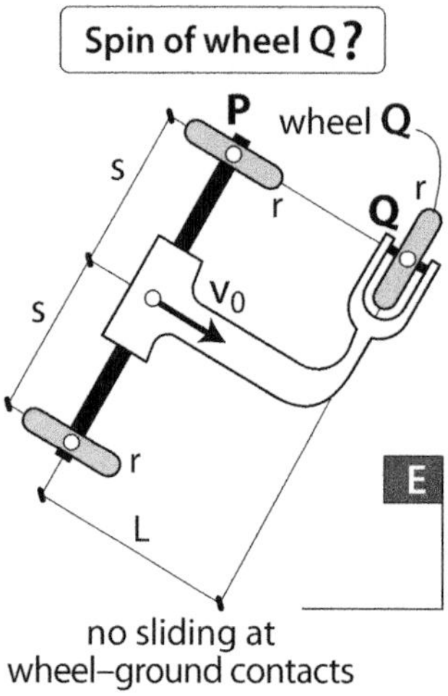

3.50 The vehicle has three identical wheels with radius r. What is the angular velocity of wheel Q relative to the chassis?

A $(v_0L)/(sr)$, with direction $\overline{\mathbf{PQ}}$
B v_0/r, with direction $\overline{\mathbf{PQ}}$
C 0, because the orientation of the front wheel blocks the vehicle movement
D $(v_0L)/(sr)$, with direction $\overline{\mathbf{QP}}$
E v_0/r, with direction $\overline{\mathbf{QP}}$

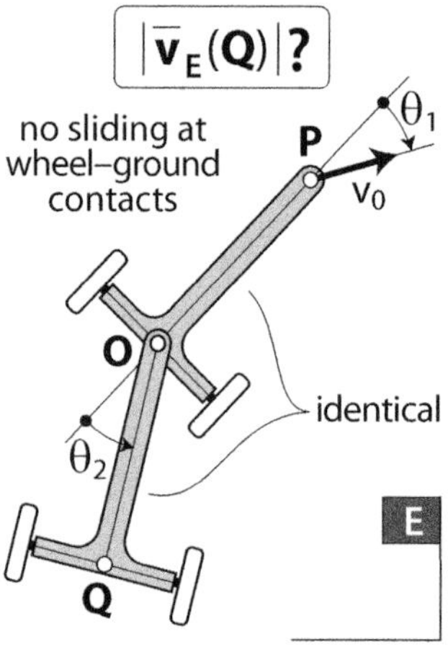

3.51 The two articulated frames are identical. What is the speed of point **Q** relative to the ground?

A v_0
B $v_0 \cos\theta_1$
C $v_0 \cos\theta_1 \cos\theta_2$
D $v_0 \cos\theta_1 / \cos\theta_2$
E $v_0 \cos\theta_2 / \cos\theta_1$

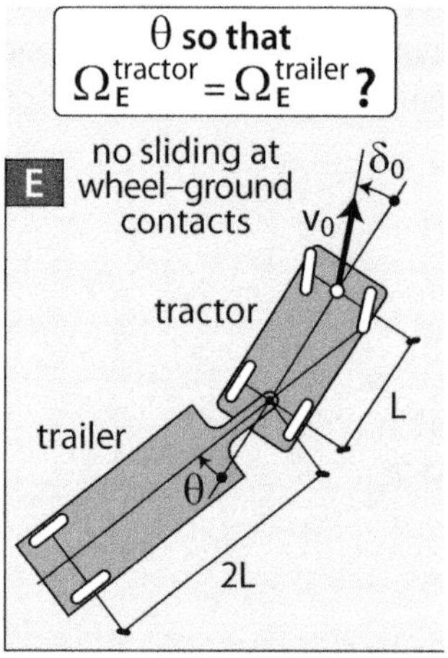

3.52 The tractor drags an articulated trailer in a cornering motion. If the speed v_0 and the steering angle δ_0 are constant, for what value of the relative angle θ do the tractor and the trailer have the same angular velocity relative to the ground?

A $\theta = 0$
B $\theta = \delta_0$
C $\theta = 2\delta_0$
D $\sin\theta = 2\tan\delta_0$
E $\cos\theta = 2\cos\delta_0$

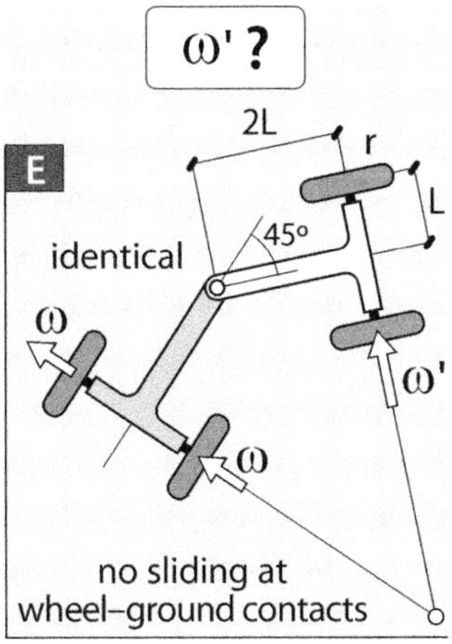

3.53 The vehicle with articulated chassis moves on the ground. If the two wheels of the rear subchassis have the same spin velocity ω, what is the value of the spin ω′ of the inner wheel of the front subchassis for the given configuration?

A ω
B $\omega/\sqrt{2}$
C $\omega/(2\sqrt{2})$
D $\omega\sqrt{2}$
E $\omega 2\sqrt{2}$

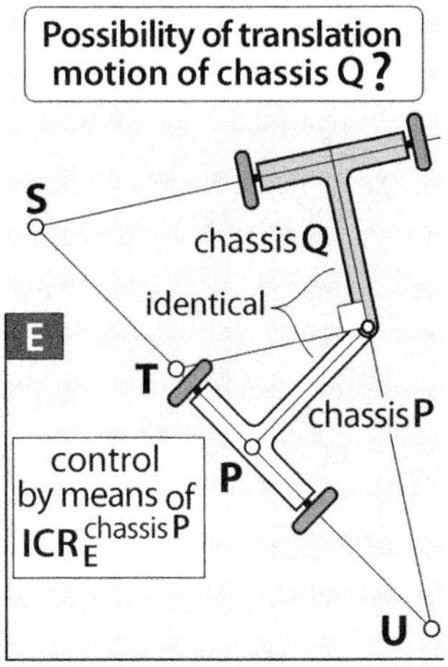

3.54 The driver of the vehicle with articulated chassis controls the position of the $\mathbf{ICR}_E^{chassis\ P}$. Is it possible to achieve a translation motion relative to the ground for chassis Q?

A Only if chassis P and chassis Q are aligned
B Only if $\mathbf{ICR}_E^{chassis\ P} = \mathbf{P}$
C Only if $\mathbf{ICR}_E^{chassis\ P} = \mathbf{S}$
D Only if $\mathbf{ICR}_E^{chassis\ P} = \mathbf{T}$
E Only if $\mathbf{ICR}_E^{chassis\ P} = \mathbf{U}$

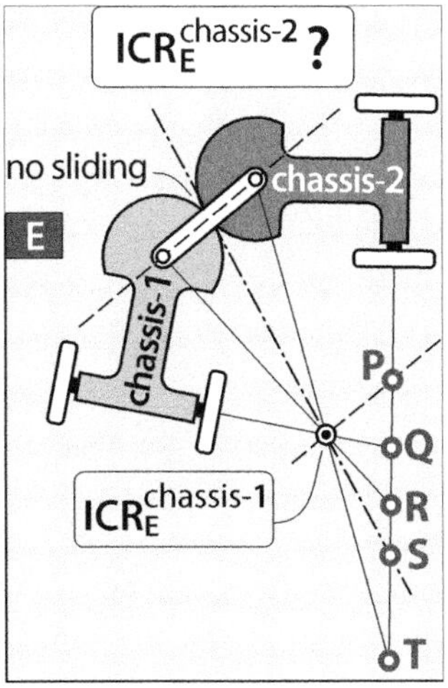

3.55 The two chassis of the vehicle have a nonsliding contact. Which point is the $\mathbf{ICR}_E^{chassis-2}$ for the given $\mathbf{ICR}_E^{chassis-1}$ and given configuration?

A **P**
B **Q**
C **R**
D **S**
E **T**

3.56 All the omnidirectional wheels of the vehicle are identical. What motorized angular velocity $\dot{\varphi}_1$ is needed for a longitudinal translation motion v_0 of the chassis relative to the ground?

A $v_0/\left(\sqrt{2}r\right)$

B $-v_0/\left(\sqrt{2}r\right)$

C $v_0/\left(\sqrt{2}R\right)$

D $-v_0/\left(\sqrt{2}R\right)$

E $v_0r/\left(2R^2\right)$

no sliding at sphere–ground
and sphere–wheel contacts

motors and sphere center fixed to the platform

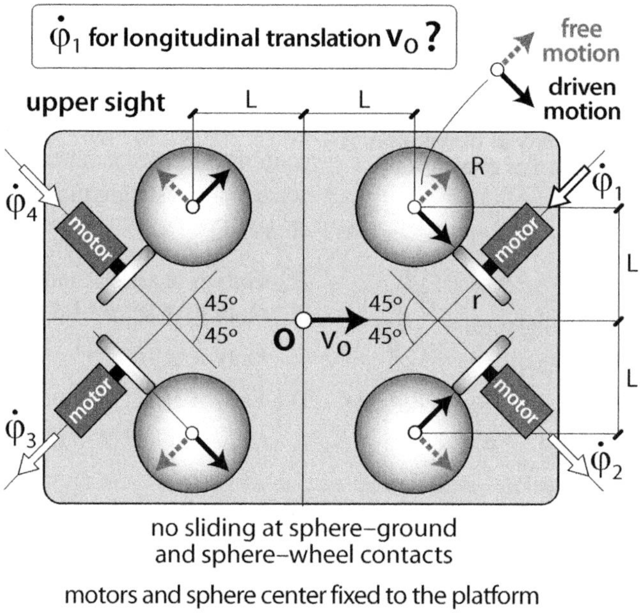

3.57 What motorized angular velocities $\{\dot{\varphi}_P, \dot{\varphi}_Q \dot{\beta}\}$ lead to a transverse translation motion v_O of the chassis relative to the ground?

A $v_O\left\{\frac{\sqrt{2}}{R}, -\frac{\sqrt{2}}{R}, \frac{1}{r}\right\}$

B $v_O\left\{-\frac{\sqrt{2}}{R}, \frac{\sqrt{2}}{R}, \frac{1}{r}\right\}$

C $v_O\left\{\frac{1}{R}, -\frac{1}{R}, 0\right\}$

D $v_O\left\{-\frac{1}{R}, \frac{1}{R}, \frac{1}{r}\right\}$

E $v_O\left\{\frac{1}{R}, -\frac{1}{R}, \frac{1}{r}\right\}$

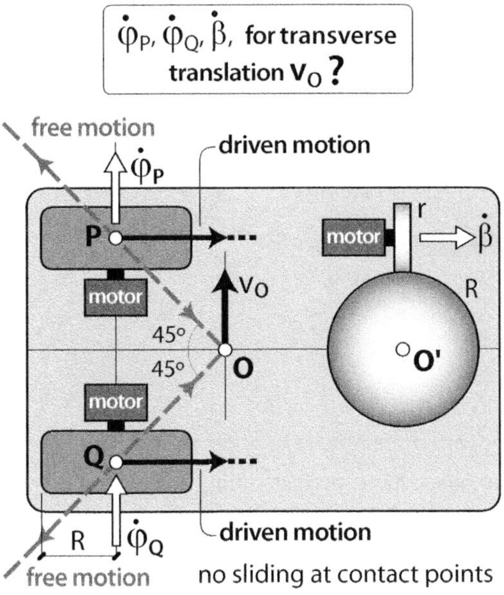

3.58 The vehicle is initially at rest on the ground. If we activate the motor of wheel **P** and those of wheels **Q** and **O** are locked, what point is the $\mathbf{ICR}_E^{\text{chassis}}$?

A **O″**
B **S**
C **S′**
D **K**
E It is not defined because the vehicle does not move under those conditions

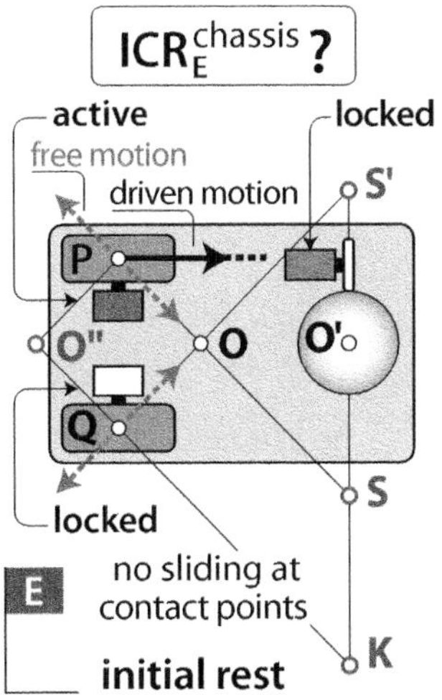

Problems

3.1 The wheel has a perfect rolling motion on a horizontal ground.

(a) Find $\bar{v}_E(\mathbf{P})$, $\bar{a}_E(\mathbf{P})$

(b) Prove that the curvature center **C** of the trajectory of **P** relative to the ground is always the symmetric point of **P** with respect to the contact point **J**

(c) The **P** trajectory (relative to the ground) is a curve named *cycloid*. Find the expression of its coordinates (x, y) as a function of θ, if $x = 0$ for $\theta = 0$

Note: the **C** trajectory relative to the ground is also a cycloid.

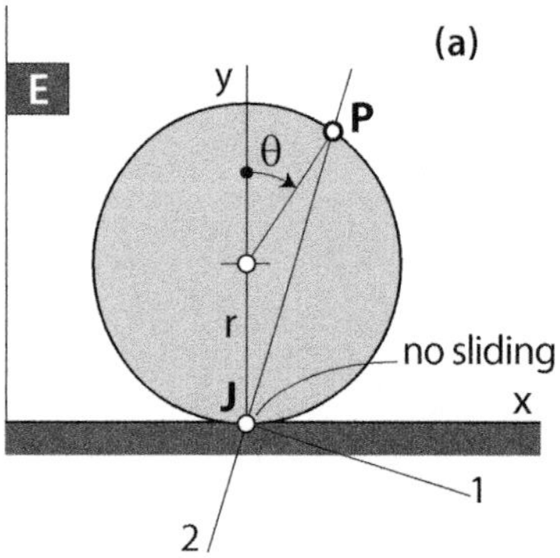

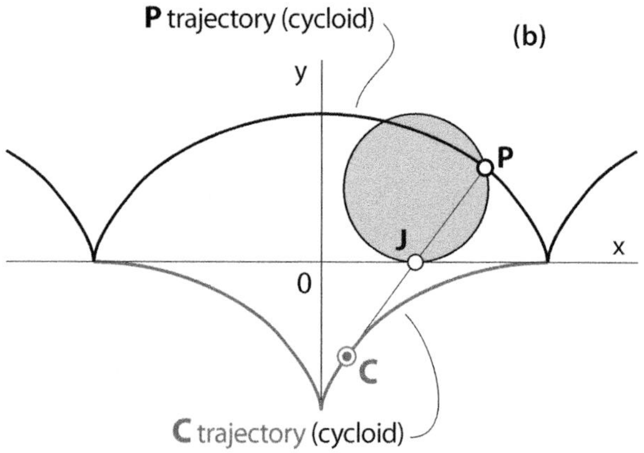

3.2 The block has a perfect rolling motion on a semicircular support. For $\theta = 0$, the central point **C** of the block is on the vertical line going through **O**.

(a) Find $\bar{v}_E(\mathbf{C})$

(b) Find $\bar{a}_E(\mathbf{C})$

(c) The position of the curvature center of the trajectory of **C** (relative to the ground) for $\theta = 0$

(d) Investigate the possibility of the **C** trajectory to have inflection points (points with zero curvature)

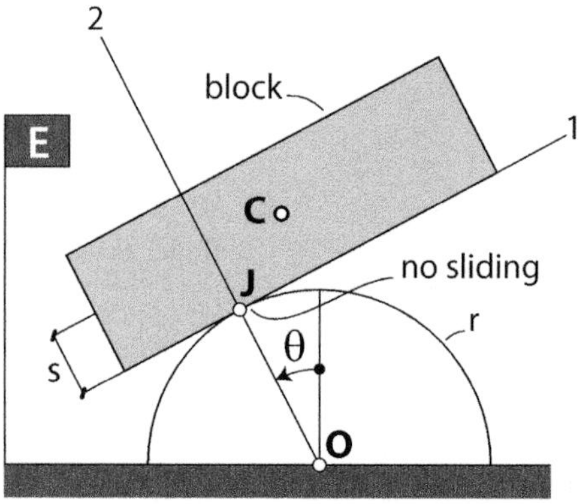

3.3 The disk has a perfect rolling motion on the rotating support. The arm, actuated by a motor, provokes a constant angular velocity $\dot{\theta}$. Find $\bar{\mathbf{a}}_{\mathrm{E}}(\mathbf{J}_{\mathrm{disc}})$.

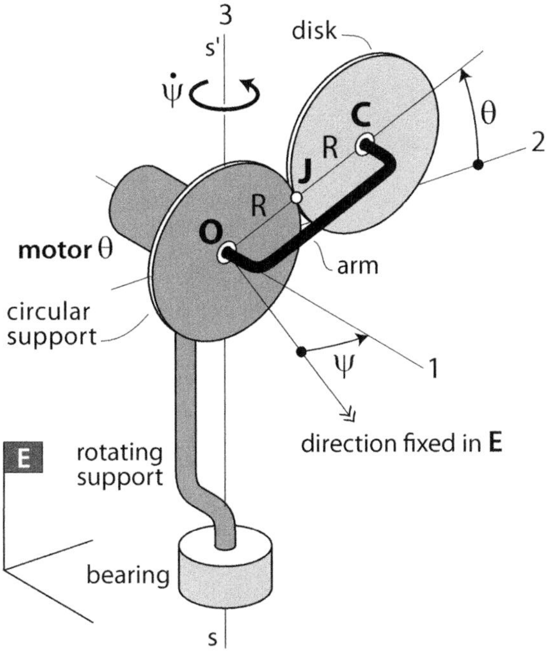

3.4 The rectangular plate can rotate about the axis r–r′ of the swingarm, which is articulated to a support rotating freely around the vertical axis s–s′. A hydraulic cylinder located between the rotating support and point **P** of the plate controls the plate inclination θ so that the angular velocity $\dot{\theta}$ is constant. Take the ground as AB reference frame and the support as REL reference frame, and find $\bar{\mathbf{a}}_{\mathrm{REL}}(\mathbf{C})$, $\bar{\mathbf{a}}_{\mathrm{tr}}(\mathbf{C})$, $\bar{\mathbf{a}}_{\mathrm{Cor}}(\mathbf{C})$.

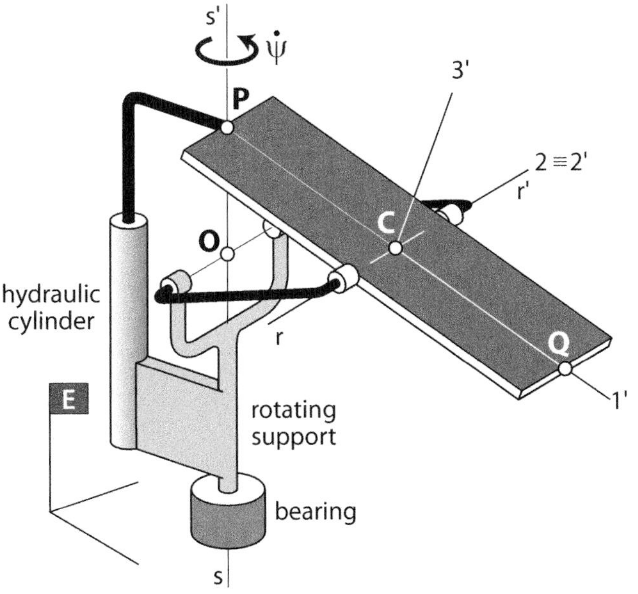

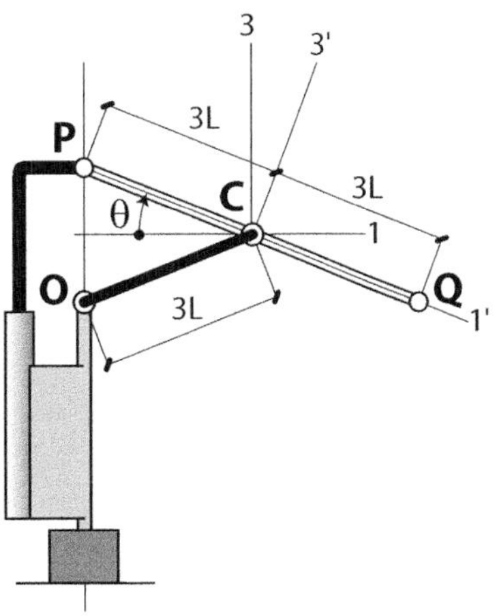

3.5 The motor Ω controls the constant rotation Ω_0 of the circular support relative to the ground. The arm rotates with angular velocity ω and angular acceleration $\dot{\omega}$ relative to the support. The motor ω acts between the arm and the wheel with center **Q**. The wheel–support contact is a nonsliding one. The bar with length 2R is fixed to the wheel.

(a) Find $\bar{\boldsymbol{\Omega}}_{\mathrm{E}}^{\mathrm{bar}}$

(b) Find $\bar{\mathbf{v}}_{\mathrm{E}}(\mathbf{P})$ and $\bar{\mathbf{a}}_{\mathrm{E}}(\mathbf{P})$ when **P** is on the vertical line through **O**

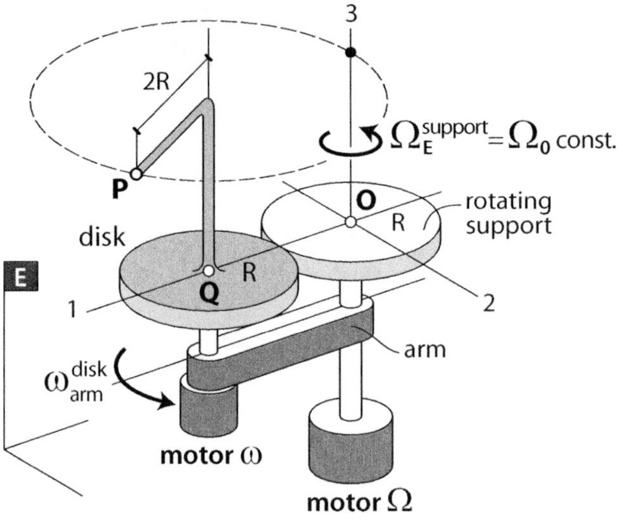

3.6 In the positioning robot shown in the figure, the bar **CP** is fixed to the wheel with center **C**. That wheel is linked to the wheel with center **O** (fixed to the support) through an inextensible chain transmission. An actuator controls the constant angular velocity Ω_0 of the support relative to the ground. A second actuator controls the variation of the arm **OC** inclination. Take the ground as AB reference frame and the support as REL reference frame.

(a) Find $\dot{\varphi}$

(b) Find $\bar{\mathbf{v}}_{\mathrm{REL}}(\mathbf{P})$, $\bar{\mathbf{v}}_{\mathrm{tr}}(\mathbf{P})$

(c) Find $\bar{\mathbf{a}}_{\mathrm{REL}}(\mathbf{P})$, $\bar{\mathbf{a}}_{\mathrm{tr}}(\mathbf{P})$, $\bar{\mathbf{a}}_{\mathrm{Cor}}(\mathbf{P})$ for $\varphi = 90^{\mathrm{o}}$

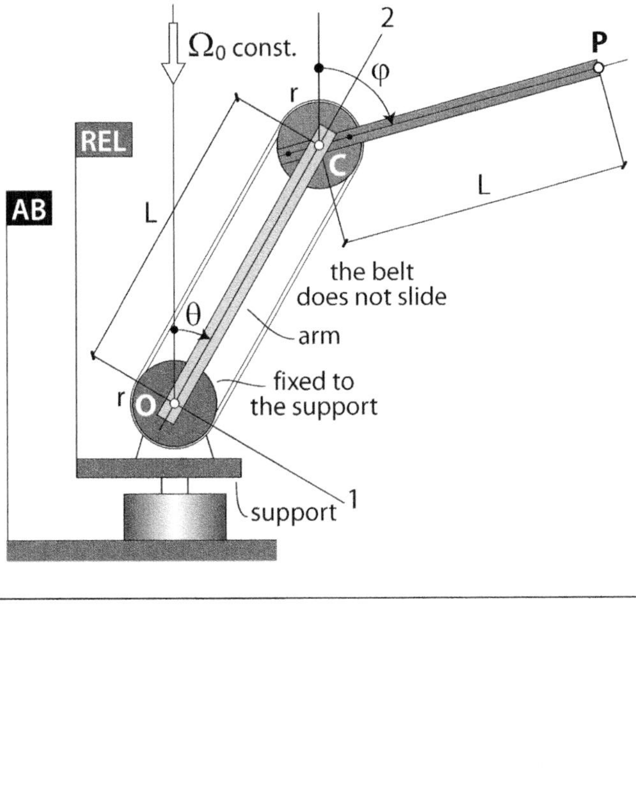

3.7 The rigid body S consists of two truncated cones joined by the base and has a perfect rolling motion relative to the ground. The value of $\dot{\varphi} = \dot{\varphi}_0$ is constant. Axis 3 of the vector basis B is vertical, and axis 2 is parallel to the contact edge of the cone with the ground.

(a) Find the fixed and moving axodes associated with the S motion relative to the ground

(b) Find $\bar{\boldsymbol{\Omega}}^{\mathrm{S}}_{\mathrm{E}}, \bar{\boldsymbol{\alpha}}^{\mathrm{S}}_{\mathrm{E}}$

(c) Find $\bar{\mathbf{v}}_{\mathrm{E}}(\mathbf{P}'), \bar{\mathbf{a}}_{\mathrm{E}}(\mathbf{P}')$, where $\mathbf{P}'$ is the highest point of the rigid body

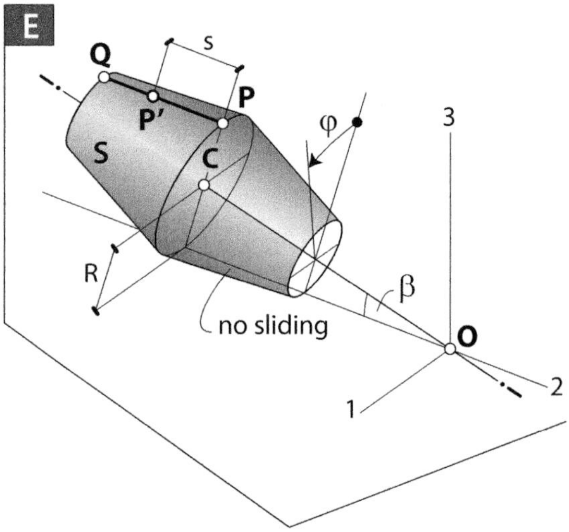

3.8 The ball of the bearing has nonsliding contacts at **P** and **Q** with the outer ring fixed to the ground, and at **S** with the inner ring. The inner ring rotates with constant ω around the bearing axis $\mathbf{OO'}$. Axis 1 of the vector basis B is vertical, and axis 2 is parallel to the $\overline{\mathbf{O}_{\mathrm{E}}\mathbf{Q}_{\mathrm{g}}}$ line.

(a) Find $\bar{\boldsymbol{\Omega}}_{\mathrm{E}}^{\mathrm{B}}$
(b) Find $\bar{\boldsymbol{\Omega}}_{\mathrm{E}}^{\mathrm{ball}}, \bar{\boldsymbol{\alpha}}_{\mathrm{E}}^{\mathrm{ball}}$
(c) Find the fixed and moving axodes associated with the ball movement relative to the ground
(d) Find $\bar{\mathbf{a}}_{\mathrm{E}}(\mathbf{Q})$
(e) Find the set of ball points that come into contact with the rings

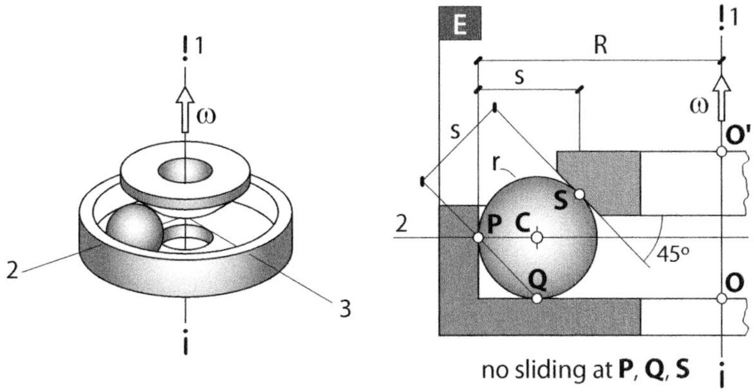

3.9 The ball of radius R is fixed to the radial rod whose end point **O** is fixed to the ground by means of a ball-and-socket joint. The ball has two nonsliding contacts with the rotating supports SP and SQ at points **P** and **Q**. SP and SQ rotate with constant angular velocity ω in opposite directions. Axis 2 of the vector basis B is vertical, and axis 1 is parallel to the $\overline{\mathbf{P}_g\mathbf{Q}_g}$ line.

(a) Find $\bar{\boldsymbol{\Omega}}_{E}^{ball}, \bar{\boldsymbol{\alpha}}_{E}^{ball}$

(b) Find the fixed and moving axodes associated with the ball movement relative to the ground

(c) Find the angular velocities $(\dot{\psi}, \dot{\varphi})$ associated with the ball's Euler angles (ψ, φ) $(\theta = 45^o)$

(d) Find $\bar{\mathbf{a}}_E(\mathbf{P}), \bar{\mathbf{a}}_E(\mathbf{Q})$

(e) Find $\bar{\mathbf{v}}_{SP}(\mathbf{P}_g), \bar{\mathbf{v}}_{SQ}(\mathbf{Q}_g)$

(f) Find $\bar{\mathbf{a}}_{SP}(\mathbf{P}), \bar{\mathbf{a}}_{SQ}(\mathbf{Q})$

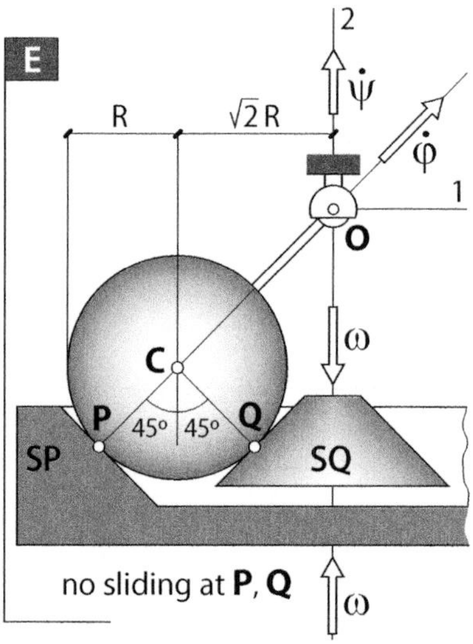

3.10 The ball with radius r has three nonsliding contacts with the rotors Q, C, and S at points **Q**, **C**, and **S**, respectively. The rotors' angular velocities ω_Q, ω_C and ω_S are constant. The speed of points **Q**, **C** and **S** relative to the ground are $|\bar{\mathbf{v}}_E(\mathbf{Q})| = 4v_0, |\bar{\mathbf{v}}_E(\mathbf{C})| = |\bar{\mathbf{v}}_E(\mathbf{S})| = v_0$. Axis 2 of the vector basis B is vertical, and axis 1 is parallel to the $\overline{\mathbf{C}_g\mathbf{S}_g}$ line.

(a) Find $\bar{\boldsymbol{\Omega}}_E^B, \bar{\boldsymbol{\Omega}}_E^{ball}, \bar{\boldsymbol{\alpha}}_E^{ball}$

(b) Find $\bar{\mathbf{v}}_E(\mathbf{P}), \bar{\mathbf{a}}_E(\mathbf{P})$

(c) Find the ISA and the fixed and moving axodes associated with the ball movement relative to the ground

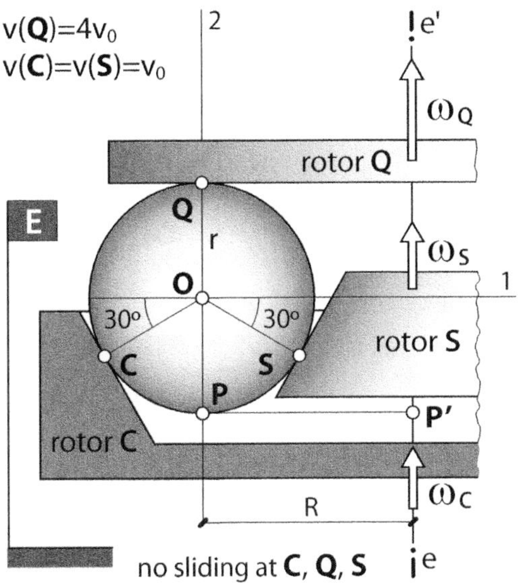

3.11 The motor 1 drives the semispherical shell through the fork, which rotates with a variable angular velocity ω relative to the ground around the vertical axis. The shell has a nonsliding contact at **J** with a platform whose angular velocity, with value ω and direction opposite to that of the fork, is controlled by motor 2.

(a) Find the screw axis of the shell relative to the ground, and the corresponding fixed and moving axodes

(b) Find $\bar{\boldsymbol{\Omega}}_{\mathrm{E}}^{\mathrm{shell}}, \bar{\boldsymbol{\alpha}}_{\mathrm{E}}^{\mathrm{shell}}$

(c) Find $\bar{\mathbf{v}}_{\mathrm{E}}(\mathbf{P}), \bar{\mathbf{a}}_{\mathrm{E}}(\mathbf{P})$, and $\mathfrak{R}_{\mathrm{E}}(\mathbf{P})$ when **P** reaches the highest position

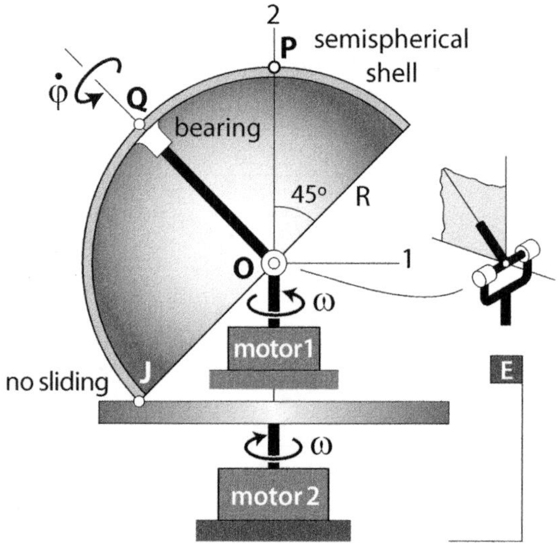

3.12 The orientation of the semispherical shell relative to the ground is described through the Euler angles ψ (first) and φ (third). The motor φ, actuating between the shell and the axis articulated to the fork, controls the rotation $\dot{\varphi} = \omega\sqrt{2}$ (where ω **is variable**). The rotation $\dot{\psi}$ is a consequence both of that $\dot{\varphi}$ and the nonsliding contact at **J** between the shell and the platform. The platform rotates with 2ω (also **variable**) relative to the ground under the action of motor 2.

(a) Find $\dot{\psi}$

(b) Find the screw axis of the shell relative to the ground, and the corresponding fixed and moving axodes

(c) Find $\bar{\boldsymbol{\Omega}}_{E}^{shell}, \bar{\boldsymbol{\alpha}}_{E}^{shell}$

(d) Find $\bar{\mathbf{v}}_{E}(\mathbf{Q}_{shell}), \bar{\mathbf{a}}_{E}(\mathbf{Q}_{shell})$, and $\mathfrak{R}_{E}(\mathbf{Q}_{shell})$ when ω is constant

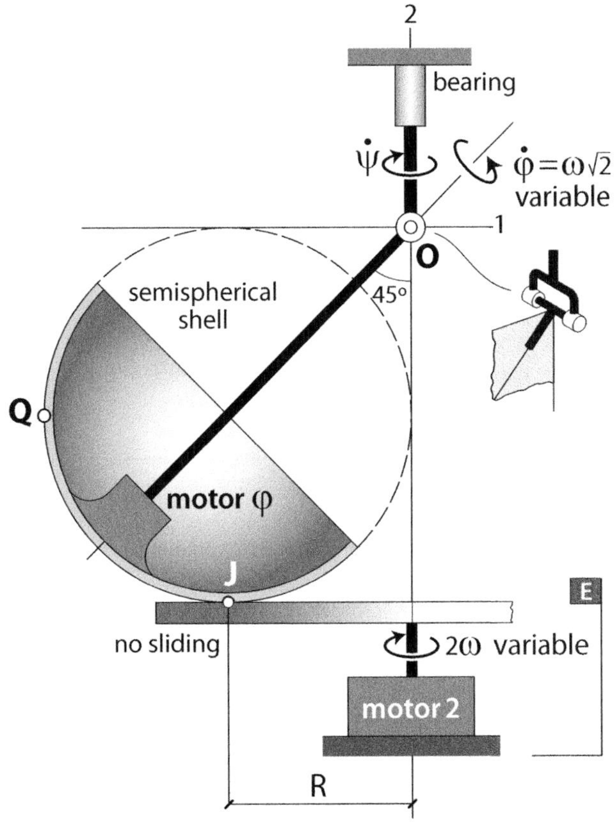

3.13 The ring is articulated to the platform through two revolute joints and has a perfect rolling motion on it. The platform rotates about the vertical axis going through the ground-fixed point **O** with constant angular velocity Ω_0. For $\theta = \pi$, the $\dot{\theta}$ value is minimum and the ring center **C** is instantaneously located on the platform rotation axis.

(a) Find $\bar{\mathbf{a}}_{\mathrm{E}}(\mathbf{G})$ for the particular configuration $\theta = \pi$

(b) Find $\bar{\mathbf{v}}_{\mathrm{E}}(\mathbf{G})$, $\bar{\boldsymbol{\Omega}}^{\mathrm{ring}}_{\mathrm{plat}}$, $\bar{\boldsymbol{\Omega}}^{\mathrm{ring}}_{\mathrm{E}}$ for a general configuration

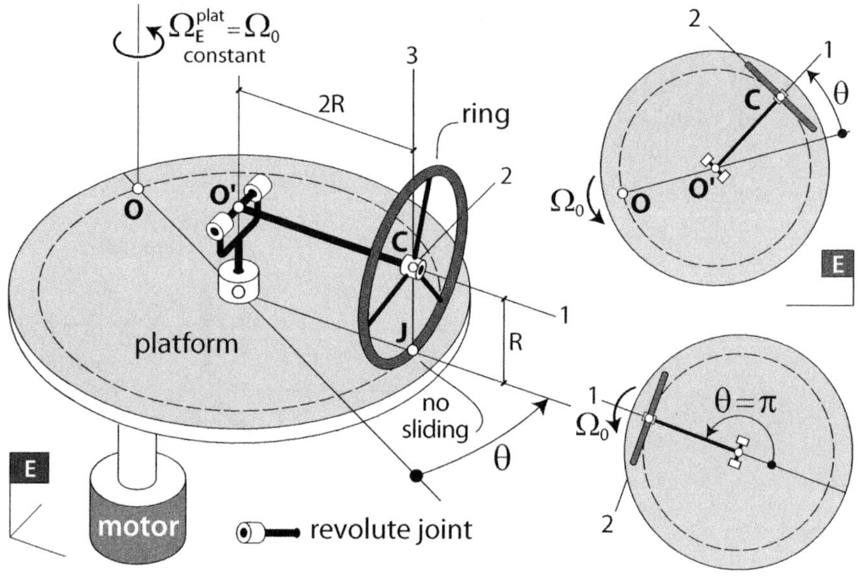

3.14 The wheel has a perfect rolling motion on the ground and is always perpendicular to it.

(a) Find $\bar{\mathbf{v}}_E(\mathbf{C}), \bar{\mathbf{a}}_E(\mathbf{C})$ and the location of the curvature center $\mathbf{CC}_E(\mathbf{C})(\equiv \mathbf{O})$ from point $\mathbf{C}$ $(\overline{\mathbf{CO}})$

(b) Find $\bar{\mathbf{v}}_E(\mathbf{P}), \bar{\mathbf{a}}_E(\mathbf{P})$

(c) Find $\bar{\mathbf{a}}_E(\mathbf{J}_{\text{wheel}})$

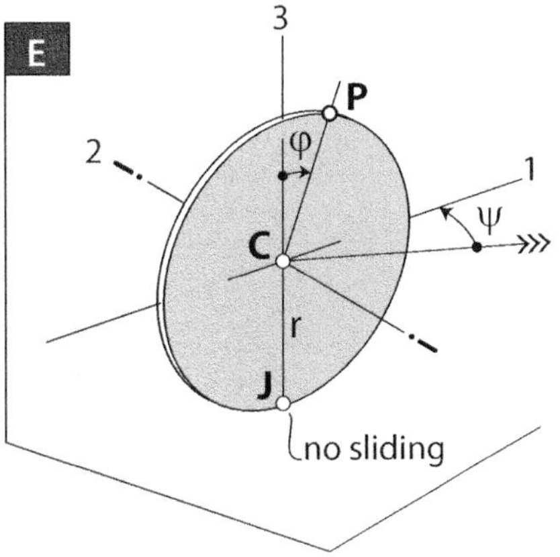

3.15 The wheel moves under the action of motor φ, which controls the constant angular velocity $\dot{\varphi}_0$ of the wheel relative to the arm. The wheel slides transversally on the ground but not longitudinally (that is, it skids but does not skate).

(a) Find $\bar{\mathbf{\Omega}}_E^{\text{wheel}}, \bar{\boldsymbol{\alpha}}_E^{\text{wheel}}$

(b) Take the ground as AB reference frame and the arm as REL reference frame, and find $\bar{\mathbf{a}}_{\text{REL}}(\mathbf{J}),\ \bar{\mathbf{a}}_{\text{tr}}(\mathbf{J}),\ \bar{\mathbf{a}}_{\text{Cor}}(\mathbf{J})$

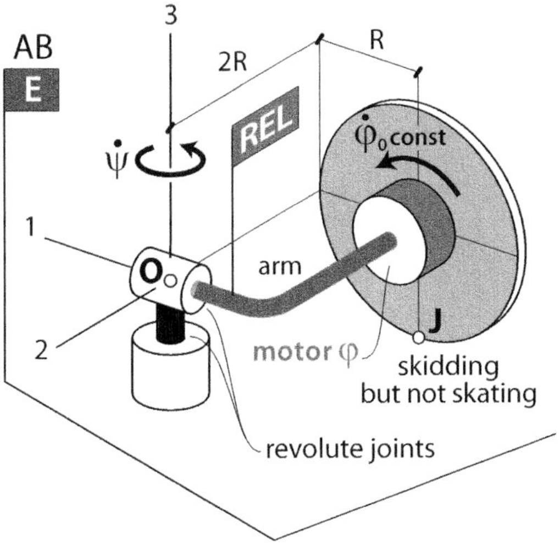

3.16 The driving element arm–fork–wheel drags the block along the rectilinear slot (Fig. a). This element is equipped with the propulsion motor φ, which drives the wheel from the fork, and the steering motor θ, which orientates the fork with respect to the arm. The wheel has a single-point nonsliding contact with the ground.

(a) Find $v_E(\text{block}), \Omega_E^{\text{arm}}$, and $\Re_E(\mathbf{C})$

(b) We replace the steering motor θ by a device formed by a pulley, with radius r_1 and fixed to the block, and another pulley, with radius r_2 and fixed to the fork, both linked through a chain (Fig. b). Find the new angular velocities $\Omega_{\text{arm}}^{\text{fork}}, \Omega_E^{\text{fork}}$ as a function of $\dot{\psi}$

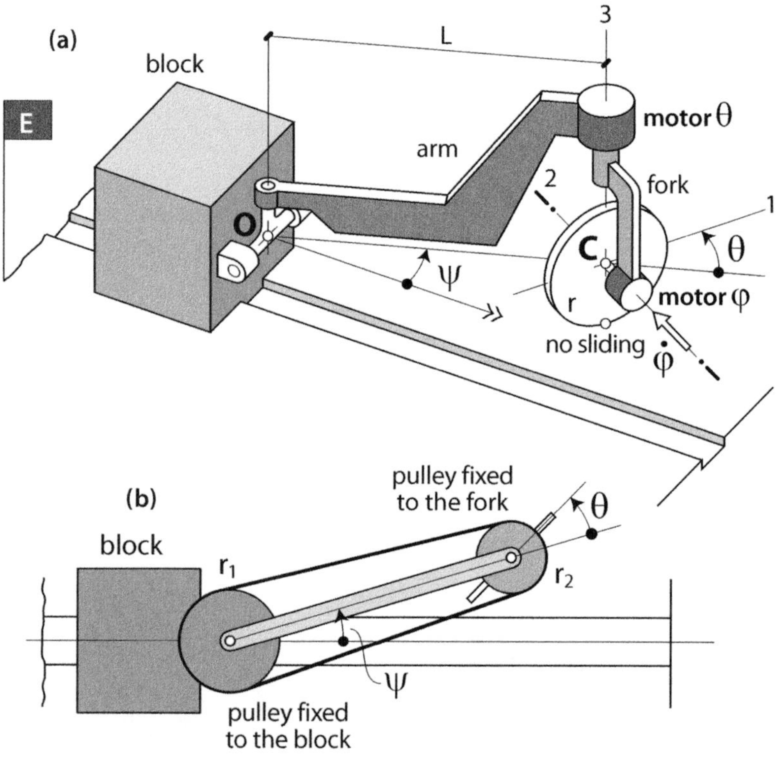

3.17 The mobile support (Fig. a) has a rectilinear and uniform translation motion with speed v_0 in the horizontal direction u–u′ (which is fixed to the ground). The arm rotates with respect to the support around the vertical axis passing through **O**′ and which defines the angle ψ with the direction u–u′. The wheel fork rotates in the opposite direction relative to the arm, so that the wheel plane defines an angle $\theta = \theta_0 + \lambda\psi$ with it. The wheel does not slide on the ground.

(a) Find $\Omega_{E}^{\text{wheel}}, \dot{\theta}$

(b) Find the equilibrium values of ψ. Are they stable or unstable?

(c) If we introduce the rotation θ of the fork relative to the arm through the wheels (Fig. b) (which do not slide at their contacts), what values of R and R′ yield to the same law $\theta = \theta_0 + \lambda\psi$?

(a)

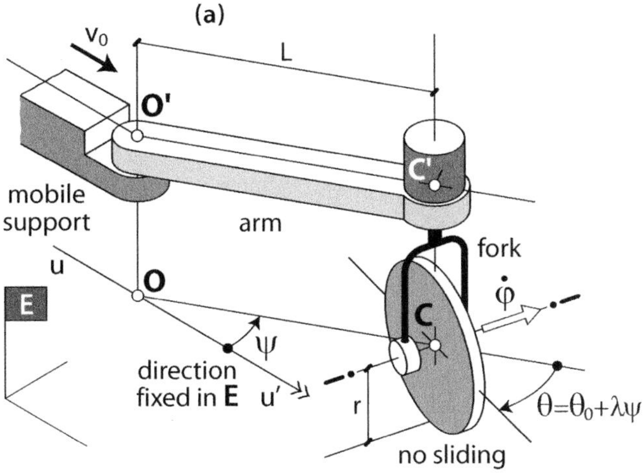

(b)

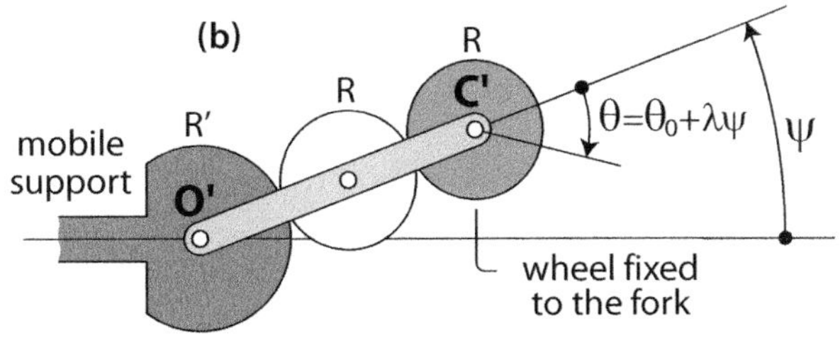

3.18 In the articulated vehicle, the motor θ controls the angle θ between the two frames, while the motor φ controls the rotation $\dot{\varphi}$ of the wheel with center **P**. The wheels do not slide on the ground.

(a) Find $v_E(\mathbf{Q}), \dot{\psi}, \dot{\psi}'$

(b) Find $\mathfrak{R}_E(\mathbf{C})$

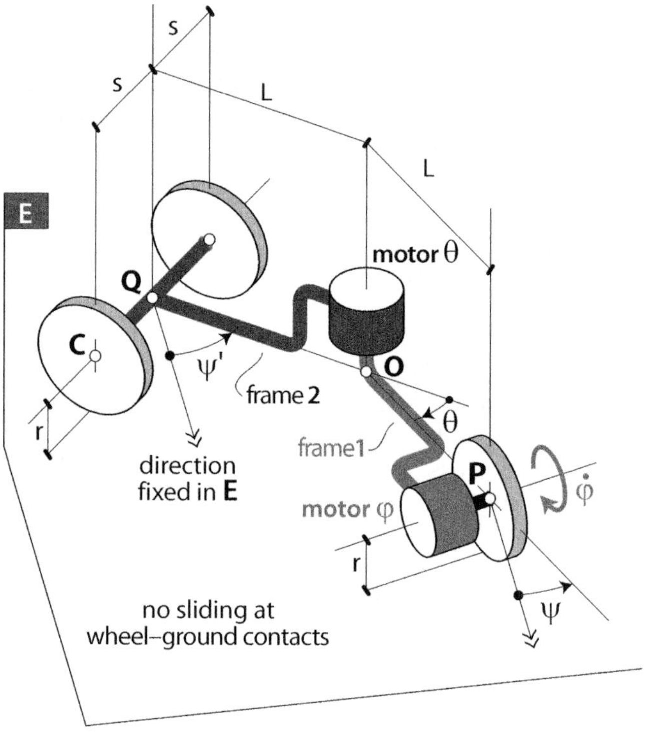

3.19 The public works machine, which consists of two chassis articulated at point **O** and four nonsteered wheels, moves on a horizontal ground without sliding. A single motor controls a rotation φ that is distributed between the two drive wheels A and B (both articulated to chassis 1) through a differential ($\dot{\varphi} = (\dot{\varphi}_A + \dot{\varphi}_B)/2$). The wheels articulated to chassis 2 can rotate freely with respect to their axes. The steering control is carried out by the hydraulic cylinder articulated between **P** and **Q**.

(a) Find $\dot{\theta}$, $v_E(\mathbf{O}_2)$

(b) Find the separation velocity between points **P** and **Q** ($\dot{\rho} = d|\mathbf{PQ}|/dt$)

(c) Find $\bar{\mathbf{a}}_E(\mathbf{G}_1)$

(d) Find $\bar{\boldsymbol{\Omega}}_E^{\text{wheel C}}$

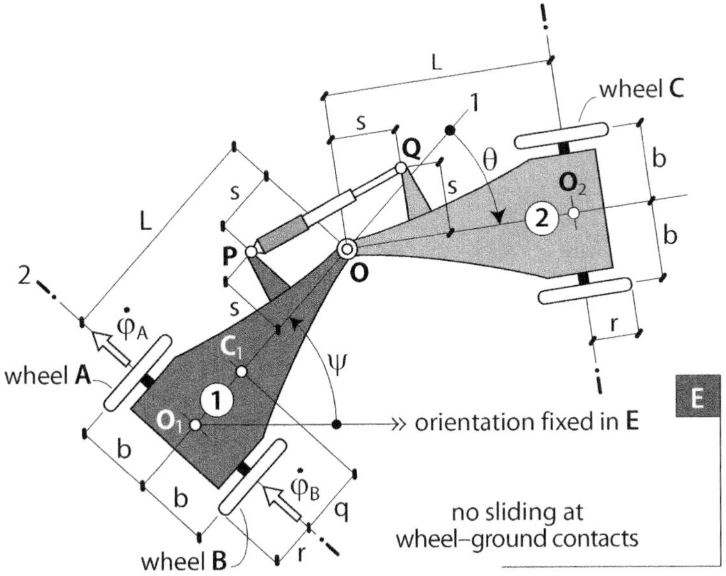

3.20 In an airport finger, the end section can move with respect to the initial section, which has the point **O** fixed to the terminal. The motor θ orientates the wheels with respect to the end section, and the motor φ imposes the rotation $\dot{\varphi}$ that is distributed between the wheels through a differential ($\dot{\varphi} = (\dot{\varphi}_A + \dot{\varphi}_B)/2$). The wheels do not slide on the ground.

(a) Find the separation velocity $\dot{x}$ and the angular velocity $\dot{\psi}$

(b) Find $\bar{\mathbf{a}}_E(\mathbf{C})$

(c) Find $\dot{\varphi}_A, \dot{\varphi}_B$

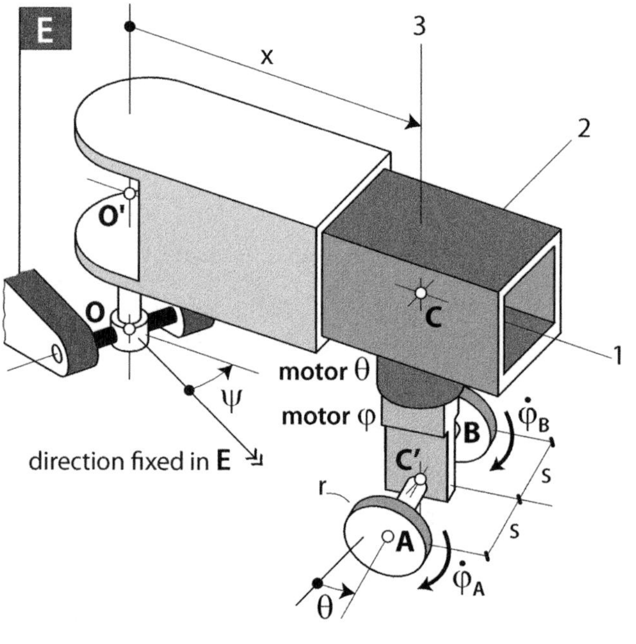

3.21 The roller has a perfect rolling motion on the rotating telescopic gangway. The actuators of the gangway cause the point **Q** to move relative to the ground with a velocity with constant module v that forms a constant angle β with the longitudinal axis of the gangway.

(a) Find the separation velocity $\dot{x}$ and the angular velocity $\dot{\psi}$

(b) Find $\Re_E(\mathbf{C})$. Prove that the $\mathbf{CC}_E(\mathbf{C})$ coincides with the $\mathbf{ICR}_E^{\text{gangway}}(\equiv \mathbf{I})$ in this case

(d) Take the ground as AB reference frame and the arm as REL reference frame, and find $\bar{\mathbf{v}}_{\text{REL}}(\mathbf{G}), \bar{\mathbf{v}}_{\text{tr}}(\mathbf{G}), \bar{\mathbf{a}}_{\text{tr}}(\mathbf{G})$

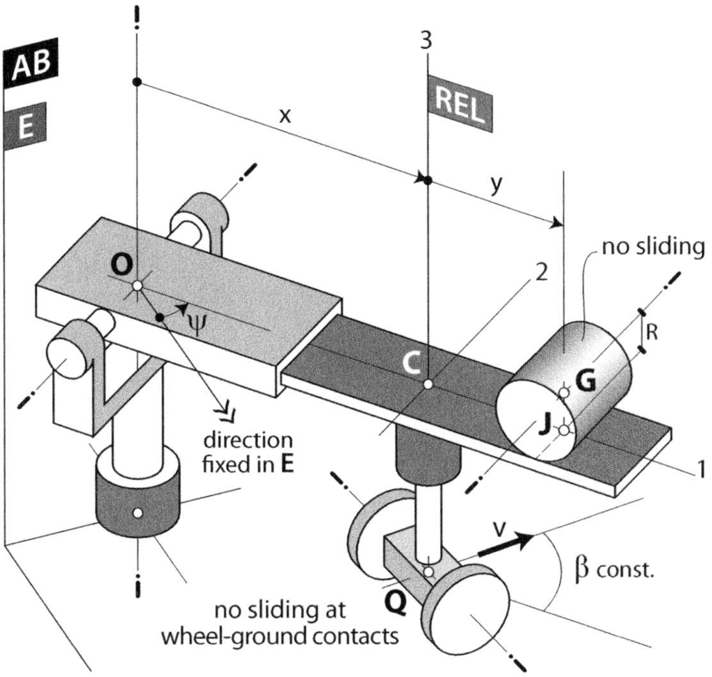

3.22 The wheel with center **C** rolls without sliding on the ground towed by a vehicle that runs along a circular track. Point **Q** describes a circular path of radius R and center **O** relative to the ground, with a variable speed v(t). The fork can rotate around the vertical axis going through **Q**.

(a) Find $\dot{\psi}, \dot{\theta}, \dot{\varphi}$

(b) Find $\mathfrak{R}_E(\mathbf{J}_g)$. Prove that, in this case, the curvature center (≡**I**) of the $\mathbf{J}_g$ trajectory coincides with the **ICR** of the fork (relative to the ground)

(c) Find the equilibrium values of θ. Are they stable or unstable?

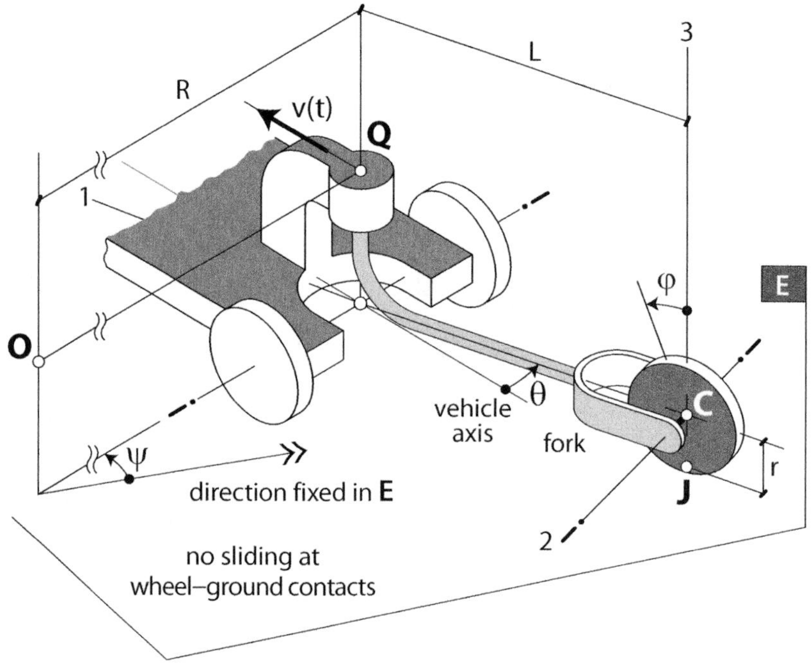

3.23 The wheel with radius R of the paving roller does not slide on the ground. Motor φ controls its angular velocity $\dot{\varphi}$, while motor θ controls the wheel plane orientation relative to the chassis (angle θ). If we assume that the roller midpoint **P** does not slide on the ground.

(a) Find $\dot{\varphi}'$ (roller spin about its axis) and $\dot{\psi}'$ (angular velocity of the roller axis relative to the ground)

(b) Find $v_E(\mathbf{Q}), \bar{\mathbf{a}}_E(\mathbf{G})$

(c) Find $\mathfrak{R}_E(\mathbf{J}_g)$ (curvature radius of the trajectory relative to the ground of the geometric point $\mathbf{J}_g$ associated with the wheel–ground contact)

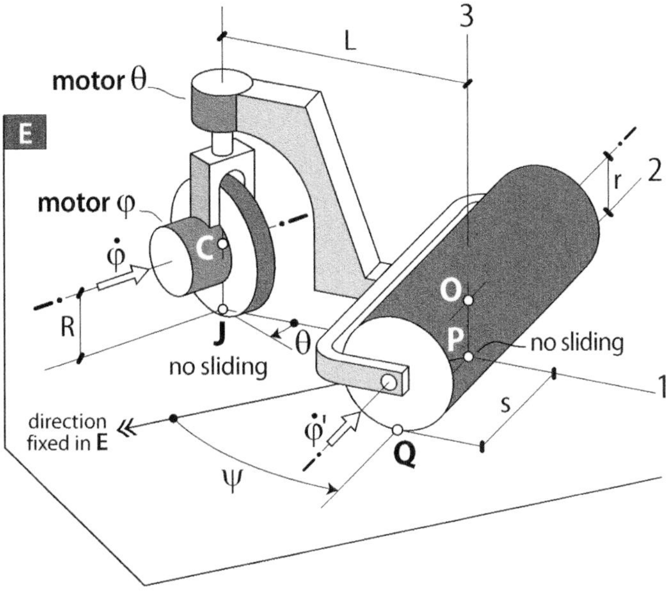

3.24 The sled is powered by a driving wheel with radius r and constant angular velocity $\dot{\varphi}_0$ (spin). The orientation of the wheel and of the rear skate determine the change of orientation $\dot{\psi}$ of the sled relative to the ground. The wheel does not slide on the ground, while the velocity of point $\mathbf{J}_2$ relative to the ground is permanently parallel to the skate longitudinal direction. The angles θ_1 and θ_2 (which give the orientation of the wheel and the skate relative to the sled) are constantly related by $\tan\theta_1 = 2\tan\theta_2$.

(a) Find $v_E(\mathbf{J}_2), \dot{\psi}$

(b) Find the instantaneous center of rotation of the sled relative to the ground $\left(\mathbf{ICR}_E^{\text{sled}} \equiv \mathbf{I}\right)$

(c) Find $\Re_E(\mathbf{J}_{1g})$ (curvature radius of the trajectory relative to the ground of the geometric point $\mathbf{J}_{1g}$ associated with the wheel–ground contact)

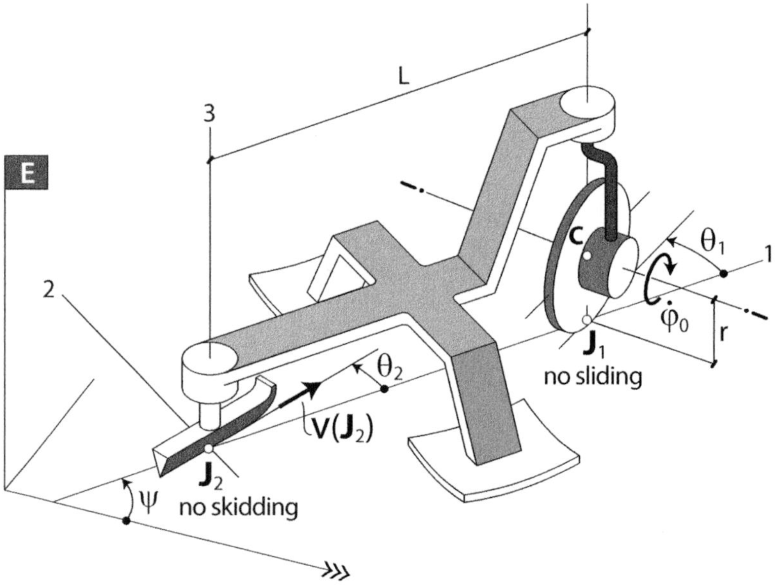

3.25 The vehicle moves on a horizontal ground. The three wheels do not slide on the ground. The wheel with center **C** rotates around the axis p–p′ parallel to $\mathbf{O}' - \mathbf{O}''$. The speed $v_E(\mathbf{O})$ and the change of orientation ψ are variable.

(a) Find the velocities $\dot{x}$ and $\dot{\varphi}$ of the wheel relative to the chassis

(b) Find the curvature center **C**′ of the **C** trajectory relative to the ground

(c) Take the ground as AB reference frame and the chassis as REL reference frame, and find $\bar{\mathbf{a}}_{REL}(\mathbf{C})$, $\bar{\mathbf{a}}_{tr}(\mathbf{C})$, $\bar{\mathbf{a}}_{Cor}(\mathbf{C})$

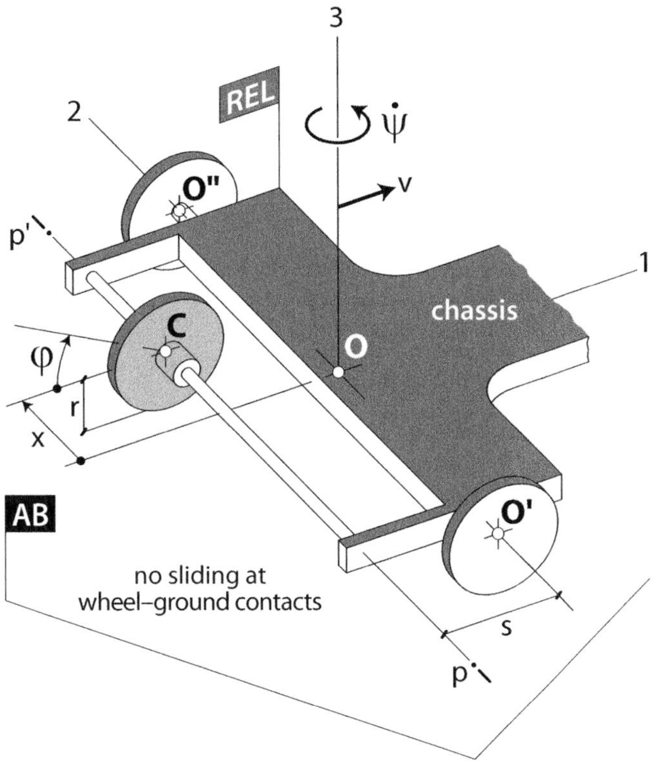

3.26 The wheels of the vehicle do not slide on the ground. The rear wheel (with center **C**) rotates around the axis **P–P′**. The orientation of the articulated parallelogram **PP′QQ′** relative to the chassis is given by angle θ. The speed $v_E(\mathbf{O})$ and the angular velocity $\dot{\psi}$ are variable.

(a) Find the angular velocities $\dot{\theta}$ and $\dot{\varphi}$ that determine the wheel relative motion

(b) Find the curvature center **C′** of the **C** trajectory relative to the ground

(c) Take the ground as AB reference frame and the chassis as REL reference frame, and find $\bar{\mathbf{a}}_{\mathrm{REL}}(\mathbf{C})$, $\bar{\mathbf{a}}_{\mathrm{tr}}(\mathbf{C})$, $\bar{\mathbf{a}}_{\mathrm{Cor}}(\mathbf{C})$

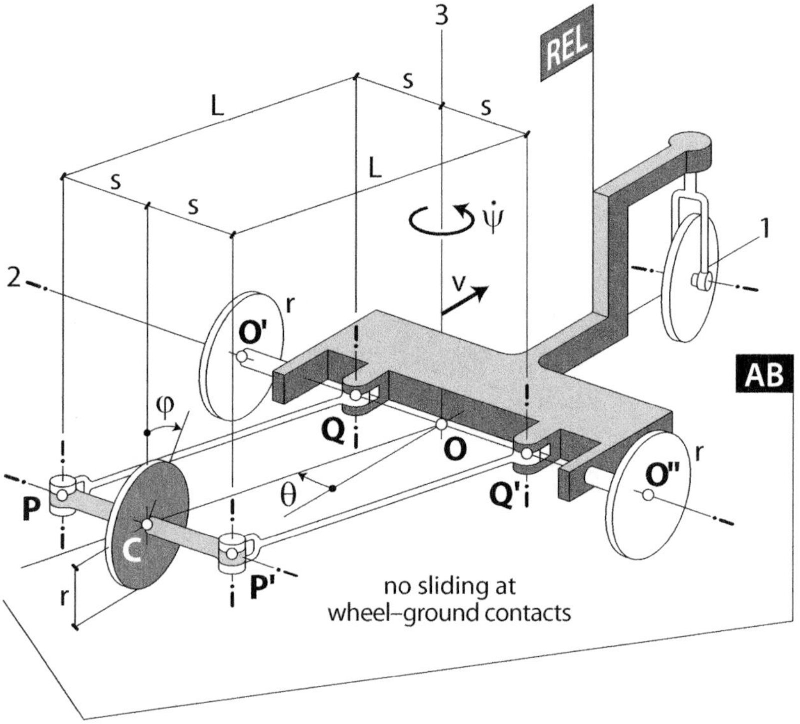

3.27 The spheres with radius R do not slide on the ground and are in contact with the frame through a frictionless spherical bearing. The small wheels with radius r have a nonsliding contact with the spheres, and their rotation $\dot{\beta}$ is governed by the corresponding motor β. The frame has three DoF described through the variables $v_1, v_2, \dot{\psi}$.

(a) Find $\dot{\beta}_1, \dot{\beta}_2, \dot{\beta}_3$ as a function of $v_1, v_2, \dot{\psi}$

(b) For the particular case $\dot{\beta}_2 = 0$ and $\dot{\beta}_1 = \dot{\beta}_3 = \dot{\beta}$ with $\ddot{\beta}_1 \neq 0$ and $\ddot{\beta}_3 = 0$, find the $\mathbf{ICR}_E^{\text{frame}}$ and the curvature center of the **O** trajectory relative to the ground

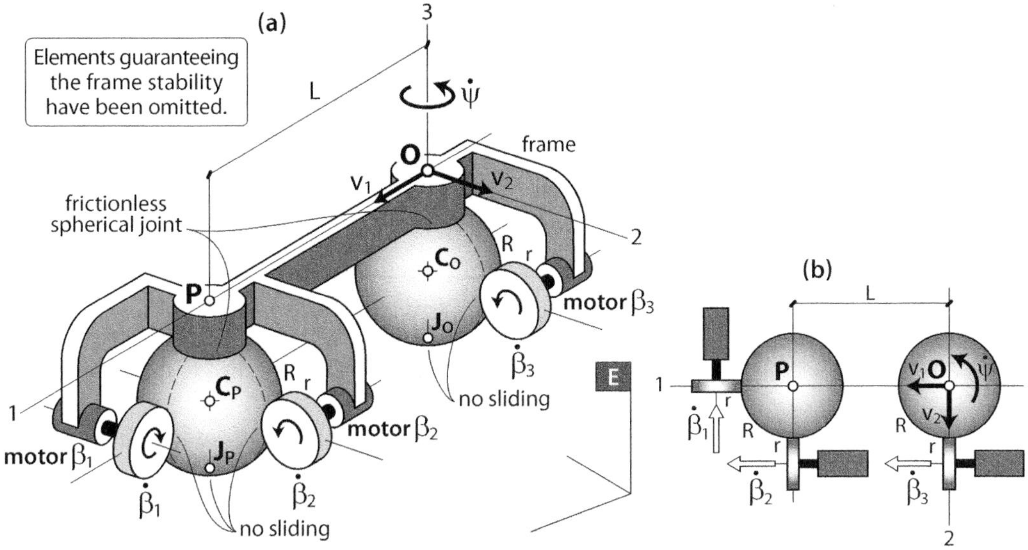

Puzzles

3.1 Rolling of a beer barrel

A ramp consisting on three parallel straight wooden planks is used to unload a ship of beer barrels.

Is it possible that the barrels go down the unloading ramp with a perfect rolling motion?

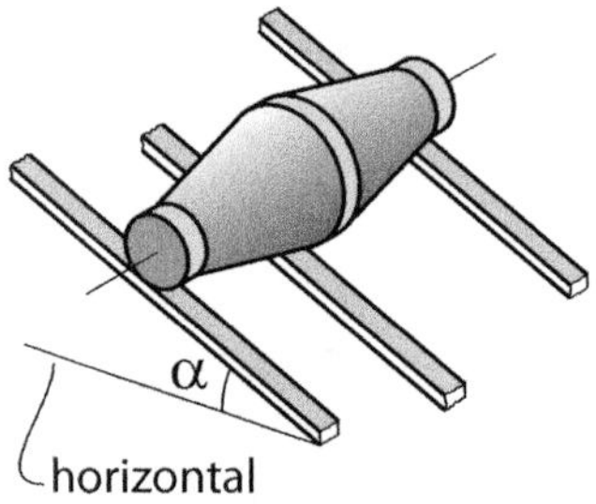

3.2 Moving a bicycle by pulling a thread

From the ground, we pull backward an inextensible thread attached to the lowest pedal of a bicycle.

Does the bicycle move forward or backward? What is the rotation direction of the pedals?

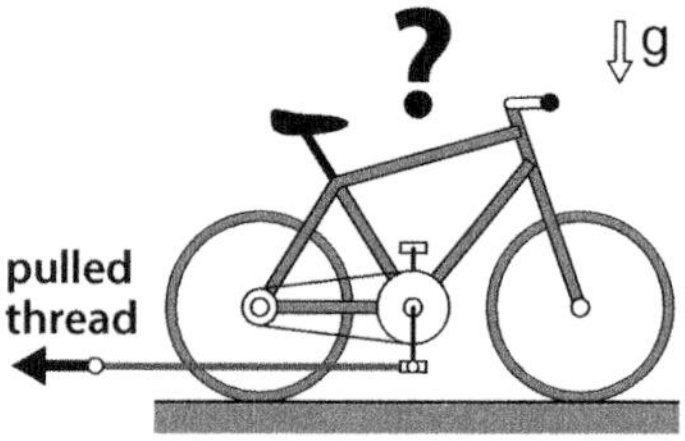

Quiz Questions: Answers

	1	2	3	4	5	6	7	8	9	10
	B	A	C	C	C	E	C	D	E	D
+10	B	B	D	E	A	D	C	E	E	E
+20	C	D	B	E	C	E	D	B	C	C
+30	E	A	A	D	A	C	D	B	B	B
+40	B	B	D	C	D	D	B	C	D	D
+50	C	D	C	C	D	A	A	D		

Problems: Answers

3.1 (a) $\{\bar{\mathbf{v}}_E(\mathbf{P})\}_B = \begin{Bmatrix} 2r\dot{\theta}\cos(\theta/2) \\ 0 \end{Bmatrix}$, $\{\bar{\mathbf{a}}_E(\mathbf{P})\}_B = \begin{Bmatrix} 2r\ddot{\theta}\cos(\theta/2) - r\dot{\theta}^2\sin(\theta/2) \\ r\dot{\theta}^2\cos(\theta/2) \end{Bmatrix}$

(b) The curvature center **C** is located on the **PJ** line; the distance **PC** is the curvature radius $\Re_E(\mathbf{P})$, which can be calculated from the previous results: $\Re_E(\mathbf{P}) = \dfrac{v_E^2(\mathbf{P})}{|a_E^n(\mathbf{P})|} = 4r\cos(\theta/2)$. Finally: $\dfrac{\Re_E(\mathbf{P})}{|\bar{\mathbf{PJ}}|} = \dfrac{4r\cos(\theta/2)}{2r\cos(\theta/2)} = 2$, which proves that **C** is the symmetric point of **P** with respect to **J**

(c) $x = r(\theta + \sin\theta)$, $y = r(1 + \cos\theta)$

3.2 (a) $\{\bar{\mathbf{v}}_E(\mathbf{C})\}_B = \begin{Bmatrix} -s\dot{\theta} \\ r\theta\dot{\theta} \end{Bmatrix}$

(b) $\{\bar{\mathbf{a}}_E(\mathbf{C})\}_B = \begin{Bmatrix} -s\ddot{\theta} - r\theta\dot{\theta}^2 \\ r\theta\ddot{\theta} + (r-s)\dot{\theta}^2 \end{Bmatrix}$

(c) $\{\bar{\mathbf{OC}}\}_B = \dfrac{s^2}{r-s}\begin{Bmatrix} 0 \\ 1 \end{Bmatrix}$

(d) $\theta_{\inf\text{lex}} = \pm\sqrt{\frac{s(r-s)}{r^2}}$ (condition of existence : $r > s$)

3.3 $\{\bar{\mathbf{a}}_E(\mathbf{J}_{\text{disc}})\}_B = \begin{Bmatrix} -R\ddot{\psi}\cos\theta \\ R(2\dot{\theta}^2 - \dot{\psi}^2)\cos\theta \\ 2R\dot{\theta}^2\sin\theta \end{Bmatrix}$

3.4 $\{\bar{\mathbf{a}}_{\text{REL}}(\mathbf{C})\}_B = \begin{Bmatrix} -3L\dot{\theta}^2\cos\theta \\ 0 \\ -3L\dot{\theta}^2\sin\theta \end{Bmatrix}$, $\{\bar{\mathbf{a}}_{\text{tr}}(\mathbf{C})\}_B = \begin{Bmatrix} -3L\dot{\psi}^2\cos\theta \\ 3L\ddot{\psi}\cos\theta \\ 0 \end{Bmatrix}$,

$\{\bar{\mathbf{a}}_{\text{Cor}}(\mathbf{C})\}_B = \begin{Bmatrix} 0 \\ -6L\dot{\theta}\dot{\psi}\sin\theta \\ 0 \end{Bmatrix}$

3.5 (a) $\left\{\bar{\mathbf{\Omega}}_E^{\text{bar}}\right\}_B = \begin{Bmatrix} 0 \\ 0 \\ \Omega_0 + 2\omega \end{Bmatrix}$

(b) $\{\bar{\mathbf{v}}_E(\mathbf{P})\}_B = \begin{Bmatrix} 0 \\ -2R\omega \\ 0 \end{Bmatrix}$, $\{\bar{\mathbf{a}}_E(\mathbf{P})\}_B = \begin{Bmatrix} 6R\omega^2 + 4R\omega\Omega_0 \\ -2R\dot{\omega} \\ 0 \end{Bmatrix}$

3.6 (a) $\dot{\varphi} = 0$

(b) $\{\bar{\mathbf{v}}_{\text{REL}}(\mathbf{P})\}_B = \begin{Bmatrix} L\dot{\theta} \\ 0 \\ 0 \end{Bmatrix}$, $\{\bar{\mathbf{v}}_{\text{tr}}(\mathbf{P})\}_B = \begin{Bmatrix} 0 \\ 0 \\ L\Omega_0(\sin\theta + \sin\varphi) \end{Bmatrix}$

(c) $\{\bar{\mathbf{a}}_{REL}(\mathbf{P})\}_B = \begin{Bmatrix} L\ddot{\theta} \\ -L\dot{\theta}^2 \\ 0 \end{Bmatrix}$, $\{\bar{\mathbf{a}}_{tr}(\mathbf{P})\}_B = -L\Omega_0^2(1+\sin\theta)\begin{Bmatrix} \cos\theta \\ \sin\theta \\ 0 \end{Bmatrix}$,

$\{\bar{\mathbf{a}}_{Cor}(\mathbf{P})\}_B = \begin{Bmatrix} 0 \\ 0 \\ 2L\Omega_0\dot{\theta}\cos\theta \end{Bmatrix}$

3.7 (a) **Fixed axode:** conical surface with vertical axis, half aperture 90°, and vertex $\mathbf{O}'$

Moving axode: conical surface with axis $\mathbf{O}'\mathbf{C}$, vertex $\mathbf{O}'$, and containing $\mathbf{P}$ and $\mathbf{Q}$

(b) $\left\{\bar{\boldsymbol{\Omega}}_E^S\right\}_B = \begin{Bmatrix} 0 \\ \dot{\varphi}_0\cos\beta \\ 0 \end{Bmatrix}$, $\left\{\bar{\boldsymbol{\alpha}}_E^S\right\}_B = \begin{Bmatrix} -\dot{\varphi}_0^2\sin\beta\cos\beta \\ 0 \\ 0 \end{Bmatrix}$

(c) $\{\bar{\mathbf{v}}_E(\mathbf{P}')\}_B = \begin{Bmatrix} 2R\dot{\varphi}_0\cos^2\beta \\ 0 \\ 0 \end{Bmatrix}$, $\{\bar{\mathbf{a}}_E(\mathbf{P}')\}_B = \dot{\varphi}_0^2\cos\beta\begin{Bmatrix} 0 \\ R\sin(2\beta) \\ -R+s\cdot\sin\beta \end{Bmatrix}$

3.8 (a) $\left\{\bar{\boldsymbol{\Omega}}_E^B\right\}_B = -\omega\dfrac{R-s}{s\sqrt{2}}\begin{Bmatrix} 1 \\ 1 \\ 0 \end{Bmatrix}$

(b) $\left\{\bar{\boldsymbol{\Omega}}_E^{ball}\right\}_B = \omega\dfrac{R-s}{R-r}\dfrac{r}{s\sqrt{2}}\begin{Bmatrix} 1 \\ 0 \\ 0 \end{Bmatrix}$, $\left\{\bar{\boldsymbol{\alpha}}_E^{ball}\right\}_B = -\omega^2\dfrac{r}{2(R-r)}\left(\dfrac{R-s}{s}\right)^2\begin{Bmatrix} 0 \\ 0 \\ 1 \end{Bmatrix}$

(c) **Fixed axode:** conical surface with axis $\mathbf{O}''\mathbf{O}$, vertex $\mathbf{O}''$, and half aperture 45° ($\mathbf{O}''$ is the intersection of lines $\mathbf{OO}'$ and $\mathbf{PQ}$)

Moving axode: conical surface with axis $\mathbf{O}''\mathbf{C}$, vertex $\mathbf{O}''$, and containing $\mathbf{P}$ and $\mathbf{Q}$

(d) $\{\bar{\mathbf{a}}_E(\mathbf{Q})\}_B = \omega^2\dfrac{r}{2}\left(\dfrac{R-s}{s}\right)^2\begin{Bmatrix} 1 \\ -1 \\ 0 \end{Bmatrix}$

(e) The three ball intersections with the planes perpendicular to $\mathbf{O}''\mathbf{C}$ and going through points $\mathbf{P}$, $\mathbf{Q}$, and $\mathbf{S}$

3.9 (a) $\left\{\bar{\boldsymbol{\Omega}}_E^{ball}\right\}_B = \omega\begin{Bmatrix} 1 \\ 2 \\ 0 \end{Bmatrix}$, $\left\{\bar{\boldsymbol{\alpha}}_E^{ball}\right\}_B = \begin{Bmatrix} 0 \\ 0 \\ -\omega^2 \end{Bmatrix}$

(b) **Fixed axode:** conical surface with vertical axis, vertex $\mathbf{O}$, and half aperture 26.6°

Moving axode: conical surface with axis $\mathbf{OC}$, vertex $\mathbf{O}$, and half aperture 18.4°

(c) $\dot{\psi} = \omega, \dot{\varphi} = \omega\sqrt{2}$

(d) $\{\bar{\mathbf{a}}_E(\mathbf{P})\}_B = \omega^2 R \frac{3}{\sqrt{2}} \begin{Bmatrix} 1 \\ 0 \\ 0 \end{Bmatrix}, \{\bar{\mathbf{a}}_E(\mathbf{Q})\}_B = \omega^2 R \frac{5}{\sqrt{2}} \begin{Bmatrix} -1 \\ 1 \\ 0 \end{Bmatrix}$

(e) $\bar{\mathbf{v}}_{SP}(\mathbf{P}_g) = \bar{\mathbf{0}}, \{\bar{\mathbf{v}}_{SQ}(\mathbf{Q}_g)\}_B = \begin{Bmatrix} 0 \\ 0 \\ \omega R\sqrt{2} \end{Bmatrix}$

(f) $\bar{\mathbf{a}}_{SP}(\mathbf{P}) = \bar{\mathbf{0}}, \{\bar{\mathbf{a}}_{SQ}(\mathbf{Q})\}_B = \omega^2 R 3\sqrt{2} \begin{Bmatrix} -1 \\ 1 \\ 0 \end{Bmatrix}$

3.10 (a) $\left\{\bar{\mathbf{\Omega}}_E^B\right\}_B = 2\frac{v_0}{r} \begin{Bmatrix} 1 \\ 0 \\ 0 \end{Bmatrix}, \left\{\bar{\mathbf{\Omega}}_E^{ball}\right\}_B = 2\frac{v_0}{R} \begin{Bmatrix} 0 \\ 1 \\ 0 \end{Bmatrix}, \left\{\bar{\boldsymbol{\alpha}}_E^{ball}\right\}_B = 4\frac{v_0^2}{Rr} \begin{Bmatrix} 0 \\ 0 \\ -1 \end{Bmatrix}$

(b) $\bar{\mathbf{v}}_E(\mathbf{P}) = \bar{\mathbf{0}}, \{\bar{\mathbf{a}}_E(\mathbf{P})\}_B = 4\frac{v_0^2}{r} \begin{Bmatrix} 0 \\ 1 \\ 0 \end{Bmatrix}$

(c) **Screw axis:** line through **P** perpendicular to axis e–e′

Fixed axode: conical surface with axis e–e′, vertex **P′**, and half aperture 90°

$\left\{\overline{\mathbf{OP'}}\right\}^T = \{R \quad -r \quad 0\}$

Moving axode: conical surface with axis **P′O**, vertex **P′**, and tangent to the ball

3.11 (a) **Screw axis:** horizontal line rotating about the vertical axis and going through **O**

Fixed axode: conical surface with vertical axis, vertex **O**, and half aperture 90°

Moving axode: conical surface with axis equal to the shell revolution axis, vertex **O**, and half aperture 45°

(b) $\left\{\bar{\mathbf{\Omega}}_E^{shell}\right\}_B = \begin{Bmatrix} \dot{\varphi}/\sqrt{2} \\ \omega - \left(\dot{\varphi}/\sqrt{2}\right) \\ 0 \end{Bmatrix} = \begin{Bmatrix} \omega \\ 0 \\ 0 \end{Bmatrix}, \left\{\bar{\boldsymbol{\alpha}}_E^{shell}\right\}_B = \begin{Bmatrix} \dot{\omega} \\ 0 \\ -\omega^2 \end{Bmatrix}$

(c) $\{\bar{\mathbf{v}}_E(\mathbf{P})\}_B = \begin{Bmatrix} 0 \\ 0 \\ \omega R \end{Bmatrix}, \{\bar{\mathbf{a}}_E(\mathbf{P})\}_B = \begin{Bmatrix} \omega^2 R \\ -\omega^2 R \\ \dot{\omega} R \end{Bmatrix}, \Re_E(\mathbf{P}) = \frac{R}{\sqrt{2}}$

3.12 (a) $\dot{\psi} = \omega$

(b) **Screw axis:** horizontal line rotating about the vertical axis and going through **O**

Fixed axode: conical surface with vertical axis, vertex **O**, and half aperture 90°

Moving axode: conical surface with axis equal to the shell revolution axis, vertex **O**, and half aperture 45°

(c) $\left\{\bar{\mathbf{\Omega}}_E^{shell}\right\}_B = \begin{Bmatrix} \dot{\varphi}/\sqrt{2} \\ \left(\dot{\varphi}/\sqrt{2}\right) - \omega \\ 0 \end{Bmatrix} = \begin{Bmatrix} \omega \\ 0 \\ 0 \end{Bmatrix}, \left\{\bar{\boldsymbol{\alpha}}_E^{shell}\right\}_B = \begin{Bmatrix} \dot{\omega} \\ 0 \\ \omega^2 \end{Bmatrix}$

(d) $\{\bar{\mathbf{v}}_E(\mathbf{Q}_{shell})\}_B = \begin{Bmatrix} 0 \\ 0 \\ -\omega R \end{Bmatrix}, \{\bar{\mathbf{a}}_E(\mathbf{Q}_{shell})\}_B = \begin{Bmatrix} \omega^2 R \\ -\omega^2 R \\ 0 \end{Bmatrix}, \Re_E(\mathbf{Q}_{shell}) = \dfrac{R}{\sqrt{2}}$

3.13 (a) $\{\bar{\mathbf{a}}_E(\mathbf{G})\}_B = \begin{Bmatrix} -2R(\dot\theta^2 + 2\Omega_0\dot\theta) \\ 0 \\ 0 \end{Bmatrix}$

(b) $\{\bar{\mathbf{v}}_E(\mathbf{G})\}_B = 2R\begin{Bmatrix} \Omega_0 \sin\theta \\ \Omega_0(1+\cos\theta) + \dot\theta \\ 0 \end{Bmatrix},$

$\left\{\bar{\boldsymbol{\Omega}}^{ring}_{plat}\right\}B = \begin{Bmatrix} -2\dot\theta \\ 0 \\ \dot\theta \end{Bmatrix}, \left\{\bar{\boldsymbol{\Omega}}^{ring}_{E}\right\}_B = \begin{Bmatrix} -2\dot\theta \\ 0 \\ \dot\theta + \Omega_0 \end{Bmatrix}$

3.14 (a) $\{\bar{\mathbf{v}}_E(\mathbf{C})\}_B = \begin{Bmatrix} r\dot\varphi \\ 0 \\ 0 \end{Bmatrix}, \quad \{\bar{\mathbf{a}}_E(\mathbf{C})\}_B = \begin{Bmatrix} r\ddot\varphi \\ r\dot\psi\dot\varphi \\ 0 \end{Bmatrix}, \quad \{\bar{\mathbf{CO}}\}_B = \begin{Bmatrix} 0 \\ r(\dot\varphi/\dot\psi) \\ 0 \end{Bmatrix}$

(b) $\{\bar{\mathbf{v}}_E(\mathbf{P})\}_B = \begin{Bmatrix} r\dot\varphi(1+\cos\varphi) \\ r\dot\psi\sin\varphi \\ -r\dot\varphi\sin\varphi \end{Bmatrix},$

$\{\bar{\mathbf{a}}_E(\mathbf{P})\}_B = \begin{Bmatrix} r\ddot\varphi(1+\cos\varphi) - r(\dot\psi^2 + \dot\varphi^2)\sin\varphi \\ r\ddot\psi\sin\varphi + r\dot\psi\dot\varphi(1+2\cos\varphi) \\ -r\ddot\varphi\sin\varphi - r\dot\varphi^2\cos\varphi \end{Bmatrix}$

(c) $\{\bar{\mathbf{a}}_E(\mathbf{J}_{wheel})\}_B = \begin{Bmatrix} 0 \\ -r\dot\psi\dot\varphi \\ r\dot\varphi^2 \end{Bmatrix}$

3.15 (a) $\left\{\bar{\boldsymbol{\Omega}}^{wheel}_{E}\right\}_B = \begin{Bmatrix} 0 \\ \dot\varphi_0 \\ \dot\varphi_0/2 \end{Bmatrix}, \left\{\bar{\boldsymbol{\alpha}}^{wheel}_{E}\right\}_B = \begin{Bmatrix} -\dot\varphi_0^2/2 \\ 0 \\ 0 \end{Bmatrix}$

(b) $\{\bar{\mathbf{a}}_{REL}(\mathbf{J})\}_B = \begin{Bmatrix} 0 \\ 0 \\ R\ddot\varphi_0 \end{Bmatrix}, \{\bar{\mathbf{a}}_{tr}(\mathbf{J})\}_B = \dfrac{R\dot\varphi_0^2}{4}\begin{Bmatrix} 1 \\ 2 \\ 0 \end{Bmatrix}, \{\bar{\mathbf{a}}_{Cor}(\mathbf{J})\}_B = \begin{Bmatrix} 0 \\ -R\dot\varphi_0^2 \\ 0 \end{Bmatrix}$

3.16 (a) $v_E(\text{block}) = \dot\varphi r \dfrac{\cos\theta}{\cos\psi},$

$\Omega^{arm}_E (= \dot\psi) = \dot\varphi \dfrac{r}{L}\dfrac{\sin(\psi+\theta)}{\cos\psi}, \quad \Re_E(\mathbf{C}) = L\left(\dfrac{\sin(\psi+\theta)}{\cos\psi} + \dfrac{L}{r}\dfrac{\dot\theta}{\dot\varphi}\right)^{-1}$

(b) $\Omega^{arm}_{fork} = -\dot\psi\dfrac{r_1}{r_2}, \quad \Omega^{arm}_E = \dot\psi\dfrac{r_2 - r_1}{r_2}$

3.17 (a) $\Omega_E^{wheel} = \frac{v_0}{r}\frac{\cos\psi}{\cos(\theta_0+\lambda\psi)}, \dot{\theta} = \frac{v_0}{L}[\sin\psi - \cos\psi\tan(\theta_0+\lambda\psi)]$

(b) $\psi_{eq} = \frac{\theta_0}{1-\lambda}$; stable if $(\lambda > 1)$, unstable if $(\lambda < 1)$

(c) $R = \frac{L}{3+\lambda}, R' = \frac{\lambda L}{3+\lambda}$

3.18 (a) $v_E(\mathbf{Q}) = \dot{\varphi}r + \dot{\theta}L\frac{\sin\theta}{1+\cos\theta}, \dot{\psi} = -\frac{\dot{\theta}L + \dot{\varphi}r\sin\theta}{L(1+\cos\theta)}, \dot{\psi}' = -\frac{\dot{\theta}L\cos\theta - \dot{\varphi}r\sin\theta}{L(1+\cos\theta)}$

(b) $\Re_E(\mathbf{C}) = \frac{L(1+\cos\theta)}{\sin\theta + (\dot{\theta}/\dot{\varphi})(L/r)}$

3.19 (a) $\dot{\theta} = \dot{\psi}(1+\cos\theta) + \dot{\varphi}\frac{r}{L}\sin\theta, v_T(\mathbf{O}_2) = \dot{\varphi}r\cos\theta - \dot{\psi}L\sin\theta$

(b) $\dot{\rho} = s\dot{\theta}\left(\cos\frac{\theta}{2} - \sin\frac{\theta}{2}\right)$

(c) $\{\bar{\mathbf{a}}_E(\mathbf{G}_1)\}_B = \begin{Bmatrix} \ddot{\varphi}r - \dot{\psi}^2 q \\ \ddot{\psi}q + \dot{\psi}\dot{\varphi}r \\ 0 \end{Bmatrix}$

(d) $\left\{\bar{\mathbf{\Omega}}_E^{wheel\ C}\right\}_B = \begin{Bmatrix} 0 \\ \dot{\varphi}(\cos\theta + (b/L)\sin\theta) + (\dot{\psi}/r)(b\cos\theta - L\sin\theta) \\ -\dot{\varphi}(r/L)\sin\theta - \dot{\psi}\cos\theta \end{Bmatrix}$

3.20 (a) $\dot{x} = \dot{\varphi}r\cos\theta, \dot{\psi} = \dot{\varphi}(r/x)\sin\theta$

(b) $\{\bar{\mathbf{a}}_E(\mathbf{G})\}_B = \begin{Bmatrix} \ddot{\varphi}r\cos\theta - \dot{\varphi}\dot{\theta}\sin\theta - \dot{\varphi}^2(r^2/x)\sin^2\theta \\ \ddot{\varphi}r\sin\theta + \dot{\varphi}\dot{\theta}r\cos\theta + \dot{\varphi}^2(r^2/x)\sin\theta\cos\theta \\ 0 \end{Bmatrix}$

(c) $\dot{\varphi}_A = \dot{\varphi}\left(1 + \frac{s}{x}\sin\theta\right) + \dot{\theta}\frac{s}{r}, \dot{\varphi}_B = \dot{\varphi}\left(1 - \frac{s}{x}\sin\theta\right) - \dot{\theta}\frac{s}{r}$

3.21 (a) $\dot{x} = v\cos\beta, \dot{\psi} = (v/x)\sin\beta$

(b) $\Re_E(\mathbf{C}) = \frac{x}{\sin\beta}, \{\mathbf{OI}\}_B = \begin{Bmatrix} -x \\ x\cot\beta \\ 0 \end{Bmatrix}$

(c) $\{\bar{\mathbf{v}}_{REL}(\mathbf{G})\}_B = \begin{Bmatrix} \dot{y} \\ 0 \\ 0 \end{Bmatrix}, \{\bar{\mathbf{v}}_{tr}(\mathbf{G})\}_B = \begin{Bmatrix} v\cos\beta \\ (v/x)(x+y)\sin\beta \\ 0 \end{Bmatrix},$

$\{\bar{\mathbf{a}}_{tr}(\mathbf{G})\}_B = \left(\frac{v}{x}\right)^2\sin\beta\begin{Bmatrix} -(x+y)\sin\beta \\ (x-y)\cos\beta \\ 0 \end{Bmatrix}$

3.22 (a) $\dot{\psi} = \frac{v}{R}, \dot{\theta} = -\frac{v}{L}\left(\frac{L}{r} + \sin\theta\right), \dot{\varphi} = \frac{v}{R}\cos\theta$

(b) $\Re_E(J_g) = L\cot\theta, \{\mathbf{CI}\}_B = \begin{Bmatrix} 0 \\ -L\cot\theta \\ 0 \end{Bmatrix}$

(c) $\theta_{eq} = -\arcsin\left(\frac{L}{R}\right)$; stable if $(\cos\theta_{eq} > 0)$, unstable if $(\cos\theta_{eq} < 0)$

3.23 (a) $\dot{\varphi}' = \dot{\varphi}\frac{R}{r}\cos\theta, \dot{\psi} = \dot{\varphi}\frac{R}{L}\sin\theta$

(b) $v_E(\mathbf{Q}) = \dot{\varphi}\frac{Rs}{L}\sin\theta, \{\bar{\mathbf{a}}_E(\mathbf{G})\}_B = \begin{Bmatrix} \ddot{\varphi}R\cos\theta - \dot{\varphi}\dot{\theta}R\sin\theta \\ \dot{\varphi}^2(R^2/L)\sin\theta\cos\theta \\ 0 \end{Bmatrix}$

(c) $\Re_E(J_g) = L\left(\sin\theta - \frac{L}{R}\frac{\dot{\theta}}{\dot{\varphi}}\right)^{-1}$

3.24 (a) $v_E(\mathbf{J}_2) = r\dot{\varphi}\frac{\sin\theta_1}{2\sin\theta_2}, \dot{\psi} = \dot{\varphi}\frac{r}{2L}\sin\theta_1$

(b) $\{\mathbf{J}_2\mathbf{I}\}_B = \begin{Bmatrix} -L \\ L\cot\theta_2 \\ 0 \end{Bmatrix} = \begin{Bmatrix} -L \\ 2L\cot\theta_1 \\ 0 \end{Bmatrix}$

(c) $\Re_E(J_{1g}) = 2L\left(\sin\theta_1 + \frac{2L}{r}\frac{\dot{\theta}_1}{\dot{\varphi}}\right)^{-1}$

(d) With $\theta_2 > \theta_1$, when taking a curve to the right/left, the wheel and the skate would have to turn left/right (opposite direction!), which is most counterintuitive. With $\theta_2 = \theta_1$, the chassis would never change its orientation; we could take curves, but always with translation motion

3.25 (a) $\dot{x} = s\dot{\psi}, \dot{\varphi} = \frac{v - x\dot{\psi}}{r}$

(b) $\{\mathbf{OC}'\}_B = \begin{Bmatrix} -s \\ v/\dot{\psi} \\ 0 \end{Bmatrix}$

(c) $\{\bar{\mathbf{a}}_{REL}(\mathbf{C})\}_B = \begin{Bmatrix} 0 \\ s\ddot{\psi} \\ 0 \end{Bmatrix}, \bar{\mathbf{a}}_{tr}(\mathbf{C}) = \begin{Bmatrix} \dot{v} - x\ddot{\psi} + s\dot{\psi}^2 \\ -s\ddot{\psi} + \dot{\psi}(v - x\dot{\psi}) \\ 0 \end{Bmatrix},$

$\bar{\mathbf{a}}_{Cor}(\mathbf{C}) = \begin{Bmatrix} -2s\dot{\psi}^2 \\ 0 \\ 0 \end{Bmatrix}$

3.26 (a) $\dot{\theta} = \dot{\psi}, \dot{\varphi} = \frac{v}{r}$

(b) $\{\mathbf{CC'}\}_B = \begin{Bmatrix} 0 \\ v/\dot{\psi} \\ 0 \end{Bmatrix}$

(c) $\{\bar{\mathbf{a}}_{REL}(\mathbf{C})\}_B = \begin{Bmatrix} L\ddot{\psi}\sin\theta + L\dot{\psi}^2\cos\theta \\ L\ddot{\psi}\cos\theta - L\dot{\psi}^2\sin\theta \\ 0 \end{Bmatrix}$,

$$\{\bar{\mathbf{a}}_{tr}(\mathbf{C})\}_B = \begin{Bmatrix} \dot{v} - L\ddot{\psi}\sin\theta + L\dot{\psi}^2\cos\theta \\ -L\ddot{\psi}\cos\theta + \dot{\psi}(v - L\dot{\psi}\sin\theta) \\ 0 \end{Bmatrix},$$

$$\{\bar{\mathbf{a}}_{Cor}(\mathbf{C})\}_B = 2L\dot{\psi}^2 \begin{Bmatrix} -\cos\theta \\ \sin\theta \\ 0 \end{Bmatrix}$$

3.27 (a) $\dot{\beta}_1 = v_1/r, \ \dot{\beta}_2 = (v_2 + \dot{\psi}L)/r, \ \dot{\beta}_3 = v_2/r$

(b) For $\dot{\beta}_2 = 0$, $\bar{\mathbf{v}}_E(\mathbf{P})$ is parallel to the frame longitudinal direction

For $\dot{\beta}_1 = \dot{\beta}_3 = \dot{\beta}$, the $\bar{\mathbf{v}}_E(\mathbf{O})$ direction is at 45° from the frame longitudinal direction

The $\mathbf{IC}_E^{frame}$ is the intersection of those $\bar{\mathbf{v}}_E(\mathbf{P})$ and $\bar{\mathbf{v}}_E(\mathbf{O})$ directions. $\mathfrak{R}_E(\mathbf{O}) = \frac{v_E^2(\mathbf{O})}{a_E^n(\mathbf{O})} = \frac{\dot{\beta}^2 r^2 2}{\dot{\beta}^2 r^2\sqrt{2}} L = L\sqrt{2} \Rightarrow \mathbf{CC}_E(\mathbf{O}) = \mathbf{ICR}_E^{frame}$

(note, however, that $\ddot{\beta}_3 \neq 0 \Rightarrow a_E^n(\mathbf{O}) = \dot{\beta}^2 \frac{r^2}{L}\sqrt{2} - \ddot{\beta}_3 \frac{r}{\sqrt{2}} \Rightarrow \mathbf{CC}_E(\mathbf{O}) \neq \mathbf{IC}_E^{frame}$).

Puzzles: Solutions

3.1 Rolling of a beer barrel

It is not possible.

For a perfect rolling motion, the instantaneous velocity of the three points of the barrel in contact with the ramp has to be zero. As the points are not aligned, they do not define an ISA. The only scenario compatible with the nonsliding condition (or perfect rolling) is that the barrel is at rest.

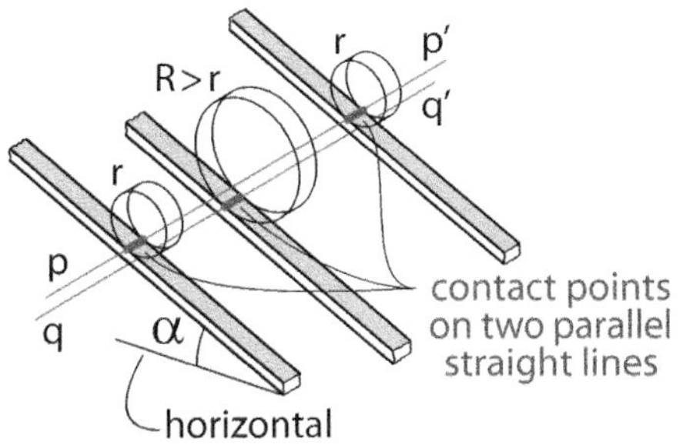

3.2 Moving a bicycle by pulling a thread

It moves backward and the pedals' rotation is counterclockwise.

Previous consideration through the energy theorem:[5] *Let's consider that the thread is pulled through a hanging weight.*

If the bicycle is on a flat ground and has no hidden motors, the weight cannot go up. If it goes down, the potential energy lost by the weight is transformed into kinetic energy of the bicycle and the weight. If it went up, where would the kinetic energy come from?

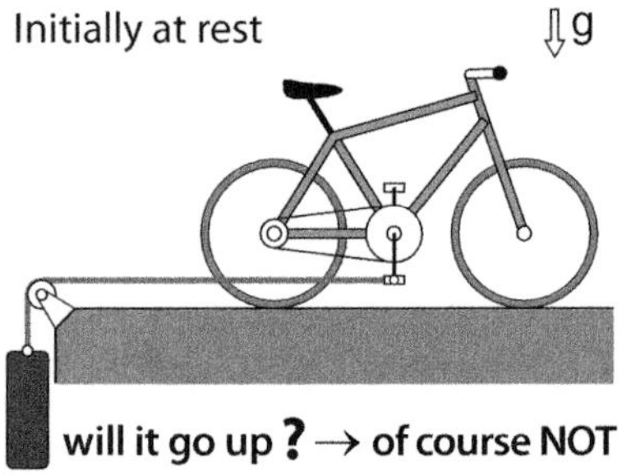

An initial (and dangerous!) intuition is that the bicycle moves forward and the pedals' rotation is clockwise. An analysis through a composition of movements (with AB = ground and REL = bicycle frame) proves that this is wrong.

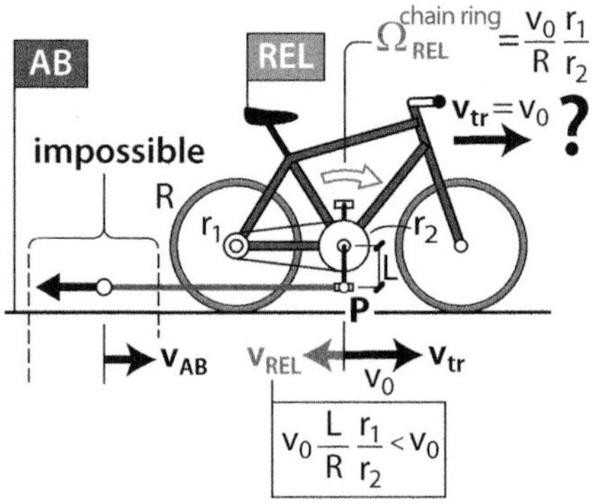

For a forward velocity v_0 of the bicycle relative to the ground ($v_{AB}(\text{bicycle}) = v_0$), the chainring (and so the pedals) would have a clockwise rotation with value $(v_0/R)(r_1/r_2)$ (both in the AB and the REL frames), and point P of the pedal would

[5] Though this theorem has not yet been presented, this previous consideration is an elementary application of that theorem that should not be a problem for undergraduate students.

have a backward velocity $v_{REL}(\mathbf{P}) = v_0(L/R)(r_1/r_2)$. Note that $v_{REL}(\mathbf{P}) < v_0$ as $(L/R) < 1$ (even if $(r_1/r_2) > 1$ in some mountain bikes, the product $(L/R)(r_1/r_2)$ remains <1).

Consequently, $\bar{\mathbf{v}}_{AB}(\mathbf{P})(= \bar{\mathbf{v}}_{REL}(\mathbf{P}) + \bar{\mathbf{v}}_{tr}(\mathbf{P}) = \bar{\mathbf{v}}_{REL}(\mathbf{P}) + \bar{\mathbf{v}}_{AB}(\text{bicycle}))$ would have to be forward, and this is impossible.

If we assume a backward motion of the bicycle relative to the ground, the previous calculations hold but with a negative sign. The resulting $\bar{\mathbf{v}}_{AB}(\mathbf{P})$ is backward because $|\bar{\mathbf{v}}_{tr}(\mathbf{P})| > |\bar{\mathbf{v}}_{REL}(\mathbf{P})|$.

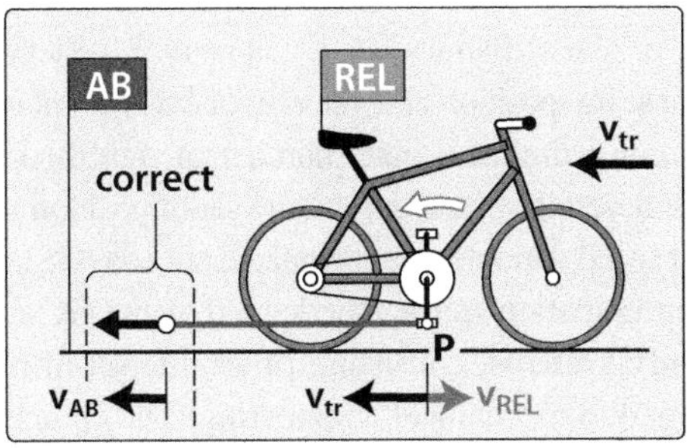

The condition $|\bar{\mathbf{v}}_{tr}(\mathbf{P})| > |\bar{\mathbf{v}}_{REL}(\mathbf{P})|$ is not accidental: it is the very reason of existence of bicycles! $|\bar{\mathbf{v}}_{REL}(\mathbf{P})|$ is the speed of the rider's feet relative to the frame and to the rider's body, and $|\bar{\mathbf{v}}_{tr}(\mathbf{P})|$ is that of the rider's body relative to the ground. When walking on the ground, $|\bar{\mathbf{v}}_{REL}(\mathbf{P})| = |\bar{\mathbf{v}}_{tr}(\mathbf{P})|$. But when riding a bicycle, $|\bar{\mathbf{v}}_{tr}(\mathbf{P})| > |\bar{\mathbf{v}}_{REL}(\mathbf{P})|$: the bicycle amplifies the speed of the walking rider. The use of sprockets with radius r_1 higher than that of the chainring r_2 (used in stiff slopes) is interesting only if $r_1 < r_2(R/L)$.

4 Introduction to Kinematics of Multibody Systems

The study of the dynamics of a mechanical system starts with the description of its **mechanical state** (position and velocity of every point in the system). For the particular case of systems consisting of a finite number of rigid bodies, though the number of points is infinite, that description calls for a finite set of position variables – **generalized coordinates** (**GC**) – and speed variables – **generalized speeds** (**GS**). The vector space defined by the GC is called **configuration space**; that defined by the GC and the GS is known as **phase space**.

Multibody systems, consisting of a finite set of rigid bodies, are the most common systems found in mechanical engineering. The choice of GC needed and GS to describe them is not unique and depends mainly on the particular application.

One possibility is to start with six coordinates and six speeds for each body (which is the number of independent GC and GS to describe the mechanical state of a rigid body evolving freely in a 3D space) and then define the relationships among them (the **configuration** and **kinematic constraint equations**).

If we want to perform analytical calculations, we will consider the actual possible configurations and motions of the system and will probably start with a number of GC and GS close to the minimum. When it is strictly the minimum, we call them **independent coordinates** (**IC**) and **degrees of freedom** (**DoF**). The DoF describe the possible independent positional changes in a short-term time scale (that is, in a differential time interval).

In many systems (for instance, in most mechanisms), there are an equal number of IC and DoF. In that case, the system is called **holonomic**. However, there are systems with more IC than DoF: though some coordinates cannot evolve independently in a differential time interval, they may show independent finite changes after a finite time interval as a result of the DoF evolution. They are **nonholonomic** systems. The chassis of an automobile is a good example.

When introducing constraints on a rigid body (through contacts with other rigid bodies), we may come across a delicate situation: **redundancy**. From the physical point of view, a new constraint is said to be redundant when some of the kinematic restrictions it would introduce on its own (if it were the only constraint acting on the rigid body) have already been guaranteed by the previous existing constraints. From the mathematical point of view, redundancy translates into linear dependency between the partial kinematic constraint equations.

The concept of redundancy is introduced and analyzed in the last sections of this chapter. It is an important issue because of the bad consequences it may have on the constraint forces (chapter 2 in *Rigid Body Dynamics* [Cambridge University Press, forthcoming]). Avoiding constraint redundancy is a basic rule in the mechanical design of multibody systems when the components' rigidity is high.

4.1 Mechanical Systems: Multibody Systems

A mechanical system is a set of particles, which can be partially grouped into subsets forming solids (rigid or deformable) and fluids. The particles may always be the same or change along time.

This text focuses on constant matter systems (without matter exchange with the environment). The mechanics of variable matter systems is very complex in the general case. In fluid mechanics, some variable matter systems are studied – like hydraulic machines – but they correspond to very particular cases where simple adaptations of the general theorems governing the mechanics of constant matter systems are appropriate.

Multibody systems, formed by a constant and finite set of rigid bodies, are the most usual mechanical systems studied in mechanical engineering and will be the ones considered in this chapter.

4.2 Generalized Coordinates: Configuration Space

The position of any point of a multibody system can be given through a finite set of variables: the **generalized coordinates** (GC).

When it comes to a **free** particle (without any contact with other elements), that set may consist of three spatial coordinates: Cartesian, cylindrical, spherical, or any other type. For a free rigid body, it can include the three spatial coordinates of one of its points and a set of orientation parameters (for instance, the three Euler angles).

The information of the position and orientation of all the elements in a multibody system is called **configuration**. This concept is broader than that of position (which refers exclusively to particles). The vector space defined by the GC is the **configuration space**. If all GC are independent, each point in the configuration space corresponds to a system configuration. Geometrically, the time evolution of the system corresponds to a trajectory in that space.

Generalized coordinates are usually distances and angles, although they do not have to conform to any conventional coordinate system. Generically they are denoted as $\mathrm{q_i}\ (\mathrm{i} = 1, 2, 3\ldots)$.

4.3 Geometric Constraints: Independent Coordinates

The m generalized coordinates used to describe the configuration of a multibody system may be mutually independent, and then their number is minimal. This is the case of the 6N generalized coordinates for a system of N rigid bodies when all of them are free.

When this is not the case (usually because of constraints associated with contact between them or with rigid bodies not belonging to the system), those 6N coordinates may be no longer independent.

If we describe the configuration through a set of n generalized coordinates of dimension greater than the minimum m (and therefore not independent), a set of (n – m) relations must be fulfilled:

$$f_j(q_1, q_2, \ldots, q_n) = 0, \quad j = 1, 2, 3 \ldots, (n - m). \tag{4.1}$$

They are called **geometric** or **configuration constraint equations** because they relate the geometric variables defining the configuration, and are usually nonlinear. The points in the configuration space corresponding to possible system configurations must be located on the **configuration manifold** (or hypersurface) defined by those equations (Eq. (4.1)).

► **Example 4.1** Points $\mathbf{P_1}$ and $\mathbf{P_2}$, belonging to two different rigid bodies, are linked by a rigid bar of length L (Fig. 4.1). Consequently, their Cartesian coordinates (x_1, y_1, z_1) and (x_2, y_2, z_2) are not independent as they must fulfill the geometric constraint equation:

$$L - \sqrt{(x_2 - x_1)^2 + (y_2 - y_1)^2 + (z_2 - z_1)^2} = 0.$$ ◄

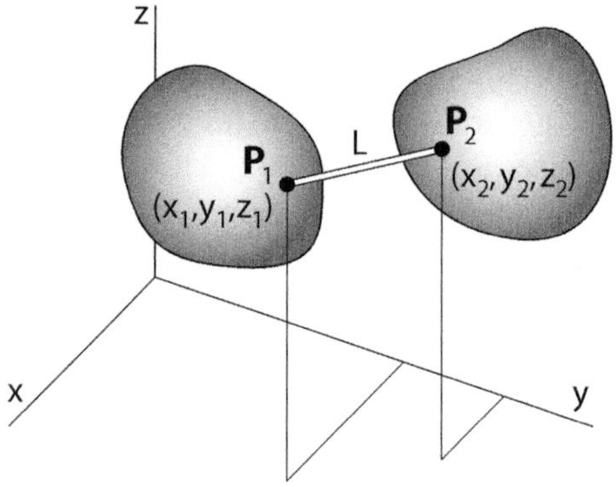

Fig. 4.1

► **Example 4.2** In a spinning top freely rotating about point **O** fixed to the ground (Fig. 4.2a), the three Euler angles (ψ, θ, φ) are independent coordinates; the configuration space is three-dimensional with axes (ψ, θ, φ).

The motion of the spinning top with angle $\theta = \theta_0$ constant and precession and spin velocities $\dot{\psi}_0$ and $\dot{\varphi}_0$ also constant corresponds to a straight line in the configuration space contained in the plane perpendicular to the θ axis through point $\theta = \theta_0$. Figure 4.2b shows the particular case of initial configuration $\psi = \varphi = 0$. ◄

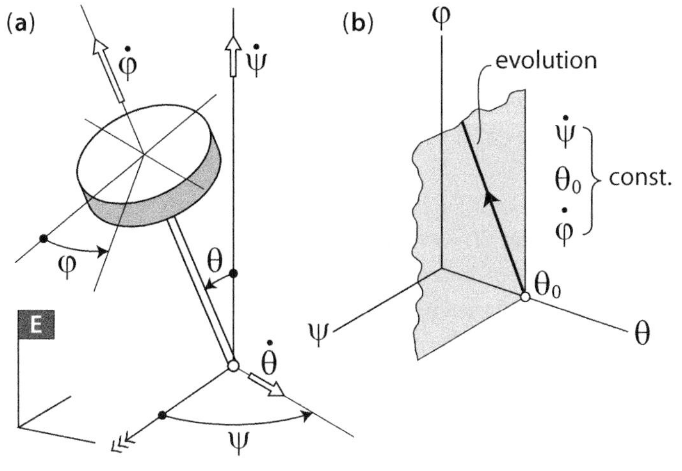

Fig. 4.2

▶ **Example 4.3** In a SCARA robot (Fig. 4.3a), which positions the clamp by means of the angles (ψ, φ) and the displacement coordinate z, the configuration space is three-dimensional with axes (ψ, φ, z).

The repetitive movement of transferring pieces from **A** to **B** corresponds to a periodic orbit in the configuration space. Figure 4.3b shows a possible example (note that an upward motion of the clamp corresponds to a decrease of the z coordinate according to the definition of z given in Fig. 4.3a). ◀

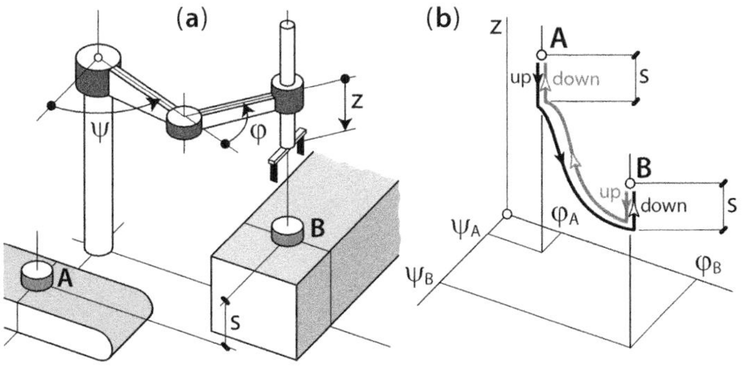

Fig. 4.3

Strictly speaking, the evolution of the generalized coordinates of a system is an unknown of the dynamical problem. However, in systems including actuators it is usual to accept that the evolution of some coordinates $q_i(t)$ (or the relationship between some of them) is totally prescribed by those actuators. In that case, the required forces or torques of the actuators that guarantee it become the unknowns. If those coordinates are explicitly substituted by their temporal function in the geometric constraint equations (Eq. (4.1)), those constraints become explicitly time dependent and are said to be **reonomic**. When there is no explicit time dependence, they are called **scleronomic** constraints.[1]

Rigid bodies not included in the system under study but constraining its motion are called **obstacles**. They may be either moving or fixed to the reference frame.

The number of **independent coordinates (IC)** of a N-bodies system can always be obtained by subtracting the number of independent geometric constraints existing among them from 6N, but this can be a long procedure. A more efficient one consists on counting the minimum number of finite increments of position variables needed to specify a new system configuration from a given configuration.

▶ **Example 4.4** The multibody system (Fig. 4.4) contains four rigid bodies:

- support ①, rotating about a vertical axis
- fork ②, articulated to the support
- prismatic slot ③, articulated to the fork
- block ④, translating along the prismatic slot

If the rigid bodies were free, there would be $4 \times 6 = 24$ independent generalized coordinates. But that number is highly reduced because of the geometric constraints between them:

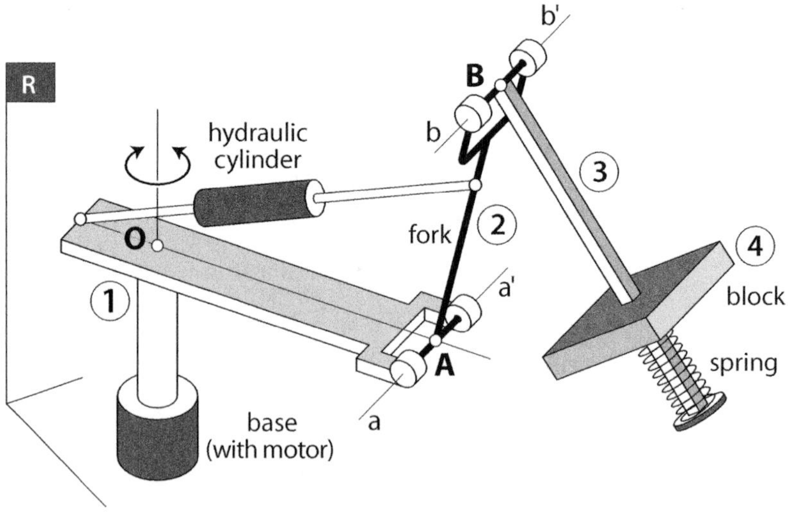

Fig. 4.4

[1] This distinction is important in analytical mechanics.

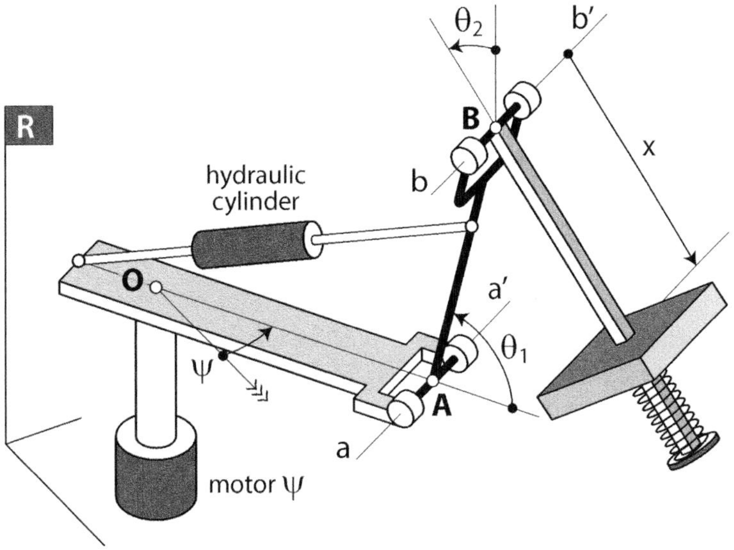

Fig. 4.5

- constraint ground →①: reduction of five coordinates because point **O** of support ① is fixed to the ground and because the support rotation is constrained to the vertical axis
- constraint support ①→ fork ②: reduction of five coordinates because point **A** of the fork is fixed to point **A** of the support and because the relative rotation between the fork and the support is constrained to axis a–a′
- constraint fork ②→prismatic slot ③: reduction of five coordinates because point **B** of the prismatic slot is fixed to point **B** of the fork and because the relative rotation between the slot and the fork is constrained to axis b–b′
- constraint prismatic slot ③→block ④: reduction of five coordinates because there is no relative rotation between the block and the prismatic slot and because the relative translation is constrained to be along the slot direction

The number of coordinates is thus reduced from 24 GC to 4 IC.

This conclusion can be attained more easily through a direct inspection of the system (Fig. 4.5): the four coordinates $(\psi, \theta_1, \theta_2, x)$ are enough to describe the system configuration, and their finite changes $(\Delta\psi, \Delta\theta_1, \Delta\theta_2, \Delta x)$ in a finite time interval are independent.

Note that ψ and θ_1 may evolve under the control of the two actuators (that is, $\dot{\psi}$ and $\dot{\theta}_1$ may be prescribed motions), whereas θ_2 and x correspond to free motions (that is, motions evolving dynamically).

If the system under study consisted just on the rigid bodies ②, ③, and ④, the rotating support would be a mobile obstacle. If it consisted only on the ③ and ④ bodies, the fork would be a mobile obstacle. ◄

4.4 Generalized Speeds: Phase Space

In a system whose configuration in a reference frame R is defined by n generalized coordinates $(q_1, q_2, \ldots, q_n)$, the components of the velocity of any point **P**, $\bar{\mathbf{v}}_R(\mathbf{P})$, and the angular velocity $\bar{\mathbf{\Omega}}_R^S$ of any rigid body in the system can be expressed as linear functions of the $\dot{q}_i$ $(i = 1, 2, \ldots, n)$:

$$\bar{\mathbf{v}}_R(\mathbf{P}) = \sum_{i=1}^{n} \bar{\mathbf{b}}_i \dot{q}_i, \bar{\mathbf{\Omega}}_R^S = \sum_{i=1}^{n} \bar{\mathbf{c}}_i \dot{q}_i, \tag{4.2}$$

where vectors $(\bar{\mathbf{b}}_i, \bar{\mathbf{c}}_i)$ are a function of $(q_1, q_2, \ldots, q_n)$ in general. Though the n generalized coordinates q_i may be independent, the n **generalized speeds** $\dot{q}_i$ may be related (as will be explained in Section 4.6).

♣ *Proof*

The position vector $\overline{\mathbf{O}_R\mathbf{P}}$ may be expressed as a function of $(q_1, q_2, \ldots, q_n)$:

$$\overline{\mathbf{O}_R\mathbf{P}} = \overline{\mathbf{O}_R\mathbf{P}}(q_1, q_2, \ldots, q_n).$$

The time derivative of this vector in R yields the velocity of **P** in R:

$$\bar{\mathbf{v}}_R(\mathbf{P}) = \sum_{i=1}^{n} \frac{\partial \overline{\mathbf{O}_R\mathbf{P}}}{\partial q_i} \dot{q}_i \equiv \sum_{i=1}^{n} \bar{\mathbf{b}}_i \dot{q}_i.$$

As the velocities of two points **P** and **Q** of a rigid body are related through $\bar{\mathbf{v}}_R(\mathbf{P}) = \bar{\mathbf{v}}_R(\mathbf{Q}) + \bar{\mathbf{\Omega}}_R^S \times \overline{\mathbf{QP}}$, the components of $\bar{\mathbf{\Omega}}_R^S$ are also linear functions of the $\dot{q}_i$. ♣

In some cases, the components of the linear and angular velocities take a simpler form if a different set of generalized speeds $(u_1, u_2, \ldots, u_n)$ is used. The new set is linearly related to the original one:

$$u_r = \sum_{i=1}^{n} A_{ri} \dot{q}_i, \quad r = 1, 2, \ldots, n, \tag{4.3}$$

where the A_{ri} are a function of $(q_1, q_2, \ldots, q_n)$ in general. The only analytical condition on these functions is that the system of equations (Eq. (4.3)) can be solved (that is, that that the $n \times n$ matrix $[A]$ with elements A_{ri} is invertible).

The velocities $(u_1, u_2, \ldots, u_n)$ may be the time derivative of n functions of the coordinates $(q_1, q_2, \ldots, q_n)$ (that is, the time derivatives of a new set of generalized coordinates), but it does not have to be so. In that case, we talk about **generalized speeds associated with pseudocoordinates**. For example, the components $(\Omega_1, \Omega_2, \Omega_3)$ of the angular velocity of a rigid body with general rotation along three orthogonal axes (which are linear functions of the time derivatives $(\dot{\psi}, \dot{\theta}, \dot{\varphi})$ of the Euler angles) can be taken as generalized speeds, but they are associated with pseudocoordinates. This is an important point in analytical mechanics, where the issues of functional dependence and partial derivatives play a prominent role.

► **Example 4.5** In a hovercraft with planar motion (Fig. 4.6), the configuration of the hull in the coast reference frame (E) can be defined by the independent coordinates $(q_1 = r, q_2 = \theta, q_3 = \psi)$, which correspond to the two polar coordinates of point **O** of the hull and the hull orientation angle ψ.

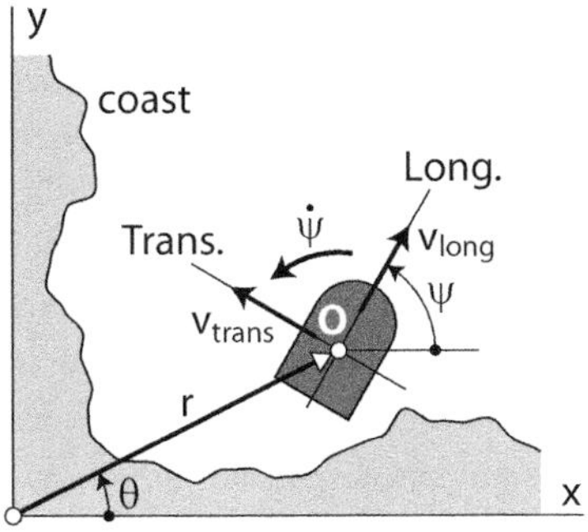

Fig. 4.6

The velocity components of any point of the hull can be expressed as linear forms of the generalized speeds $(\dot{r}, \dot{\theta}, \dot{\psi})$.

In a new set of independent coordinates as (x, y, ψ) (where x, y are the **O** Cartesian coordinates), the new generalized speeds are linearly related to the previous ones through:

$$\left\{\begin{matrix} \dot{x} \\ \dot{y} \\ \dot{\psi} \end{matrix}\right\} = [A]\left\{\begin{matrix} \dot{r} \\ \dot{\theta} \\ \dot{\psi} \end{matrix}\right\} \text{ with } [A] = \begin{bmatrix} \cos\theta & -r\sin\theta & 0 \\ \sin\theta & r\cos\theta & 0 \\ 0 & 0 & 1 \end{bmatrix} \text{ invertible.}$$

The generalized speeds corresponding to the components of $\bar{v}_E(\mathbf{O})$ in the longitudinal and transversal directions of the hull, and to the speed of orientation change of the hull, $(u_1 \equiv v_{LONG}(\mathbf{O}), u_2 \equiv v_{TRANS}(\mathbf{O}), u_3 \equiv \dot{\psi})$, may be more suitable to study the dynamics and the control of the hovercraft. They are related to $(\dot{r}, \dot{\theta}, \dot{\psi})$ through:

$$\left\{\begin{matrix} u_1 = v_{LONG}(\mathbf{O}) \\ u_2 = v_{TRANS}(\mathbf{O}) \\ u_3 = \dot{\psi} \end{matrix}\right\} = [B]\left\{\begin{matrix} \dot{r} \\ \dot{\theta} \\ \dot{\psi} \end{matrix}\right\}, \text{with } [B] = \begin{bmatrix} \cos(\psi-\theta) & r\sin(\psi-\theta) & 0 \\ -\sin(\psi-\theta) & r\cos(\psi-\theta) & 0 \\ 0 & 0 & 1 \end{bmatrix} \text{ invertible.}$$

The first two velocities (u_1, u_2) are associated with pseudocoordinates. ◄

The vector space defined by the GC and the GS is known as **phase space**.[2]

4.5 Kinematic Constraints: Degrees of Freedom

When the n generalized speeds $(u_1, u_2, \ldots, u_n)$ are not independent, they have to fulfill certain kinematic restrictions, which translate into linear equations relating them. If only m

[2] In some books, the designation **phase space** is used only when the GC and the GS are independent.

generalized speeds are independent, there are (n − m) independent **kinematic constraint equations**:

$$\sum_{i=1}^{n} C_{ji} u_i = 0, \quad j = 1, 2, \ldots, (n - m), \tag{4.4}$$

where in general the C_{ji} coefficients are functions of the generalized coordinates used to define the configuration.

Independent generalized speeds are called **degrees of freedom (DoF)**. They describe possible short-term independent positional changes (given by $u_i dt$, with $dt \to 0$).

To determine the number of DoF in a multibody system with N rigid bodies, we may start considering the 6N associated with the N free bodies, and then subtract from 6N the number of kinematic constraint equations. This procedure – very common in computer programs – is usually long, and it may be better to proceed by direct inspection of the number of independent speeds (provided the system is properly designed; bad designs are discussed in Section 4.7). In properly designed systems, an appropriate way to do this is to block one by one the possible movements of the system until total immobilization. The number of DoF is equal to that of blocked motions.

► **Example 4.6** The multibody system (Fig. 4.7), which is a simplified model of a passenger boarding bridge in an airport, contains five rigid bodies.

If they were free, there would be $5 \times 6 = 30$ DoF. But the kinematic constraints between them reduce their number:

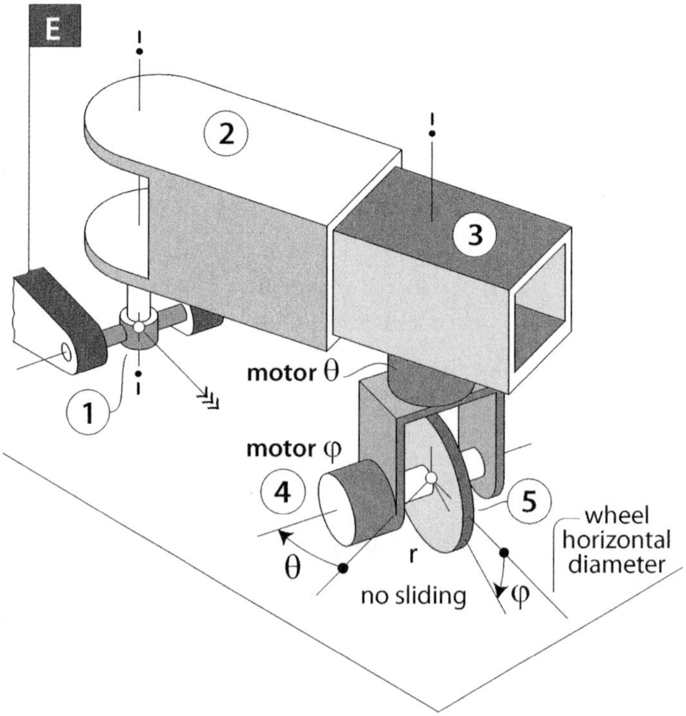

Fig. 4.7

- constraint ground →①: reduction of five DoF because this first element can only rotate about a horizontal axis fixed to the ground
- constraint ①→②: reduction of five DoF because the rigid body ② can only rotate about the vertical axis
- constraint ②→③: reduction of five DoF because the rigid body ③ can only move in and out the rigid body ②
- constraint ③→④: reduction of five because the rigid fork ④ can only rotate about the vertical axis
- constraint ④→⑤: reduction of five because the rigid wheel ⑤ can only rotate about a horizontal axis fixed to the rigid fork ④
- constraint ground →⑤: reduction of three DoF because the wheel has a single-point nonsliding contact with the ground

The total number of reductions is 28, and therefore the system has only two DoF.

We can reach the same result more easily by counting the motions that have to be blocked to bring the system to a complete stop: if we block movements φ and θ, the passenger bridge cannot move (under the nonsliding wheel–ground contact hypothesis). Thus, the system has two DoF. ◀

The constraint equations (geometric and kinematic) define a manifold in the phase space. Each point in that manifold represents a possible mechanical state of the system. That manifold is particularly interesting for **autonomous** systems (where the particles accelerations depend only on their positions and velocities): the trajectories representing the system evolution from different initial states never intersect.[3] Those trajectories bear a certain resemblance to the current lines of a fluid. Unfortunately, complete graphic visualization is only possible in systems with just one IC and one DoF (usually described by $\dot{q}$); in all other cases, the dimension of that manifold is greater than three.

► **Example 4.7** The small amplitude oscillations of a pendulum around its equilibrium position (Fig. 4.8) are sinusoidal:

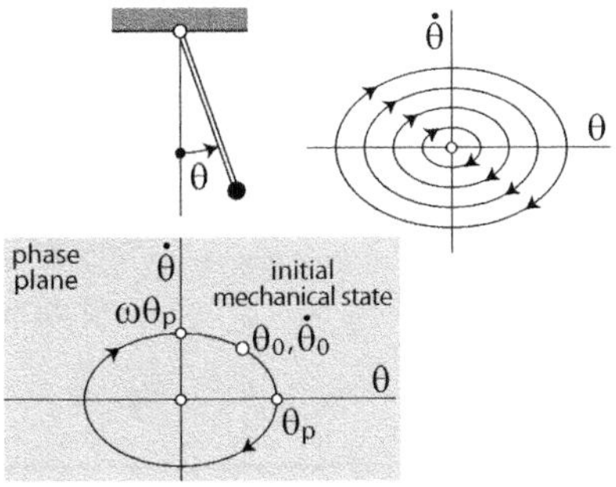

Fig. 4.8

[3] An intersection would be indicative of an indeterminacy in the movement.

$$\theta(t) = \theta_P \sin(\omega t + \varphi_0),$$
$$\dot{\theta}(t) = \theta_P \omega \cos(\omega t + \varphi_0),$$

where θ_P is the oscillation amplitude and φ_0 depends on the $(\theta, \dot{\theta})$ values for $t = 0$:

$$\tan \varphi_0 = \frac{\omega\theta(t=0)}{\dot{\theta}(t=0)} = \frac{\omega\theta_0}{\dot{\theta}_0}.$$

The set of trajectories in the phase plane $(\theta, \dot{\theta})$ for small amplitudes forms a set of concentric ellipses. ◀

4.6 Holonomy

When the number of DoF of a system coincides with that of IC, we say that the system is **holonomic**.

▶ **Example 4.8** Let's consider a free wheel in planar motion. The two velocity components $(\dot{x}(\mathbf{C}), \dot{z}(\mathbf{C}))$ of its center **C** and the angular velocity $\dot{\varphi}$ can be taken as DoF. If forced to move without sliding on a fixed line $z(\mathbf{C}) = z_0(\mathbf{C})$ (Fig. 4.9), the constraint conditions translate into two kinematic constraint equations:

$$\left.\begin{array}{ll} \bullet \text{ contact :} & \dot{z}(\mathbf{C}) = 0 \\ \bullet \text{ no sliding :} & \dot{x}(\mathbf{C}) = r\dot{\varphi} \end{array}\right\}.$$

The system now has one DoF (= 3 − 2). The two equations are integrable and yield two geometric constraint equations:

$$\left.\begin{array}{l} z(\mathbf{C}) = z_0(\mathbf{C}) = r \\ x(\mathbf{C}) = x_0(\mathbf{C}) + r(\dot{\varphi} - \dot{\varphi}_0) \end{array}\right\}.$$

The system has one IC (= 3 − 2) and is holonomic. ◀

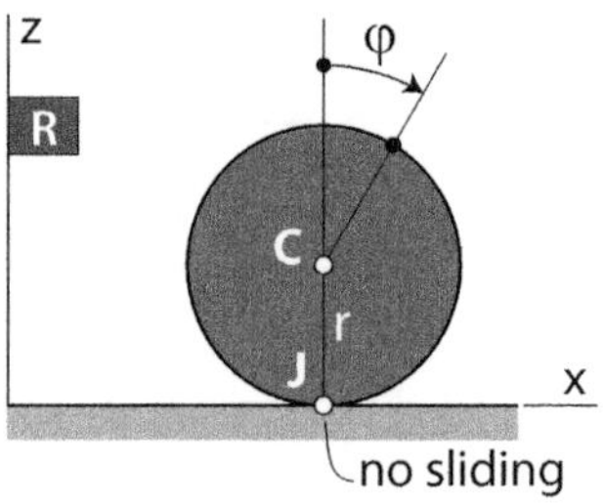

Fig. 4.9

When some kinematic constraint equations are not analytically integrable, the system is **nonholonomic**. In that case, not all kinematic constraint equations result in geometric constraint equations, and consequently the number of IC is greater than that of DoF: though some coordinates cannot evolve independently in the short-term time scale, they may show independent finite changes after a finite time interval as a result of the DoF evolution.

► **Example 4.9** For a wheel moving freely in space, the Cartesian coordinates (x, y, z) of its center **C** and the three Euler angles (ψ, θ, φ) can be taken as IC. Their time derivatives $(\dot{x}, \dot{y}, \dot{z}, \dot{\psi}, \dot{\theta}, \dot{\varphi})$ can be taken as DoF. If forced to a nonsliding motion on the fixed plane x–y while keeping a perpendicular orientation to that plane (Fig. 4.10), the constraint conditions translate into kinematic constraint equations:

$$\left.\begin{array}{ll} \bullet \text{ vertical plane}: & \dot{\theta} = 0 \\ \bullet \text{ contact}: & \dot{z} = 0 \\ \bullet \text{ nonsliding}: & \dot{x} = r\dot{\varphi}\cos\psi, \dot{y} = r\dot{\varphi}\sin\psi \end{array}\right\}.$$

The system now has two DoF (= 6 – 4) that can be described via $(\dot{\psi}, \dot{\varphi})$. The first two equations can be integrated to yield:

$$\left.\begin{array}{l} \theta = \theta_0 (= 0) \\ z = z_0 = r \end{array}\right\}.$$

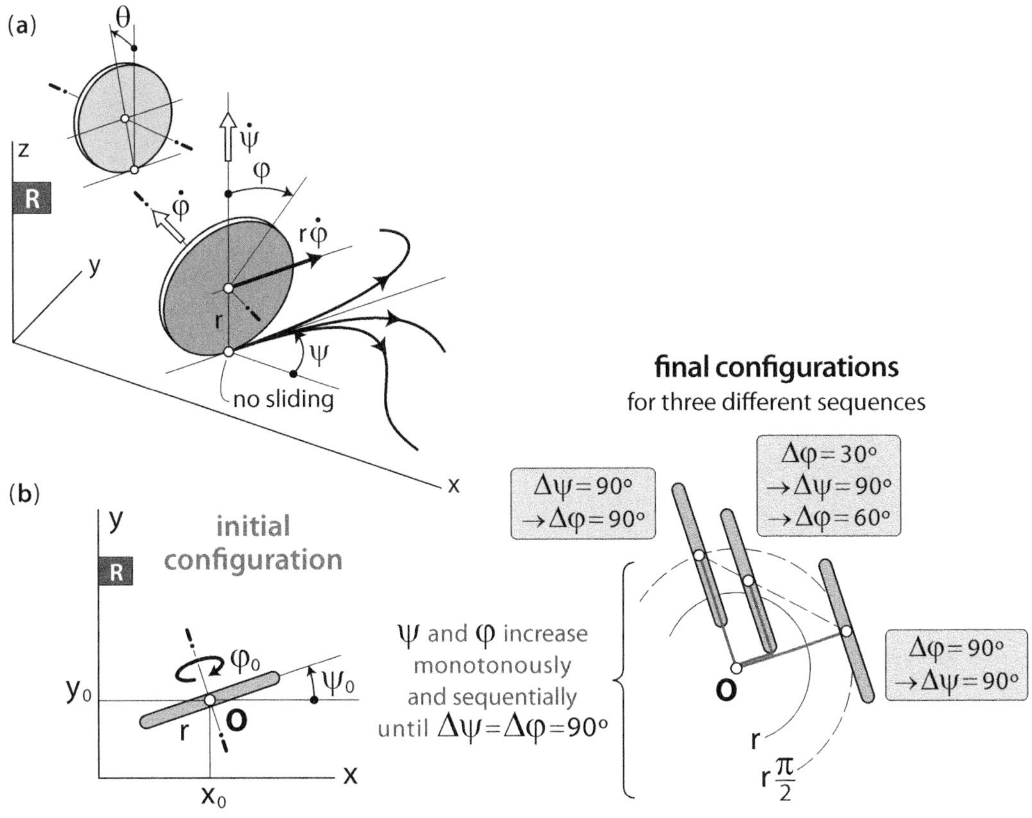

Fig. 4.10

But the other two do not have a general integral relating $(\Delta x, \Delta y)$ to $(\Delta\psi, \Delta\varphi)$ because the speed $\dot{\psi}$ is independent of the speed $r\dot{\varphi}$ of advance of its center (or advance of the geometric point $\mathbf{J}_g$ of wheel–ground contact on the wheel path).

According to the time evolution of $(\psi(t), \varphi(t))$, the wheel path will be one or another among the infinite number of possible ones for a same value of $(\Delta\psi, \Delta\varphi)$, so the $(\Delta x, \Delta y)$ increments are not univocally determined by $(\Delta\psi, \Delta\varphi)$ (Fig. 4.10a). Consequently, the number of IC is four: (ψ, φ, x, y).

Figure 4.10b illustrates this situation for $\Delta\psi = \Delta\varphi = 90^\circ$ and three different time evolutions $(\psi(t), \varphi(t))$ (where the angles increase sequentially and monotonously). In all three cases, the final orientation is the same (because we are using Euler angles), but the wheel center location is different.

The system is nonholonomic: it has two DoF and four IC (= 6 − 2) (that can be associated with (x, y, ψ, φ)). ◀

► **Example 4.10** The multibody system in Example 4.4 (Fig. 4.11), containing four rigid bodies and with four IC, has four DoF: if we block the two actuators (motor and hydraulic cylinder), the rotation of the prismatic slot about the b–b′ axis and the block translation along the slot, the system cannot move. It is a holonomic system. ◀

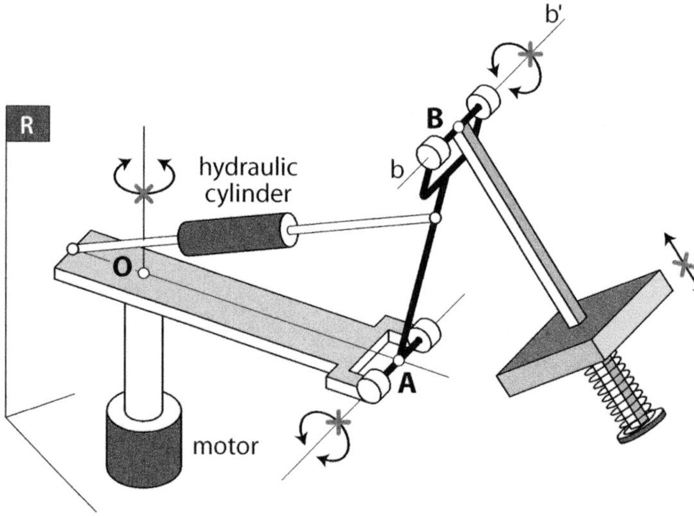

Fig. 4.11

► **Example 4.11** The passenger boarding bridge in Example 4.6 has two DoF and is nonholonomic: from an initial configuration, the finite increments of four IC (which

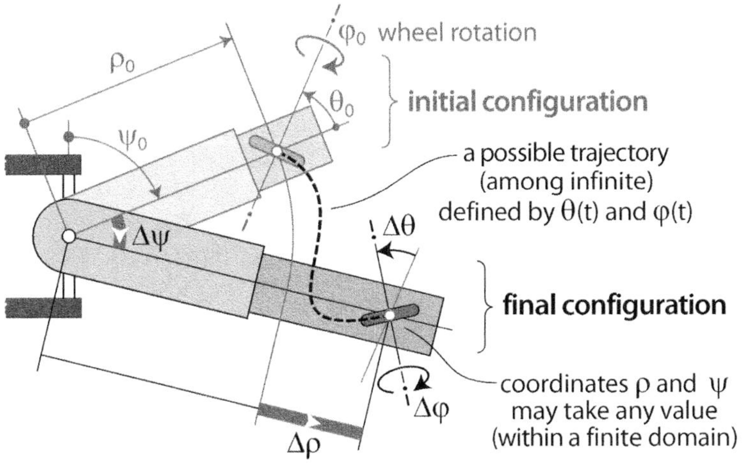

Fig. 4.12

correspond to finite increments of four generalized speeds) are necessary to define the new configuration univocally (Fig. 4.12): ($\Delta\rho, \Delta\psi, \Delta\theta, \Delta\varphi$). ◀

The chassis of many ground vehicles (conventional cars, bicycles, tricycles, etc.) is a nonholonomic system. That nonholonomic behavior is often associated with nonsliding wheels whose transverse axis can change its orientation independently from the forward displacement (as the wheel in Example 4.9). When the change of orientation of the wheel axis is univocally related to the speed of advance, the vehicle behavior is holonomic. This is the case of any vehicle moving on rails (trains, tramways, etc.).

Another element that can also be responsible for nonholonomy is the idealized skate in Fig. 4.13, which cannot skid but can slide longitudinally and change its orientation by pivoting.

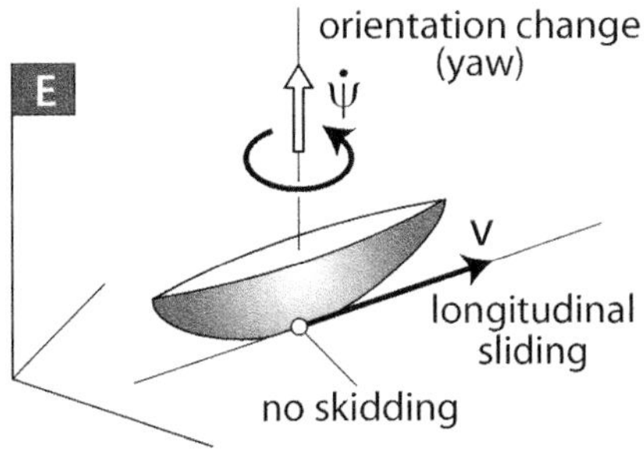

Fig. 4.13

▶ **Example 4.12** The center point **O** of a train wagon has a speed of advance v (controlled by the train machine). As the wagon has to follow the rail, that speed is proportional to the wagon's angular velocity $\dot{\psi}$ through the rail radius of curvature $\mathfrak{R}$ (Fig. 4.14). The wagon has just one DoF.

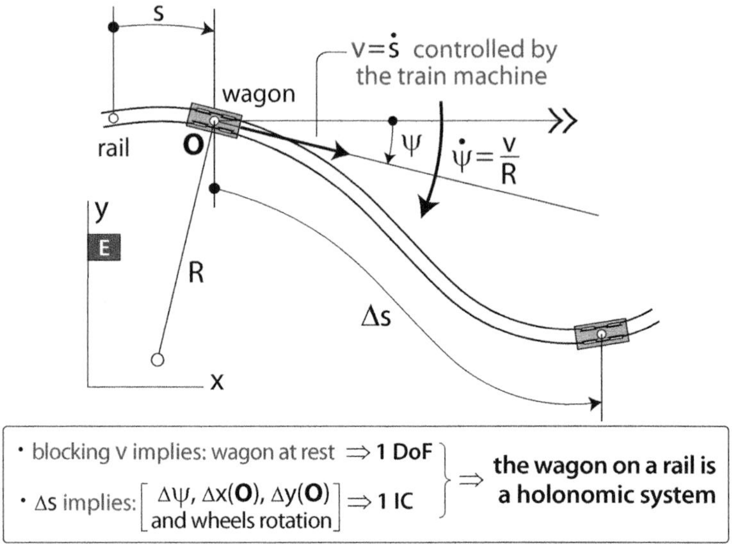

Fig. 4.14

In a finite time interval, the distance Δs traveled by the wagon center determines univocally the finite changes of the center coordinates $(\Delta x, \Delta y)$, the wagon rotation $(\Delta\psi)$, and the wheels' rotation. The wagon has just one IC, and thus it is holonomic. ◀

▶ **Example 4.13** The sledge in Fig. 4.15 has two parallel nonsteered skis and a driving steered wheel, and it moves on a flat ground. The controlled speed variables are the

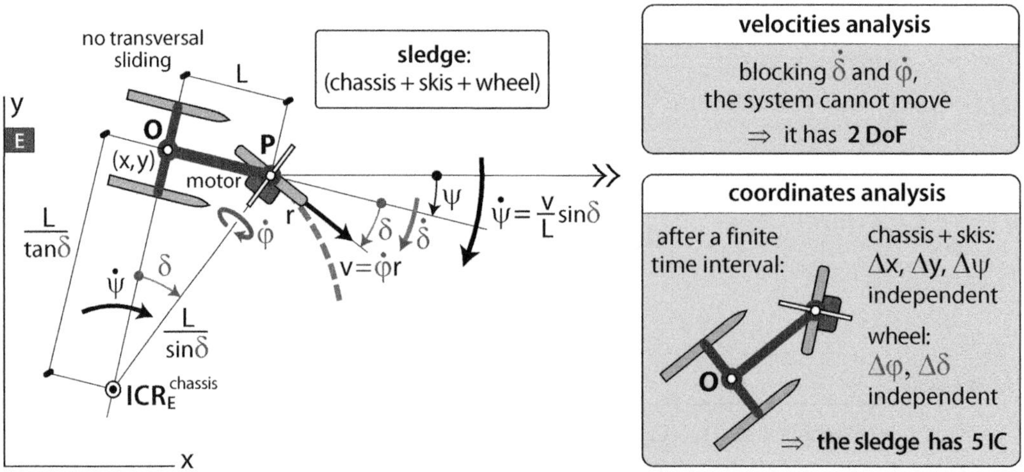

Fig. 4.15

steering velocity $\dot{\delta}$ and the wheel rotation about its axis $\dot{\varphi}$. If the skis do not skid and the wheel does not slide on the ground, the sledge is completely immobilized when we block $\dot{\delta}$ and $\dot{\varphi}$: the sledge has two DoF relative to the ground.

In a finite time interval, the chassis (and the skis, which are fixed to the chassis) can reach any configuration: the (x, y) coordinates of point **O** and the chassis orientation ψ may have any value through a suitable choice of $\dot{\delta}$ and $\dot{\varphi}$ evolutions. Thus, the chassis has three IC: (x, y, ψ). The position of the wheel center is univocally determined if we know the increments of these three coordinates. However, its orientation relative to the chassis is not: the increments of the orientation angles (δ, φ) are independent from $(\Delta x, \Delta y, \Delta\psi)$. Consequently, the sledge (as a whole) has five IC, and it is a nonholonomic system. ◀

▶ **Example 4.14** The car in Fig. 4.16 has front steering and rear traction, and it moves on a flat ground. The controlled speed variables are the steering velocity $\dot{\delta}$ and the rotation $\dot{\varphi}_m$ of the rotor relative to the stator. The steering velocities of the right and left front wheels $(\dot{\delta}_r, \dot{\delta}_l)$ are univocally determined by $\dot{\delta}$ and δ. The speed of the rear axle midpoint **O**, $v_E(\mathbf{O}) \equiv v$, is proportional to $\dot{\varphi}_m$: $v = \lambda r \dot{\varphi}_m$ (where r is the wheels radius,

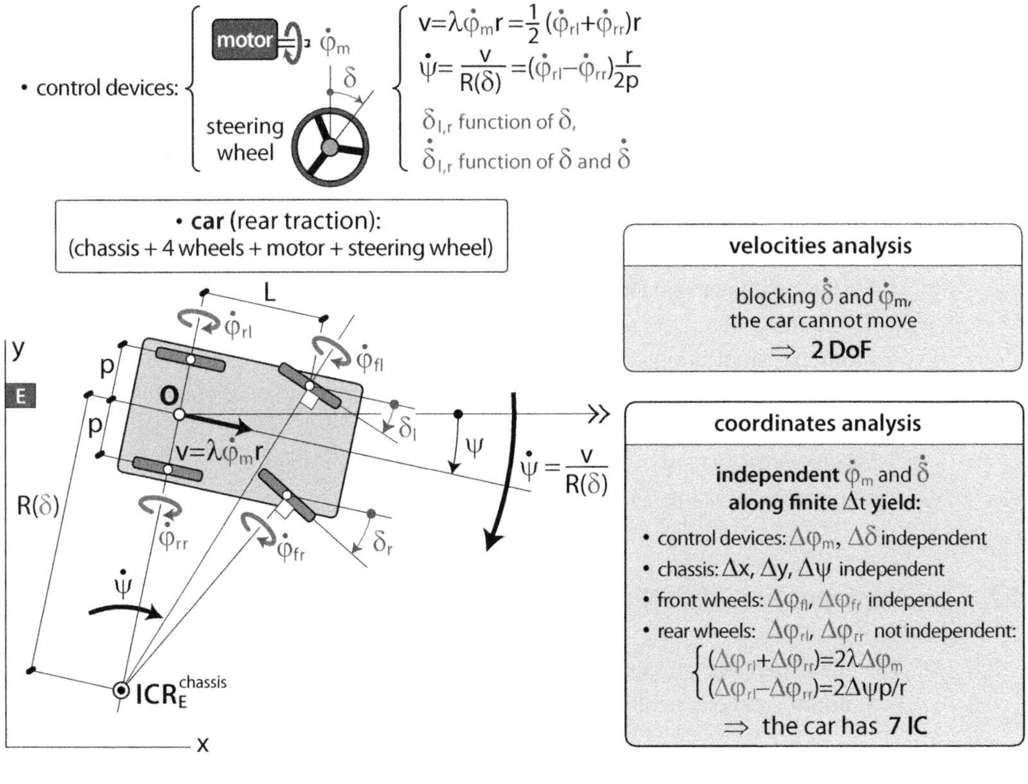

Fig. 4.16

and λ is the gear ratio). That speed is also related to the rear wheels' rotation relative to the chassis: $v = r(\dot{\varphi}_{rr} + \dot{\varphi}_{rl})/2$. The yaw angular velocity of the chassis is related to the previous variables through $\dot{\psi} = v/R(\delta) = r(\dot{\varphi}_{rr} - \dot{\varphi}_{rl})/2p$, where $R(\delta)$ is the radius of curvature of the **O** trajectory, and 2p is the rear axle track.

If the wheels are rigid and do not slide on the ground, the car is completely immobilized when we block $\dot{\delta}$ and $\dot{\varphi}_m$: the car has two DoF relative to the ground.

In a finite time interval, the control variables evolve independently, yielding finite independent changes on the corresponding angular coordinates $(\Delta\delta, \Delta\varphi_m)$. As in the previous example, the finite changes $(\Delta x, \Delta y)$ of point **O** coordinates and the finite orientation change of the chassis $\Delta\psi$ may have any value (through a suitable choice of $\dot{\delta}$ and $\dot{\varphi}_m$ evolutions). Thus, the chassis plus the control devices have five IC: $(\delta, \varphi_m, x, y, \psi)$.

As for the $\Delta\varphi$ rotations of the wheels, those corresponding to the rear ones are directly related to $(\Delta\varphi_m, \Delta\psi)$ through:

$$\Delta\varphi_{rr} + \Delta\varphi_{rl} = 2\lambda\Delta\varphi_m, \quad \Delta\varphi_{rr} - \Delta\varphi_{rl} = (2p/r)\Delta\psi.$$

However, those of the front ones are independent. As their spin angular velocities are proportional to the yaw velocity through variable coefficients:

$$\dot{\varphi}_{fr} = \dot{\psi}\frac{1}{\sin\delta_r}\frac{L}{r}, \quad \dot{\varphi}_{fl} = \dot{\psi}\frac{1}{\sin\delta_l}\frac{L}{r},$$

the time integration to obtain $(\Delta\varphi_{fr}, \Delta\varphi_{fl})$ is not possible as (δ_r, δ_l) are a function of the independent variable δ.

Consequently, the car (as a whole) has seven IC, and it is a nonholonomic system.

Note: In a front-traction car, the relationship $\Delta\varphi_{rr} + \Delta\varphi_{rl} = 2\lambda\Delta\varphi_m$ is replaced by a relationship between $\Delta\varphi_{fr}$, $\Delta\varphi_{fl}$ and φ_m. In a four-wheel traction vehicle, it is replaced by a relationship between $\Delta\varphi_{fr}$, $\Delta\varphi_{fl}$, $\Delta\varphi_{rr}$, $\Delta\varphi_{rl}$ and φ_m. ◀

Systems with just one DoF are always holonomic. Nonholonomy calls for at least two DoF and three IC.

From the engineering point of view, nonholonomic systems can be very interesting: with just a few actuators (one per DoF), the system configuration may evolve in a configuration space of high dimension.

When performing dynamical studies, nonholonomic and holonomic systems are equally treated if using a vector approach. This is not the case in analytical dynamics.[4]

4.7 Superposition of Partial Constraints: Full and Tangent Redundancy

The main purpose of this section is to introduce the concepts of **full** and **tangent redundancy** in constraints, both of paramount importance for the mechanical designer.

[4] **Ordinary Lagrange's equations** only apply to holonomic systems. Nonholonomic ones are dealt with **Lagrange's equations with multipliers**, whose application is far more laborious (see chapter 7 of *Rigid Body Dynamics*).

Common constraint joints in mechanical systems

Cylindrical joint | 3

$$\left\{ \begin{array}{c} v_1 \\ 0 \\ 0 \\ \Omega_1 \\ 0 \\ 0 \end{array} \right\}$$

1 axial   2

Revolute joint | 3

$$\left\{ \begin{array}{c} 0 \\ 0 \\ 0 \\ \Omega_1 \\ 0 \\ 0 \end{array} \right\}$$

1 axial   2

Prismatic joint | 3

$$\left\{ \begin{array}{c} v_1 \\ 0 \\ 0 \\ 0 \\ 0 \\ 0 \end{array} \right\}$$

1 axial   2

Ball-and-socket (spherical) joint | 3

$$\left\{ \begin{array}{c} 0 \\ 0 \\ 0 \\ \Omega_1 \\ \Omega_2 \\ \Omega_3 \end{array} \right\}$$

1   2

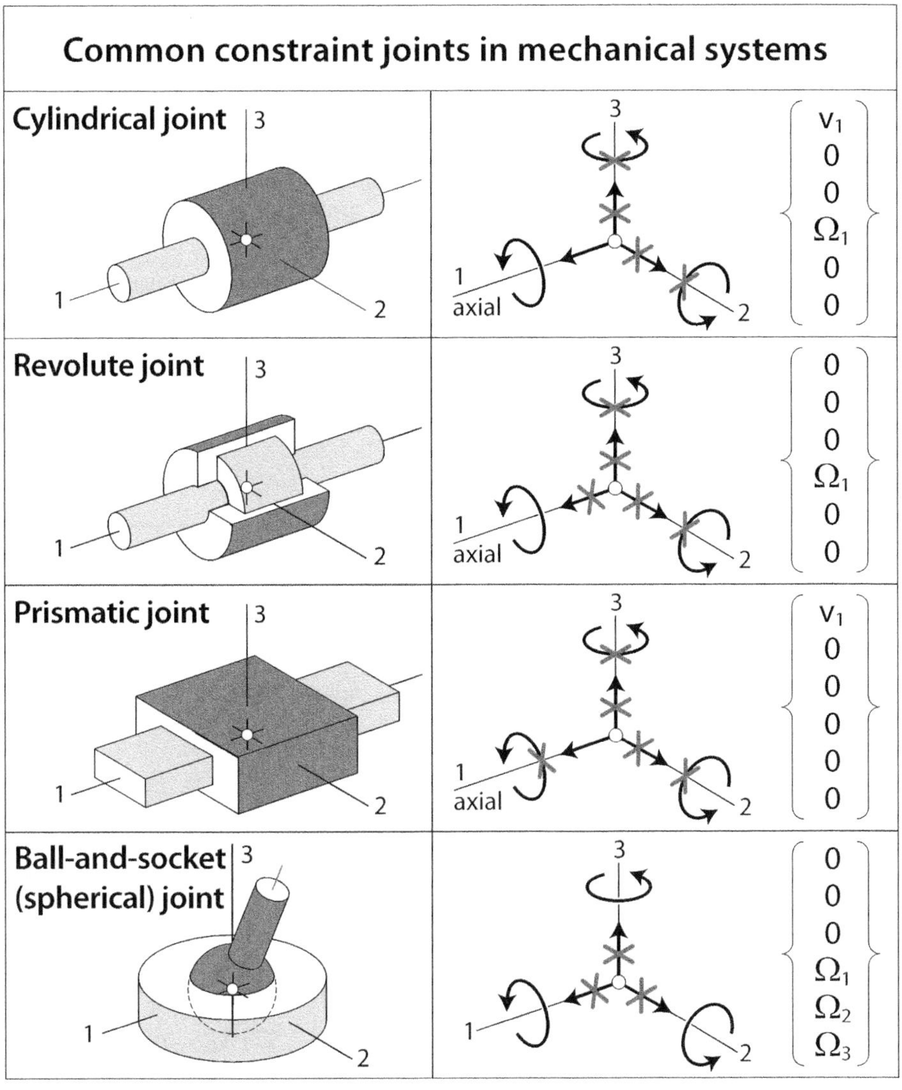

Fig. 4.17

For the sake of clarity, we will focus our analysis on the kinematic restrictions of a single rigid body S1 relative to another rigid body S2. Those restrictions are the result of **several partial constraints**. They may be either direct constraint joints between S1 and S2 (the rigid bodies have a direct contact) or indirect constraints (the rigid bodies are connected through one or several intermediate elements S_i).

Figure 4.17 shows a few usual partial constraints.[5] The right column contains the kinematic torsor $\left\{ \bar{\mathbf{v}}_{S2}(\mathbf{O}_{S1}) \ \ \bar{\boldsymbol{\Omega}}_{S2}^{S1} \right\}^{\mathrm{T}}$ at the point $\mathbf{O}$ of rigid body S1 (marked just by a

---

[5] The universal Cardan joint has not been included in this table because its kinematic analysis is less straightforward. Its kinematics and dynamics will be considered in chapter 2 of *Rigid Body Dynamics* [Cambridge University Press, forthcoming].

white point on the constraint schematics in the left column). That torsor could be referred to any other point **P** of S1:

$$\left\{ \bar{\mathbf{v}}_{\mathrm{S2}}(\mathbf{P}) \ \bar{\boldsymbol{\Omega}}_{\mathrm{S2}}^{\mathrm{S1}} \right\}^{\mathrm{T}} = \left\{ \bar{\mathbf{v}}_{\mathrm{S2}}(\mathbf{O}) + \bar{\boldsymbol{\Omega}}_{\mathrm{S2}}^{\mathrm{S1}} \times \bar{\mathbf{OP}} \ \bar{\boldsymbol{\Omega}}_{\mathrm{S2}}^{\mathrm{S1}} \right\}^{\mathrm{T}}. \tag{4.5}$$

From Eq. (4.5), it is clear that the number of nonzero components in the kinematic torsor may change when shifting to a different point. However, that of independent components is always the same.

All kinematic torsors shown in Fig. 4.17 describe the relative motion between two rigid bodies S1 and S2, which seem to be directly in contact with one another. However, most of those constraint joints include intermediate elements between S1 and S2. This is the case of the small spheres in a ball bearing. Other frequent constraints between pairs of rigid bodies are achieved through intermediate elements as taut cables, wheels, rollers, and bars.

► **Example 4.15** Rigid body S1 has initially the three DoF of planar motion (two translational and one rotational DoF). If points **P** and **Q** are constrained to move along identical circular slots on the ground (solid S2) with centers $\mathbf{O_P}$ and $\mathbf{O_Q}$ so that $\overline{\mathbf{O_P P}}$ and $\overline{\mathbf{O_Q Q}}$ are parallel (Fig. 4.18a), the number of DoF reduces to one: a circular translation.[6]

The same restriction can be achieved through two identical and parallel thin rods with ball-and-socket joints at their ends (Fig. 4.18b). ◄

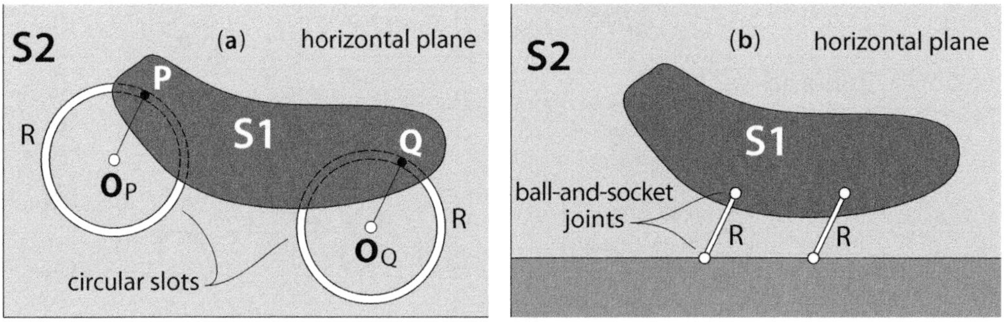

Fig. 4.18

Let's consider a rigid body S1 linked to S2 through just one partial constraint. As a result, S1 has m DoF (with $m < 6$) relative to S2.

[6] The configuration where $\overline{\mathbf{O_P P}}$, $\overline{\mathbf{O_Q Q}}$ and $\overline{\mathbf{O_P O_Q}}$ are parallel is excluded in this analysis. When the system starts moving from this configuration, there is an indeterminacy in the motion: it can be either a circular translation or a motion where $\overline{\mathbf{O_P P}}$, $\overline{\mathbf{O_Q Q}}$ rotate in opposite direction. It is called a **bifurcation configuration**.

We introduce now an additional partial kinematic constraint between S1 and S2 in a particular configuration that would be responsible for n restrictions if acting on its own. Some of those n restrictions may be independent from those introduced by the previous constraint. However, the *linear kinematic superposition* of the restrictions associated with the two partial constraints may indicate that some of the n restrictions have already been guaranteed by the previous constraint in that configuration: those restrictions are called **redundant**. If there are r redundant restrictions, the linear superposition indicates that S1 has m – (n – r) degrees of freedom relative to S2: $DoF_{lin} = m - (n - r)$.

In order to gain some insight in redundancy and for the sake of simplicity, let's consider an additional constraint between S1 and S2 that, on its own, would be responsible for just one restriction. There are three possibilities:

- The additional constraint is not redundant, and the number of DoF of S1 relative to S2 is (m – 1)
- The additional constraint is redundant ($DoF_{lin} = m$), and it has no consequences at all on the relative motion between the two rigid bodies: the configuration can evolve exactly in the same way as it did before adding that constraint. In that case, we say that the restriction is **fully redundant**
- The additional constraint is redundant ($DoF_{lin} = m$), but we discover that it does have consequences: the analysis of the motion geometry (not just the analysis of allowed velocities in the mounting configuration) shows that the actual number of relative DoF (denoted as DoF_{geom} to avoid confusion with DoF_{lin}) is lower than the one we had before we added the new constraint; the evolution of the configuration is different, and it may even be impossible to leave that configuration. We say then that there is **tangent redundancy**, and the number of DoF eliminated by the additional constraint is the **degree of tangent redundancy (DTR)**

Mechanical systems presenting redundant constraints are called **overconstrained systems**. Detecting the number of redundancies in a mechanical system through a computer program is a fairly simple matter because the associated mathematical formulation is strictly linear. However, discovering whether they are full or tangent redundancies is quite a different issue: the distinction between them calls for nonlinear formulations, and those are not so easily implemented in computers.

The analysis of tangent redundancy that we present here is just an introduction for the reader so that he or she is able to identify it.

► **Example 4.16** The three stools in Fig. 4.19 have just three DoF relative to the floor: two horizontal translations and a rotation perpendicular to the floor. The restriction, however, has been achieved differently in each of them: they have a different number of legs. Stools with more than three legs present full redundancy. ◄

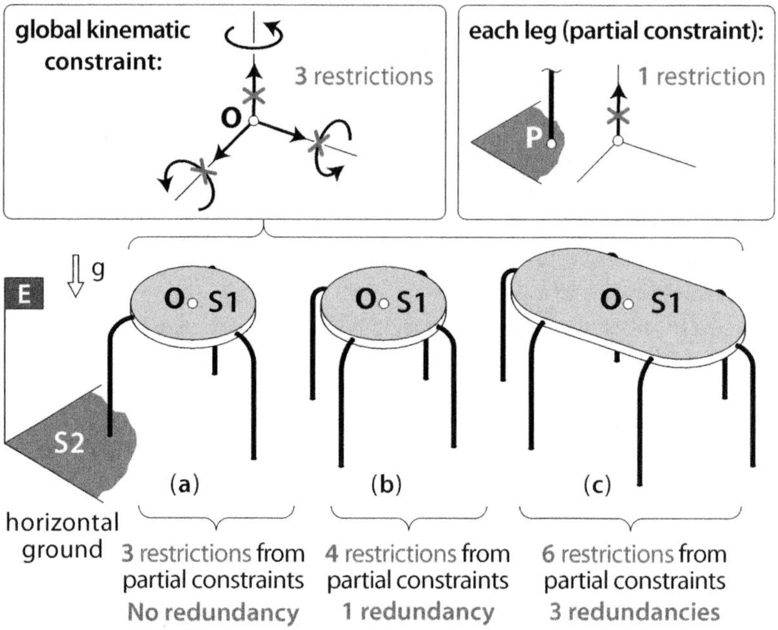

Fig. 4.19

Redundancy can be detected by introducing the partial constraints sequentially and analyzing the resulting motion or the eventual difficulties in assembling the system.

► **Example 4.17** We introduce sequentially three partial kinematic restrictions between a rigid body constrained to have planar motion (two translational DoF and one rotational DoF) and a fixed plane parallel to the plane of movement:

(a) A first skate P fixed to the rigid body (Fig. 4.20a): the velocity component of point **P** transverse to the skate is eliminated; one DoF is removed (perpendicular to the skate translation), but the number of IC remains unchanged (as the constraint is nonholonomic)

(b) A second skate Q fixed to the rigid body (Fig. 4.20b): a velocity component of point **Q** transverse to the skate is eliminated; the translation DoF is removed, and rotation is only possible around a permanent center of rotation (point **I**, fixed both to the rigid body and the reference frame)

(c) A third skate S fixed to the rigid body and with a transverse direction not going through **I** (Fig. 4.20c): a velocity component of point **S** is eliminated and the rotation is no longer possible. With these three skates, the rigid body is blocked

(d) If the transverse direction of the third skate S went through **I** (Fig. 4.20d), that constraint would be redundant as the transverse component of its velocity has already been removed by skates P and Q. It would be a full redundancy as the system would be able to change its configuration and skate S would never interfere with its motion

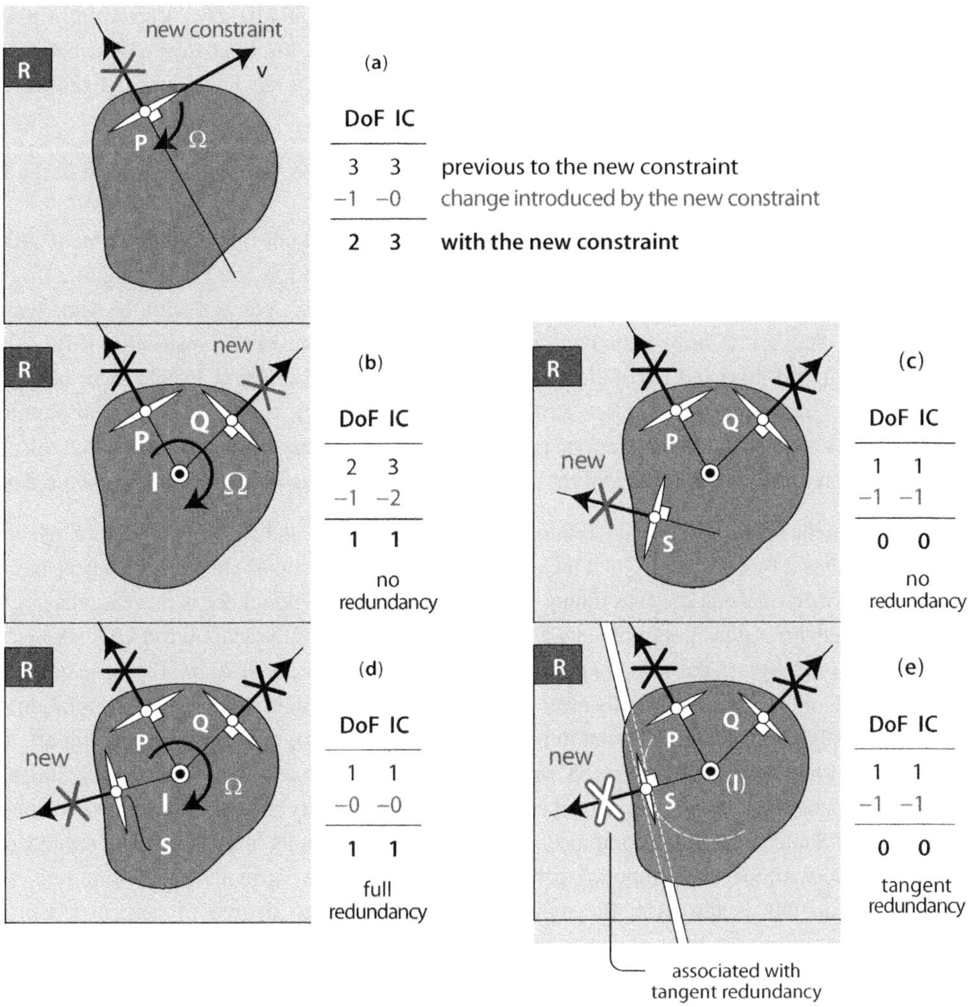

Fig. 4.20

(e) If no third skate is introduced but instead point **S** is restricted to move along the s–s′ slot perpendicular to $\overline{\mathbf{SI}}$ (Fig. 4.20e), there is tangent redundancy: in that particular configuration, the velocity component of **S** normal to the slot (that would be removed if no other constraints existed) has already been canceled by the skates P and Q; however, the rigid body is blocked because if it rotated, **S** would be forced to follow simultaneously the s–s′ line and the circumference with center **I** and radius $|\overline{\mathbf{SI}}|$. The rotation has been removed, and we say there is one degree of tangent redundancy (DTR = 1). The suppression of that DoF has not been discovered through a purely kinematic analysis (or velocity analysis): it has become evident only when considering the geometric design

If we had not introduced the constraints sequentially, we would not be able to associate the full or the tangent redundancy to a particular constraint: we would say simply that *there is* full or tangent redundancy. ◀

▶ **Example 4.18** We introduce sequentially a set of partial kinematic constraints through intermediate elements (thin rods with ball-and-socket joints at their ends) between a rigid body (RB) constrained to have planar motion (three DoF) and a fixed ground parallel to the plane of movement:

(a) A first rod between points **P** and **P′**: it eliminates the velocity component in the **PP′** direction, and removes one DoF (Fig. 4.21a)
(b) A second rod between **Q** and **Q′**, parallel to the first one and with the same length: it also eliminates a velocity component and removes a second DoF (Fig. 4.21b)
(c) A third rod between **S** and **S′** parallel to the previous ones and with the same length. This rod is fully redundant and does not add any kinematic restriction (Fig. 4.21c)
(d) If the third rod is parallel to the other two but with different length, there is tangent redundancy with DTR = 1: it blocks the motion as **S** would have to follow simultaneously the circumferences tangent to **S** with centers **S′** and **S″** (Fig. 4.21d). The reduction of another DoF has been discovered when considering the geometric design

In the sequential introduction of constraints, the eventual difficulty of assembling can be analyzed by considering that the new rod we are introducing is shorter (or longer) than the distance between the two points we want to connect. When there is no redundancy, this length error does not imply any assembling difficulties: if the second rod **QQ′** is too short (or too long), as the velocity point **Q** of the RB may have any direction (because of the two DoF of the RB), it can easily meet the rod end point. The final assembling configuration will be slightly different than that in Fig. 4.21b (the two rods will not be strictly parallel).

However, if the third rod (**SS′** or **SS″**) were too short, assembling would be difficult: point **S** of the RB can only move on a circle with center **S′**. Only if the system is flexible enough will **S** and the rod end point meet. That assembling difficulty is a symptom of redundancy.

Sometimes, the number of difficulties depends on the particular assembling sequence. When this is the case, the redundancy has a nonzero degree of tangency. But the opposite is not true: we may have tangent redundancy, but the number of elements whose assembling is difficult (from now on, *number of assembling difficulties*) may be independent of the assembling sequence.

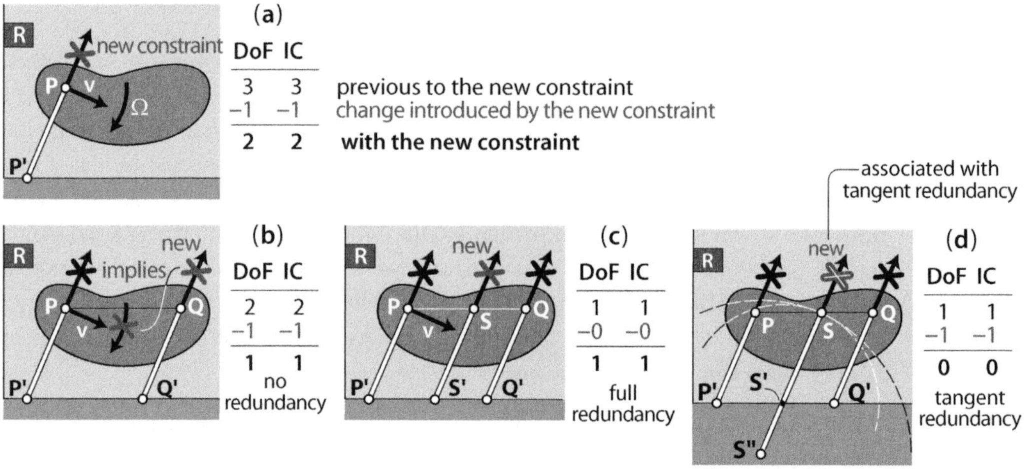

Fig. 4.21

In this example, whether you assemble the elements according to the sequence (**PP′**, **QQ′**, **SS″**) or (**PP′**, **SS″**, **QQ′**) has no consequences on the number of assembling difficulties: it is always associated to just the last rod. Hence, through this analysis we are unable to tell whether we have full or tangent redundancy. ◄

► **Example 4.19** We introduce sequentially a set of partial kinematic constraints through intermediate elements (thin rods with ball-and-socket joints at their ends) between a block on a smooth ground E (planar motion, three DoF) and four E-fixed points (Fig. 4.22).

The same sequential analysis performed in the previous example can be done on the block. If the assembling starts with the pair of rods ①②, ①③ (or ④③, ④②, which are symmetrical options that will not be discussed), the block motion is restricted to a single DoF. If it starts with rods ②③, the block may still move according to a circular translation.

The situation is not the same if we start with rods ①④: though the second rod appears to be redundant, it actually eliminates the remaining two DoF. There is tangent redundancy.

Let's proceed to the analysis of difficulties on assembling the system. If having started with rod ①, we add rod ④: if the latter is too short, assembling will be impossible if the system is totally rigid. Some flexibility will allow assembling, but as the rigid body will then be totally blocked, a length error in the remaining two rods will also provoke assembling difficulties. The total number of difficulties that may appear is three.

If rod ④ is the third one in the assembling, the difficulties will be associated with both the third and the fourth rods (two difficulties). However, if rod ④ is added once the other three have been assembled, the difficulty will be strictly associated with that last rod (one difficulty).

In this example, the number of assembling difficulties depends on the assembling sequence. We can then be sure that there is tangent redundancy, though we cannot tell its degree through this analysis. ◄

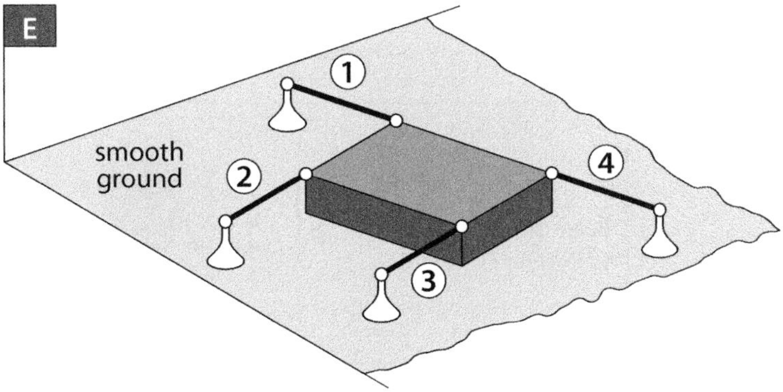

Fig. 4.22

Redundancy is a tricky condition because it may entail high tensions in the system, and should be avoided in general. Both full and tangent redundancy present drawbacks, as it will be particularly evident when studying their consequences on the constraint forces (chapter 2 of *Rigid Body Dynamics* [Cambridge University Press, forthcoming]). Full redundancy, apparently innocuous, may result in an extreme sensitivity to small geometric errors: they can prevent the assembling of the system or allow a nonsmooth movement by causing deformations.

Tangent redundancy is worse because some DoF are removed in an unclear way: the unavoidable deformability, and occasionally the presence of gaps, allow a small range of movement associated with those DoF. In addition, as will be seen in chapter 2 in *Rigid Body Dynamics* [Cambridge University Press, forthcoming], some constraint forces tend to very high values (in principle to ∞).

The existence of full or tangent redundancy can be assessed without having to impose a sequential introduction of partial constraints. It is enough to consider the partial restrictions, their linear superposition, and, if the latter indicates redundancy, the actual kinematic restriction when the geometric design is taken into account. The next two examples illustrate this procedure.

► **Example 4.20** Consider a rigid ruler with the three DoF of planar motion (Fig. 4.23a). The kinematic torsor at its midpoint **O** has three independent components: $\{\dot{x} \quad \dot{y} \quad \Omega\}^{\mathrm{T}}$. The kinematic torsors at **P** and **Q** are obtained from that torsor through RBK:

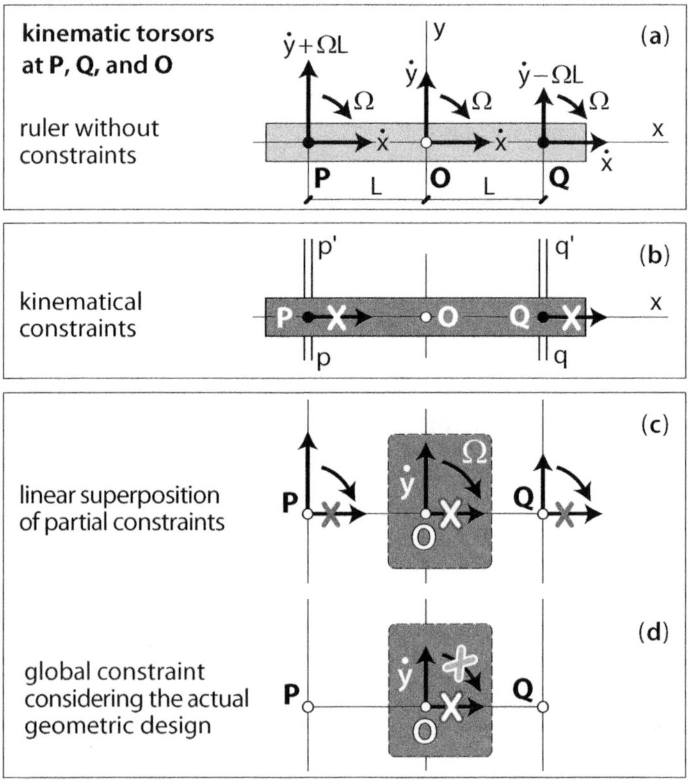

Fig. 4.23

- kinematic torsor at **P**: $\{\dot{x} \quad \dot{y} + L\Omega \quad \Omega\}^T$
- kinematic torsor at **Q**: $\{\dot{x} \quad \dot{y} - L\Omega \quad \Omega\}^T$

As expected, both torsors have three independent components.

If we force points **P** and **Q** to move along the p–p′ and q–q′ slots perpendicular to **PQ** (Fig. 4.23b), their velocity component perpendicular to the slot will be eliminated:

- partial kinematic torsor at **P**: $\{0 \quad \dot{y} + L\Omega \quad \Omega\}^T$
- partial kinematic torsor at **Q**: $\{0 \quad \dot{y} - L\Omega \quad \Omega\}^T$

Having just one point–slot contact (**P** or **Q**) or two (**P** and **Q**) has exactly the same consequence on the kinematic torsor at **O**: $\{0 \quad \dot{y} \quad \Omega\}^T$. If each point–slot contact is taken as a partial constraint, their superposition at **O** indicates that there is redundancy.

However, the analysis of the geometric design (that is, taking into account that the slots are straight) shows that the rotation Ω has been eliminated: a change in the ruler orientation would imply a separation between points **P** and **Q**, and this is not possible if both the ruler and the slots are perfectly rigid and there are no gaps. It is a tangent redundancy that eliminates one DoF (DTR = 1).

The linear superposition of partial kinematic restrictions at the same point **O** is another method to discover redundancy without having to proceed sequentially. ◀

In the previous examples, the number of redundancies was equal to the degree of tangent redundancy (and always equal to one). But this is not always the case: a single redundancy may suppress several DoF.

▶ **Example 4.21** Points **P** and **Q** of a rigid body are attached to points **P′** and **Q′** of the supports fixed to the ground by means of rigid rods with spherical ball joints at both ends. Points **P′**, **P**, **Q**, and **Q′** are aligned (Fig. 4.24a).

If these two constraints between the rigid body and the ground are taken as partial constraints, the analysis of the kinematic restrictions imposed by each of them indicates that they are redundant. Both of them have a similar effect on the kinematic torsor at **O** (Fig. 4.24b): $\{0 \quad v_x \quad v_z \quad \Omega_x \quad \Omega_y \quad \Omega_z\}^T$. However, the geometric design shows that only the Ω_x rotation is possible (Fig. 4.24c). It is a tangent redundancy with DTR = 4: it eliminates four DoF. ◀

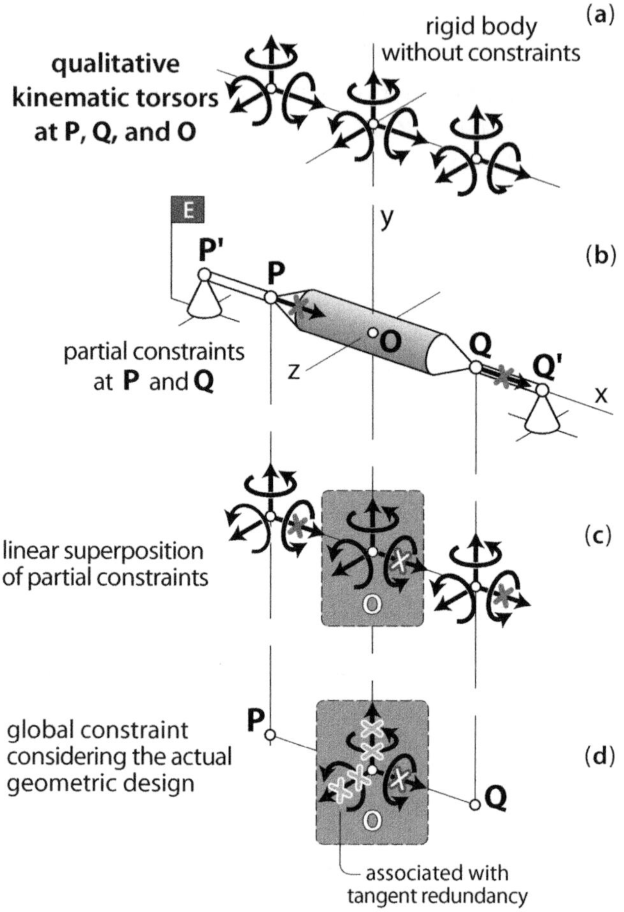

Fig. 4.24

In the following example, the partial constraints are responsible for more than one partial restriction. For the sake of clarity, we will analyze the existence of redundancy through the sequential introduction of constraints.

► **Example 4.22** We introduce sequentially two partial kinematic constraints between a rigid platform S1 and the ground S2 (Fig. 4.25):

- A first constraint consisting of a ball-and-socket joint at the platform center **O**; it removes three DoF (the three translations)
- A second constraint through a thin bar with a bearing and a ball-and-socket joint at the end points. On its own, that constraint would be responsible for the reduction of two DoF (two translations)

If the second constraint is added in the particular configuration in Fig. 4.25 (points **O**, **P**, and **O**′ not aligned), there is no redundancy: the two velocity components of $\bar{\mathbf{v}}_{S2}(\mathbf{P})$

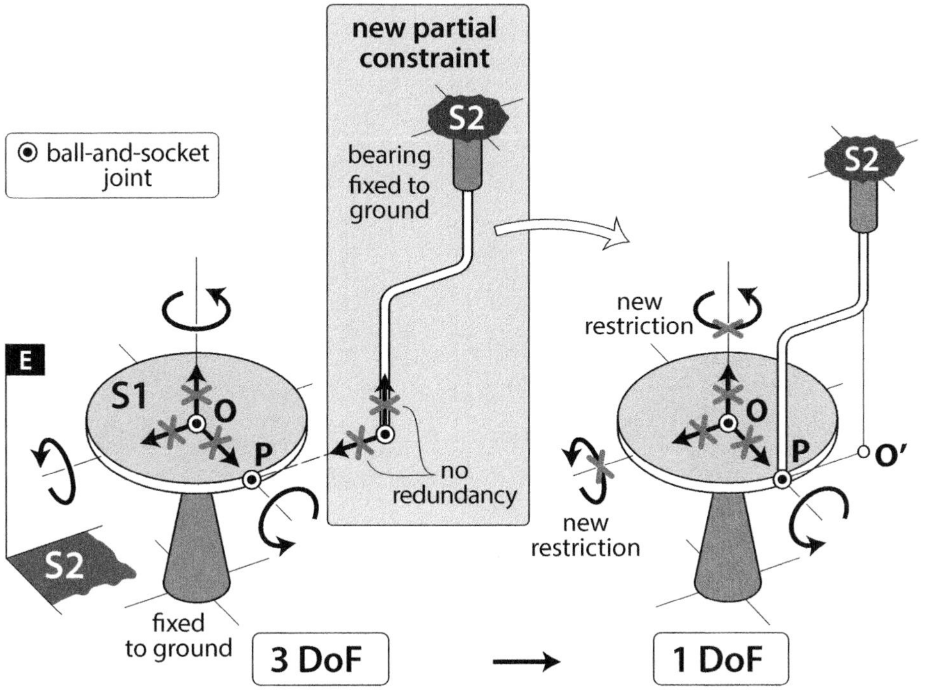

Fig. 4.25

that the bar eliminates in that particular configuration have not been eliminated by the first constraint. The platform keeps just one rotation DoF.

If the bearing is located on the vertical line through **O** (**O** = **O**′, Fig. 4.26a), there is full redundancy: the restriction of the $\bar{\mathbf{v}}_{S2}(\mathbf{P})$ radial component is redundant in all possible configurations allowed by the platform vertical rotation. The platform has two DoF.

When points **O**, **P**, and **O**′ are aligned (Fig. 4.26b), the redundancy in the radial direction is tangent: point **P** cannot leave that configuration as the circular trajectories imposed by each constraint are divergent. The vertical rotation is eliminated and the platform has just one DoF.

Finally, Fig. 4.27 shows an example where the second constraint has been changed to a rod with universal Cardan joints at its end points (each joint allows two relative rotations). When the three axes of that constraint are aligned, that partial constraint would eliminate two DoF on its own: the translation in the axes' direction and the rotation about that direction.

When the second constraint is added to the platform, the restriction of the rotation about axis **OP** is correctly suppressed. However, as the radial component of $\bar{\mathbf{v}}_{S2}(\mathbf{P})$ was already suppressed by the joint at **O**, the suppression of the translation in the axes' direction is redundant. Therefore, the platform seems to keep two rotation DoF ($\text{DoF}_{\text{lin}} = 4$, the vertical rotation and the horizontal rotation perpendicular to **OP**). But if we try to force **P** to leave that configuration either vertically or horizontally, the two partial constraints will

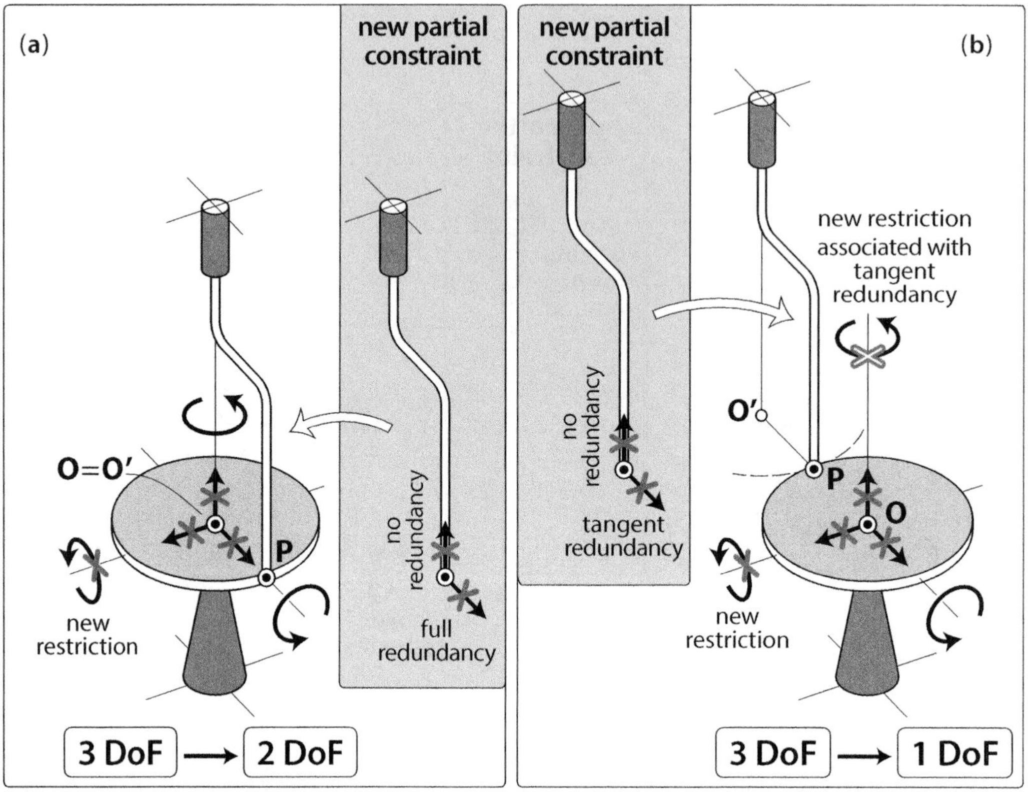

Fig. 4.26

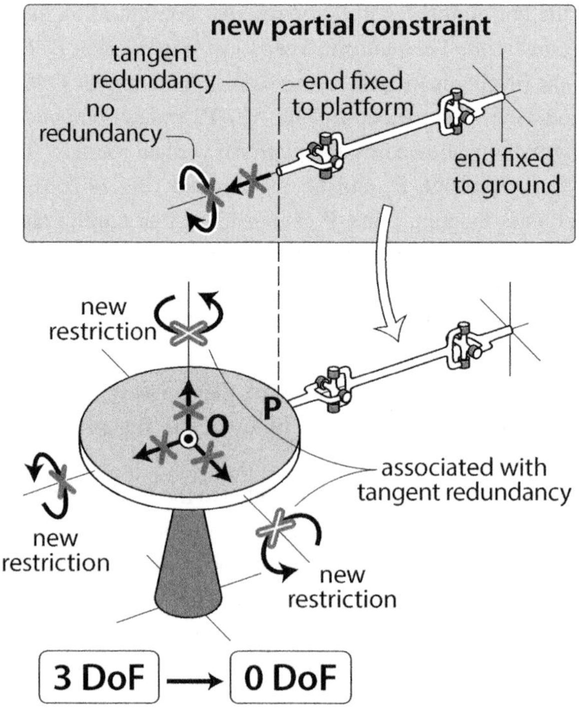

Fig. 4.27

try to impose circular divergent trajectories: the platform is blocked ($\mathrm{DoF}_{geom} = 6$). It is a tangent redundancy responsible for two restrictions (DTR = 2). ◀

4.8 Systematic Analysis of Redundancy in Partial Constraints between Rigid Bodies

The analysis of full and tangent redundancy in partial constraints between two rigid bodies – by direct contact or by means of intermediate elements – can be done in a systematic way that does not call for the consideration of a sequential introduction of constraints. The procedure is as follows:

(a) Count up the number r_p of partial kinematic restrictions (number of DoF that each partial constraint would eliminate on its own)

(b) Count up the number r_s of kinematic restrictions coming from the linear superposition of the partial constraints (number of independent components of the kinematic torsor) without taking into account the geometric design. If the sum $\sum r_p$ of partial constraints is greater than r_s, there is redundancy and the number of redundancies NR is $\mathrm{NR} = \left(\sum r_p\right) - r_s$

(c) Count up the number r_g of kinematic restrictions introduced by all constraints acting together taking into account the geometric designs. If there is redundancy ($\mathrm{NR} > 0$) and $r_g = r_s$, it is full redundancy. If $r_g > r_s$, there is tangent redundancy. The difference $\mathrm{DTR} = r_g - r_s$ is the **degree of tangent redundancy**. It is the number of DoF removed by the tangent redundancy

The following table summarizes this analysis:

Systematic analysis of redundancy

$\sum r_p \equiv$ restrictions of the partial constraints

Without taking into account the geometric design of the constraints (linear superposition)

$r_s \equiv$ restrictions from superposition of the partial constraints' kinematic restrictions

⇓ ⇓

If $\sum r_p = r_s$: **no redundancy**	If $\sum r_p > r_s$: **redundancy**
$\sum r_p = r_s = r_g$	**Number of redundancies** $\mathrm{NR} = \left(\sum r_p\right) - r_s$

⇓

Taking into account the geometric design of the constraints

$r_g \equiv$ global restrictions

⇓ ⇓

If $r_g = r_s$: *full* **redundancy**	If $r_g > r_s$: ***tangent* redundancy** **Degree of tangent redundancy:** $\mathrm{DTR} = r_g - r_s$
	(note that $\sum r_p > r_g$ or $\sum r_p < r_g$)

► **Example 4.23** For the system in Example 4.17 (Fig. 4.28), a systematic analysis of redundancy yields:

$$\left.\left.\begin{array}{l}\sum r_p = 1+1+1 = 3 \\ r_s = 2\end{array}\right\} \Rightarrow NR = \sum r_p - r_s = 1 \atop r_g = 3\right\} \Rightarrow DTR = r_g - r_s = 1 \Rightarrow 1$$

tangent redundancy. ◄

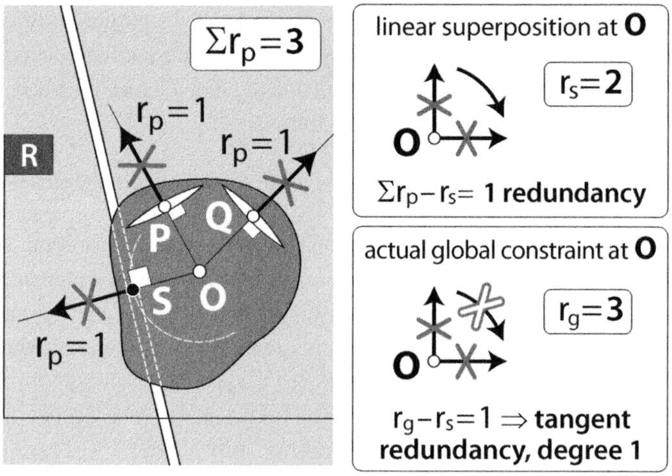

Fig. 4.28

► **Example 4.24** For the system in Example 4.18 (Fig. 4.29), each bar on its own would be responsible for one restriction $(r_p = 1)$. The superposition suggests that

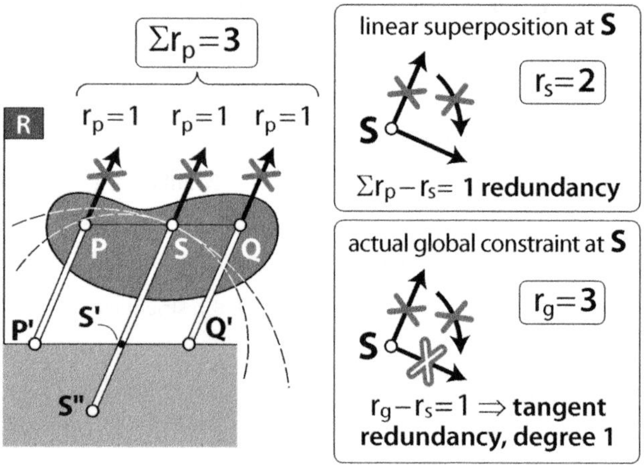

Fig. 4.29

one bar is redundant, but the rigid body is actually immobilized. So a systematic analysis yields:

$$\left.\begin{array}{l}\left.\begin{array}{l}\sum r_p = 1+1+1=3\\ r_s=2\end{array}\right\} \Rightarrow NR = \sum r_p - r_s = 1\\ r_g = 3\end{array}\right\} \Rightarrow DTR = r_g - r_s = 1 \Rightarrow 1$$

tangent redundancy. ◄

► **Example 4.25** Let's consider a door with two hinges (Fig. 4.30), where only the lower hinge restricts the vertical descending motion. If each hinge is considered as a partial constraint, a systematic analysis of redundancy yields:

$$\left.\begin{array}{l}\left.\begin{array}{l}\sum r_p = 5+4=9\\ r_s=5\end{array}\right\} \Rightarrow NR = \sum r_p - r_s = 4\\ r_g = 5\end{array}\right\} \Rightarrow DTR = r_g - r_s = 0 \Rightarrow$$

Full redundancy (4 redundancies).

(The kinematic superposition can be analyzed through the kinematic torsor at point **O** of the door.)

Note: A small number of total redundancies is acceptable provided that the door has some flexibility. If the flexibility is high, redundancies may be convenient to increase the door's rigidity. ◄

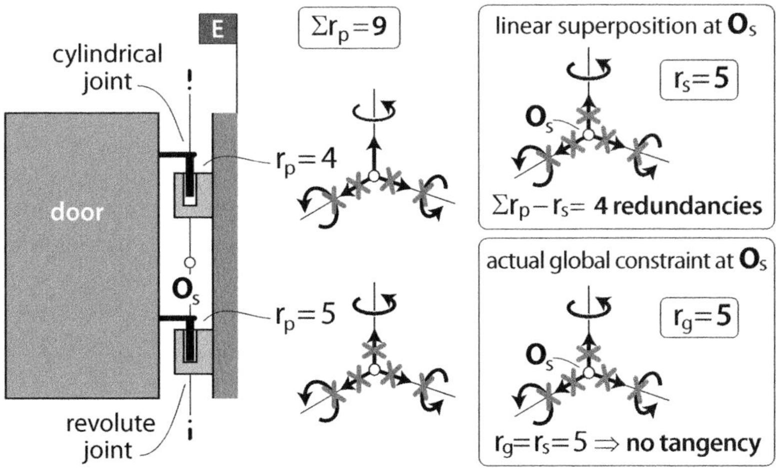

Fig. 4.30

► **Example 4.26** We want to immobilize a rigid body relative to the ground by means of two partial constraints (Fig. 4.31):

- a ball-and-socket joint at **O** between the rigid body and the ceiling $(r_p = 3)$
- a tensioned rope passing through the thin ring Q of the rigid body with ends **P** and **P′** fixed to the ground $(r_p = 1)$

The systematic analysis of redundancy yields:

$$\left.\begin{array}{l}\left.\begin{array}{l}\sum r_p = 3 + 1 = 4\\ r_s = 3\end{array}\right\} \Rightarrow NR = \sum r_p - r_s = 1\\ r_g = 5\end{array}\right\} \Rightarrow DTR = r_g - r_s = 2 \Rightarrow$$

$\Rightarrow$ 1 redundancy, degree of tangent redundancy 2.

Note: The kinematic superposition to obtain $r_s = 3$ can be studied through the kinematic torsor at point **O**. The value $r_g = 5$ comes from point **Q** being forced to move on a sphere with center **O** (due to the ball-and-socket joint) and on an ellipsoid with foci **P** and **P′** (due to the rope). Both surfaces are tangent at point **Q**. ◄

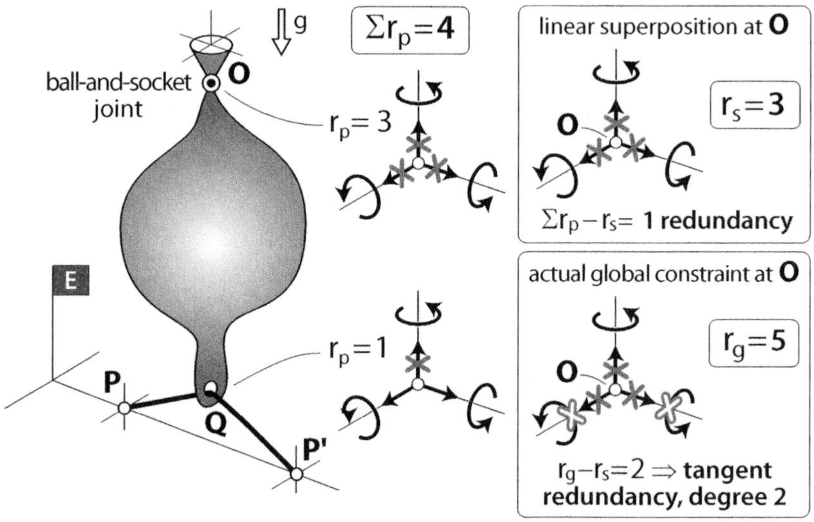

Fig. 4.31

► **Example 4.27** In a parallel robot, the end element has a constraint with the ground through a universal Cardan joint, which restricts its motion to two rotations about the two axes of the joint $(r_p = 4)$. The control of those two DoF is achieved through two hydraulic cylinders (Fig. 4.32a). In the proposed design, the cylinders act on the robot through two rods with ball-and-socket joints at the ends.

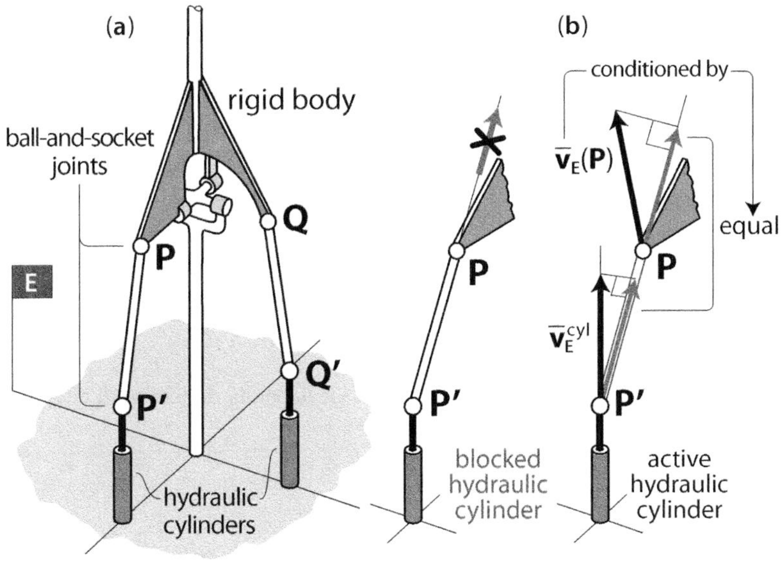

Fig. 4.32

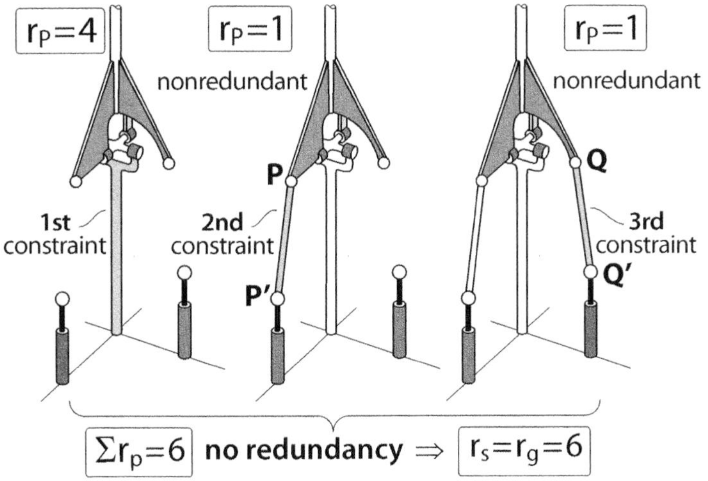

Fig. 4.33

When a cylinder is blocked, the rod eliminates the velocity component of a point of the end element along the direction of the linking rod $\left(r_p = 1\right)$; when it is activated, it controls that component (Fig. 4.32b).

Consequently (Fig. 4.33):

$$\left.\begin{array}{l}\left.\begin{array}{l}\sum r_p = 4 + 1 + 1 = 6 \\ r_s = 6\end{array}\right\} \Rightarrow NR = \sum r_p - r_s = 0 \\ r_g = 6\end{array}\right\} \Rightarrow \text{No redundancy at all!}$$

◀

Quiz Questions

Threads, cables, and ropes are considered inextensible throughout the whole collection of questions unless stated otherwise. The ground reference frame is always denoted as E. All data are declared in the figures, but only part of them are declared in the text.

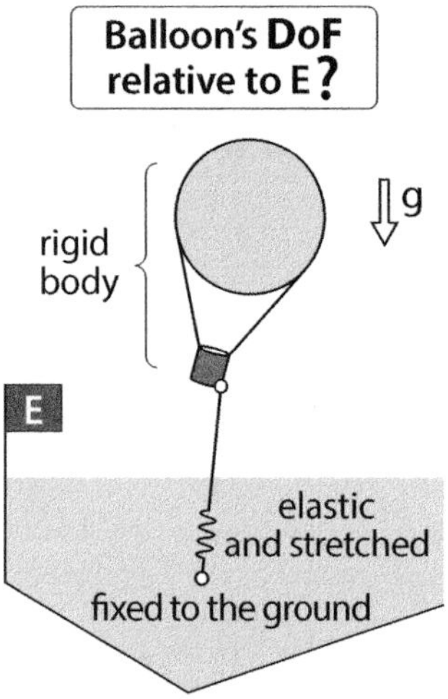

4.1 A balloon is attached to the ground through a stretched elastic cable. How many DoF does the balloon (modeled as a rigid body) have relative to the ground?

A It depends on the wind characteristics.
B 3
C 4
D 5
E 6

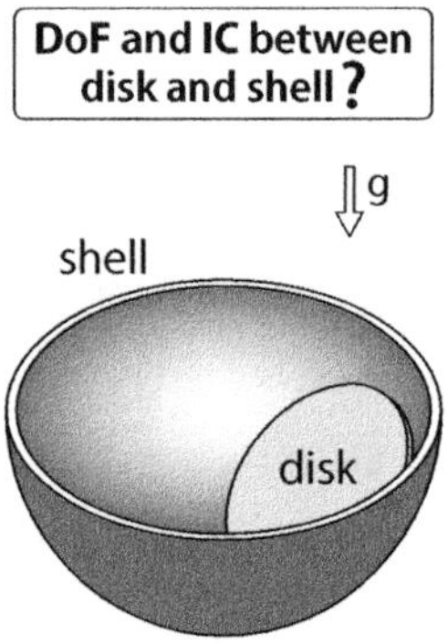

4.2 The homogenous disk slides without friction inside the spherical shell, keeping its contour in contact with the spherical surface. How many DoF and IC are there between disk and shell?

	DoF	IC
A	4	4
B	3	3
C	2	3
D	2	2
E	3	4

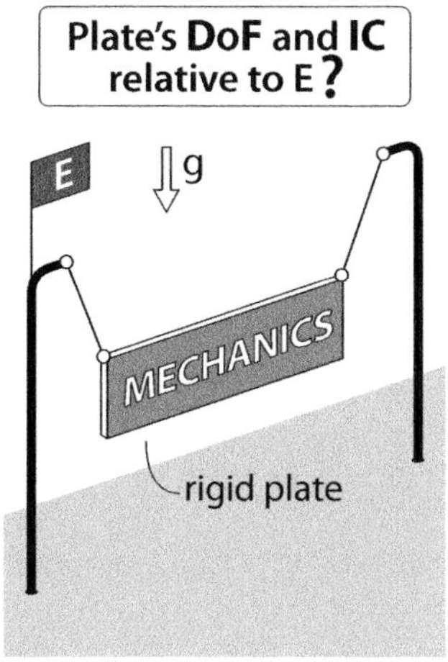

4.3 The rigid plate hangs from the ground-fixed poles through two taut cables. How many DoF and IC does it have relative to the ground?

	DoF	IC
A	2	2
B	2	3
C	3	3
D	3	4
E	4	4

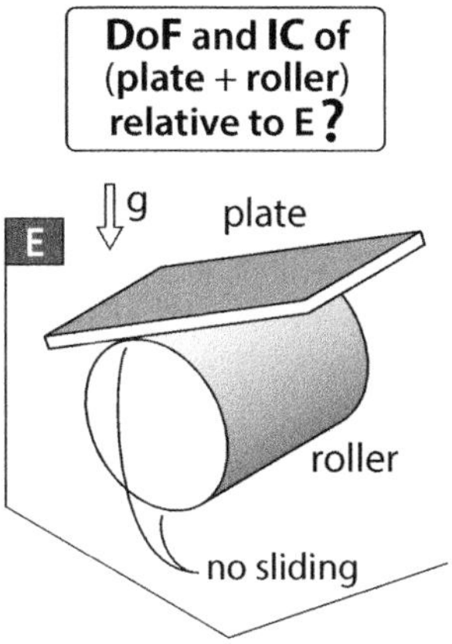

4.4 If there is no sliding at the contact points between plate, roller, and ground, how many DoF and IC does the system (plate + roller) have relative to the ground?

	DoF	IC
A	3	4
B	2	2
C	3	3
D	2	4
E	2	3

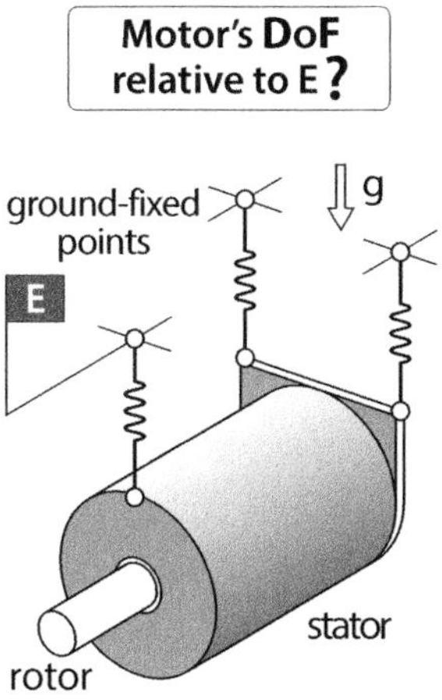

4.5 An electric motor, formed by a stator and a rotor, is suspended from the ceiling by three springs. How many DoF are there between motor and ground?

A 3
B 4
C 5
D 6
E 7

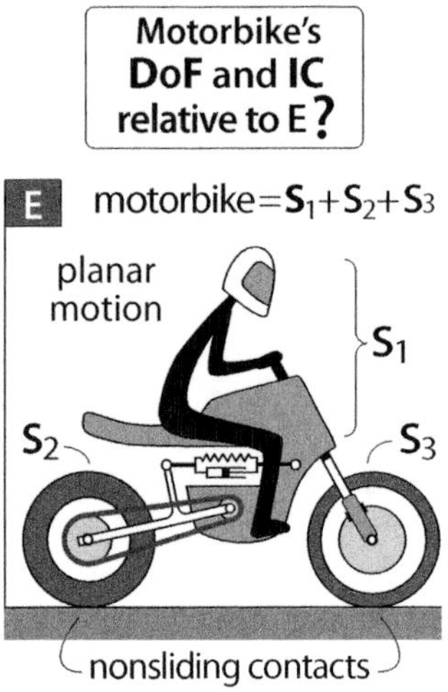

4.6 In a simplified study of a motorbike, the wheels and the assembly (frame + pilot) are modeled as rigid bodies. If the motorbike has a planar motion and the wheels do not slide on the ground, how many DoF and IC does it have relative to the ground?

	DoF	IC
A	4	4
B	3	4
C	3	3
D	2	3
E	2	2

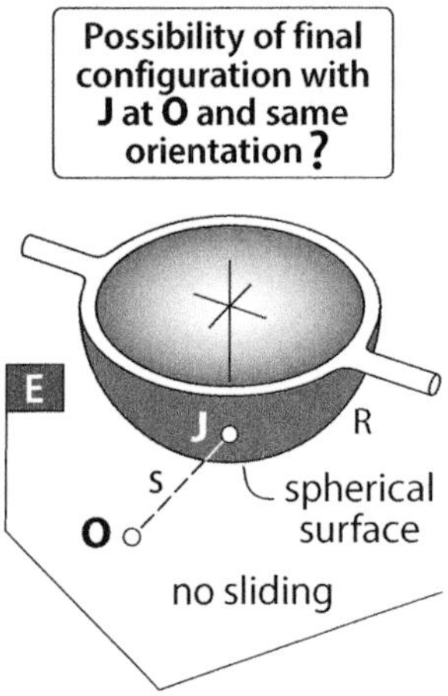

4.7 The rigid hemisphere has a single-point contact with the horizontal ground. Initially it is at rest in the shown configuration. If it does not slide on the ground, is it possible that it reaches a configuration with the same orientation but with the contact point at **O**?

A It is possible
B It is possible only if $s < R$
C It is possible only if $s > R$
D It would only be possible if it were a complete sphere
E It is not possible because the nonsliding condition leads to a nonholonomic behavior

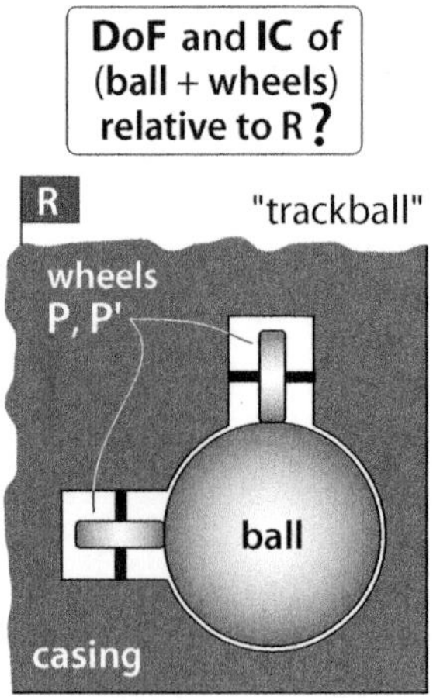

4.8 The "trackball" that is used in laptops to move the cursor on the screen consists of a ball with its center fixed to the casing (R). Its rotation is controlled with the fingers, and it has nonsliding contacts with two wheels P and P′ whose axes are fixed to the casing. The wheels' rotation provides the signals to move the cursor. How many DoF and IC does the system (ball + wheels) have relative to the casing?

	DoF	IC
A	2	>2
B	2	2
C	3	>3
D	4	4
E	3	3

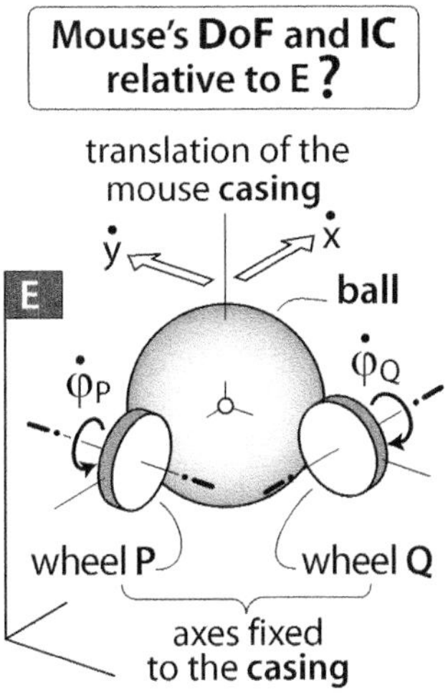

4.9 An old computer mouse consists of a casing and a ball with nonsliding contacts with the table and the wheels P and Q. The wheels' axes are fixed to the housing. If the housing has a translational motion x–y relative to the table, how many DoF and IC does the mouse have relative to the ground?

	DoF	IC
A	2	2
B	2	3
C	3	4
D	2	5
E	3	3

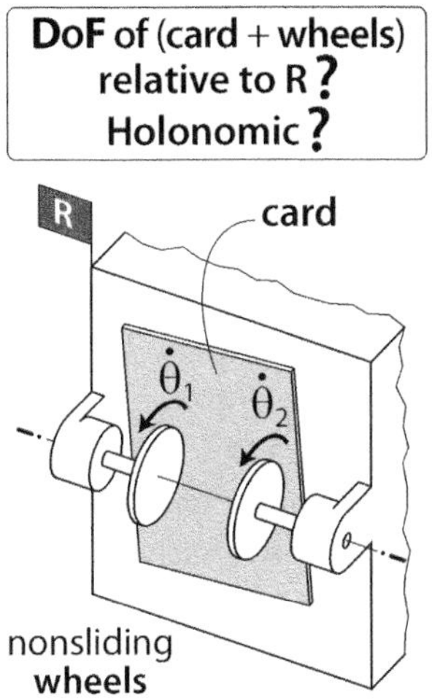

4.10 In a card printer (R), the card configuration is controlled by friction wheels driven by motors. The wheels do not slide on the card. How many DoF does the system (card + wheels) have with respect to the printer? Is it a holonomic system?

A 1 DoF, holonomic
B 2 DoF, holonomic
C 2 DoF, nonholonomic
D 3 DoF, holonomic
E 3 DoF, nonholonomic

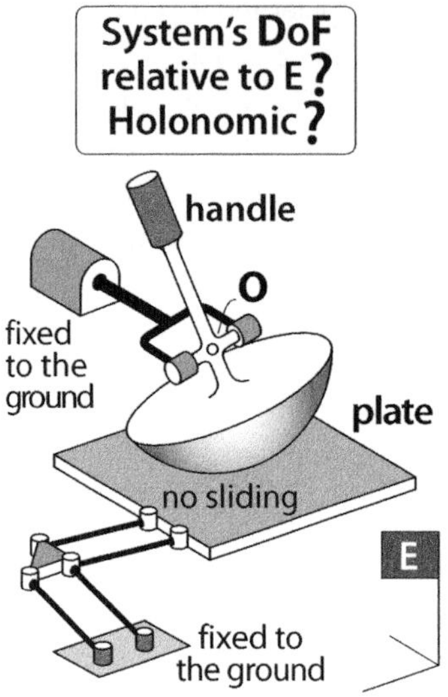

4.11 The plate has the horizontal motion constrained by the articulated bars (forming two series parallelograms). The spherical cap is articulated to the ground and has a nonsliding contact with the plate. The motion of the plate relative to the ground is a result of the cap rotation, which is controlled through the handle. How many DoF does the system formed by all these elements have relative to the ground? Is it a holonomic system?

A 2 DoF, holonomic
B 2 DoF, nonholonomic
C 3 DoF, holonomic
D 3 DoF, nonholonomic
E 4 DoF, holonomic

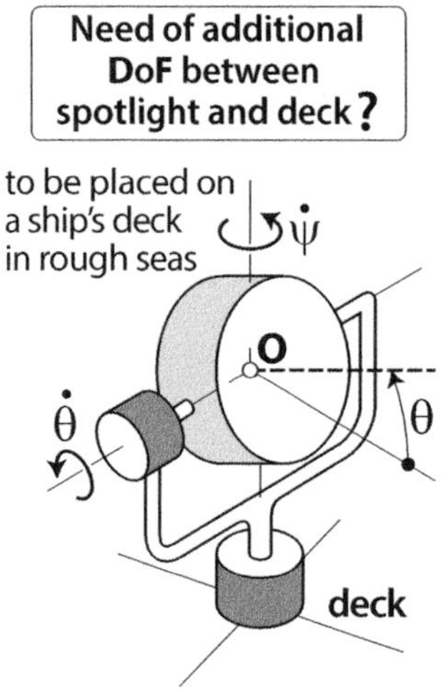

4.12 A spotlight installed on a ship has two Euler rotations $\dot{\psi}$ and $\dot{\theta}$ relative to the deck controlled by two actuators. If we neglect the displacement of point **O**, do we need any extra DoF between spotlight and deck to avoid the effect of the ship oscillations in rough seas on the spotlight performance?

A We need one extra DoF
B We need two extra DoF
C We need three extra DoF
D We do not need any extra DoF; we just need to control the rotations $\dot{\psi}$ and $\dot{\theta}$
E We need two extra rotations around ship-fixed axes

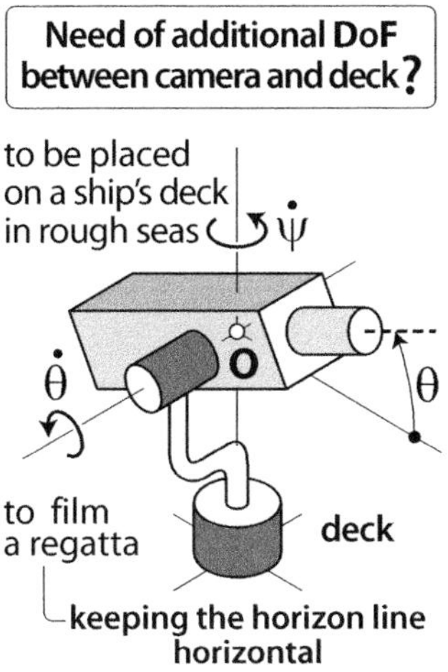

4.13 A camera installed on a ship has two Euler rotations $\dot{\psi}$ and $\dot{\theta}$ relative to the deck controlled by two actuators. If we film a regatta in rough seas and want to see the horizon line always horizontal in the film, do we need any extra DoF between the camera and the deck?

A We need one extra DoF
B We need two extra DoF
C We need three extra DoF
D We do not need any extra DoF; we just need to control the rotations $\dot{\psi}$ and $\dot{\theta}$
E We need two extra rotations around ship-fixed axes

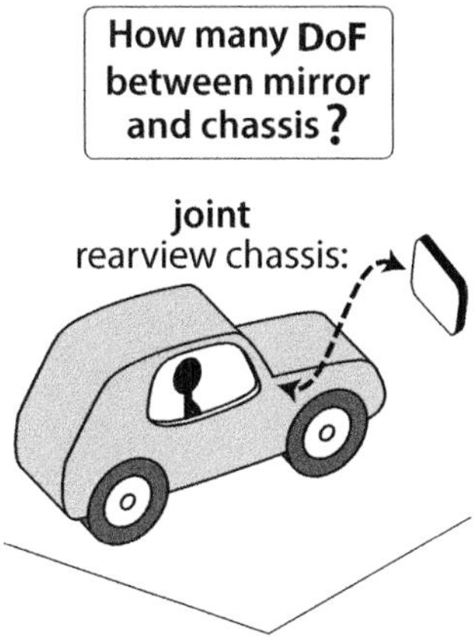

4.14 What is the minimum number of DoF between the rearview mirror and the chassis of a vehicle so that the mirror orientation can be properly adjusted?

A 1
B 2
C 3
D 4
E 5

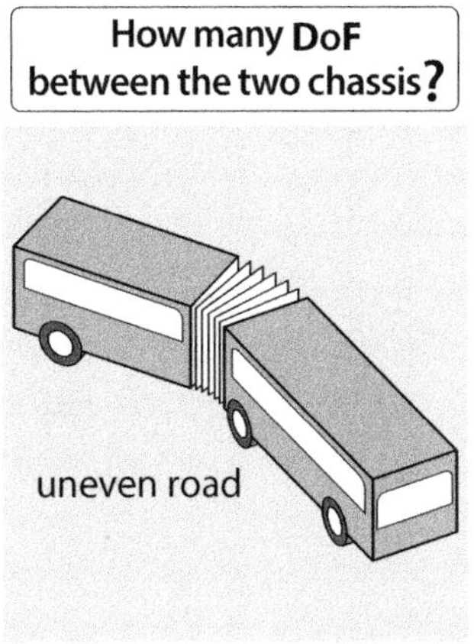

4.15 How many DoF do we need between the front and the rear chassis of an articulated bus so that it can circulate on any road without losing ground contact and without forcing the suspension?

A 1
B 2
C 3
D 4
E 5

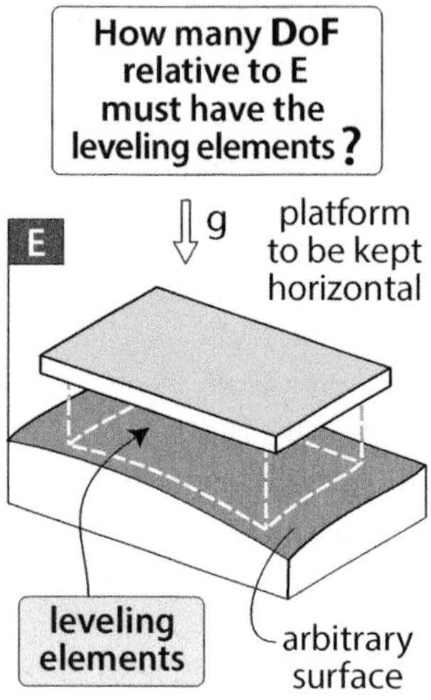

4.16 How many DoF relative to the ground do we need for the set of leveling elements of a platform so that its surface is horizontal?

A 1
B 2
C 3
D 4
E 5

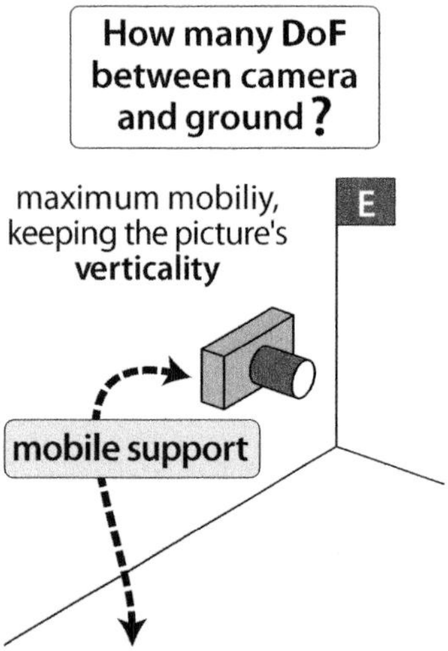

4.17 How many DoF do we need between a camera support and the ground if it has to provide maximum mobility to the camera while keeping the horizon line parallel to the base of the images?

A 1
B 2
C 3
D 4
E 5

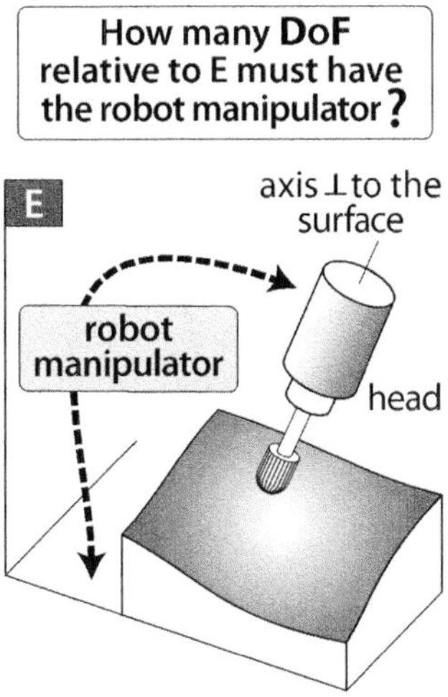

4.18 A computer-controlled machine tool has to machine a surface with a rotary tool so that its axis is perpendicular to the surface at each point of intersection. How many DoF relative to the ground must be controlled by the robot to move the toolhead?

A 5
B 3
C 4
D 6
E 7

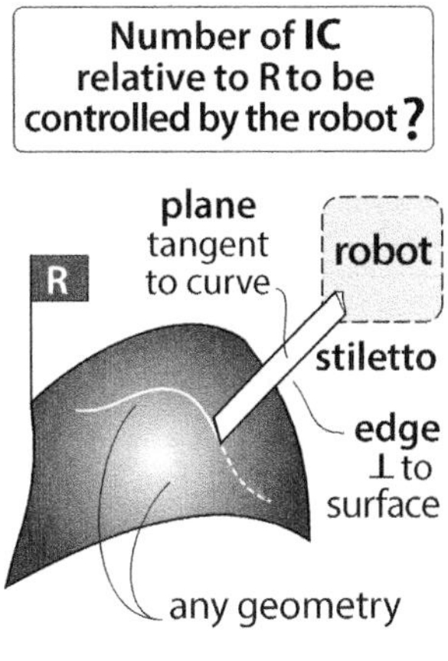

4.19 A robot manipulates a stiletto on a surface. Both the shape of the incision and the surface may have any geometry. If the stiletto edge must remain perpendicular to the surface and its plane must remain tangent to the curve, what is the minimum number of IC (position coordinates and Euler angles) to be controlled by the robot?

	Position coordinates	Euler angles
A	1	2
B	2	2
C	3	3
D	3	2
E	2	3

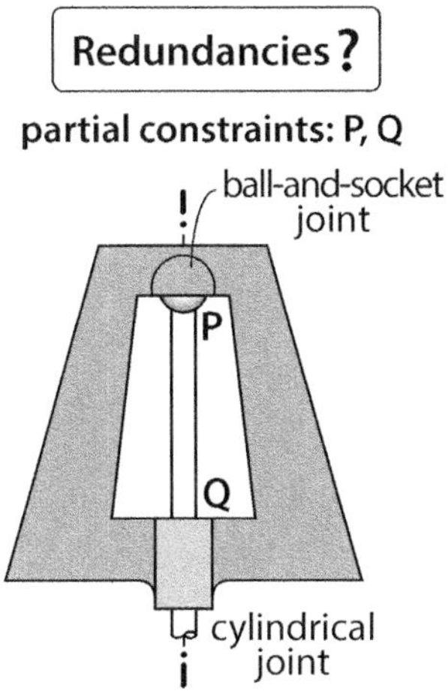

4.20 The motion of the rotor is constrained by the axis at **P** and **Q**. If the contacts at **P** and **Q** are taken as partial constraints, what is the number of redundancies?

A 0
B 1
C 2
D 3
E 4

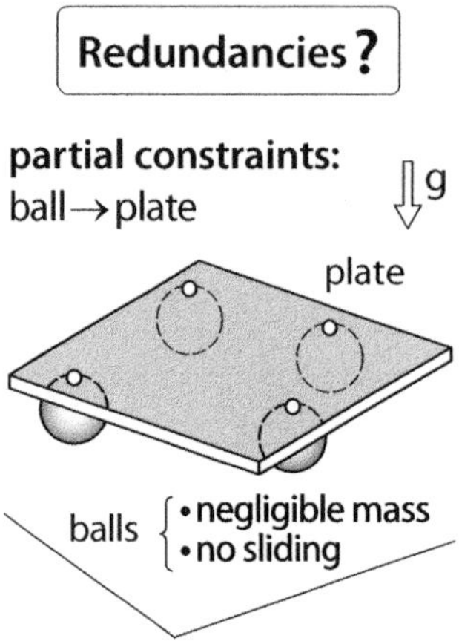

4.21 The motion of the plate relative to the ground is constrained by its contact with the four balls, which may rotate without sliding in their contacts with the ground and the plate. If each ball introduces a partial constraint, what is the number of redundancies?

A 0
B 1
C 2
D 3
E 4

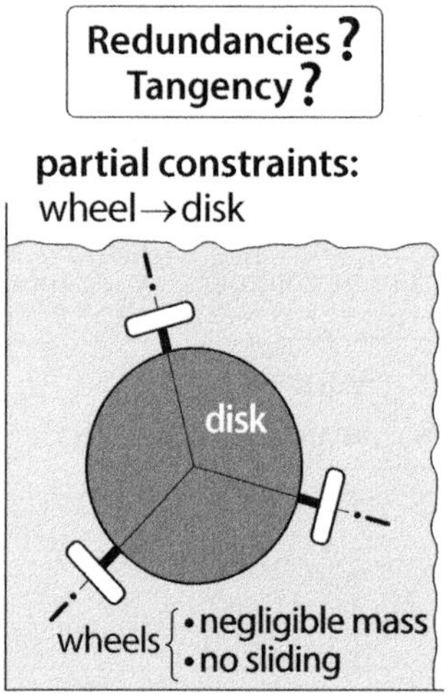

4.22 The motion of the disk relative to the ground is constrained through three wheels, which may rotate freely about their axis without sliding on the ground. If each wheel introduces a partial constraint, what is the number of redundancies? If nonzero, is there tangent redundancy?

	Redundancies	Tangent redundancy
A	1	No
B	2	Yes
C	1	Yes
D	2	No
E	0	No

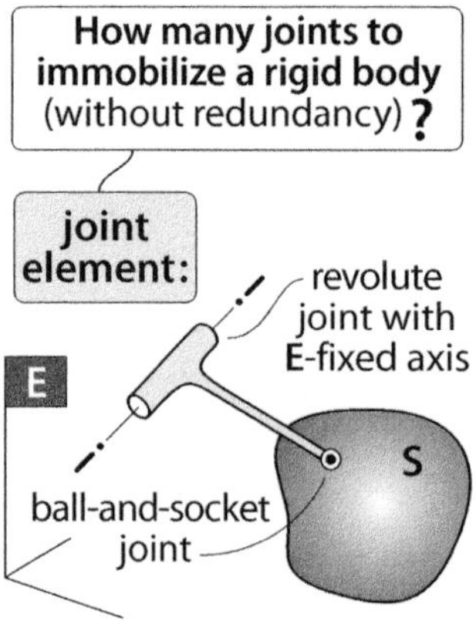

4.23 We want to immobilize a rigid body S with respect to the ground by means of bars with a ball-and-socket joint at the S-fixed end and a revolute joint with a ground-fixed axis at the other end. How many joints do we need if we want to avoid redundancy?

A 2
B 3
C 4
D 5
E 6

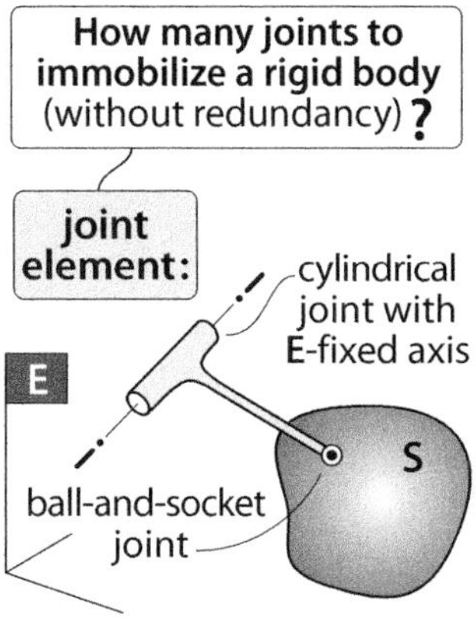

4.24 We want to immobilize a rigid body S with respect to the ground by means of bars with a ball-and-socket joint at the S-fixed end and a cylindrical joint with a ground-fixed axis at the other end. How many joints do we need if we want to avoid redundancy?

A 2
B 3
C 4
D 5
E 6

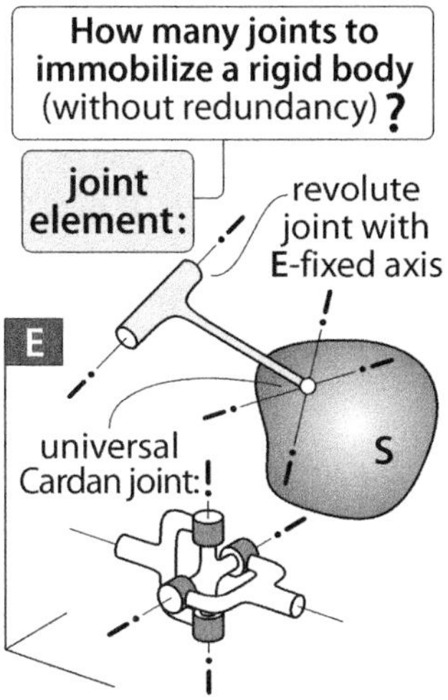

4.25 We want to immobilize a rigid body S with respect to the ground by means of bars with a universal Cardan joint at the S-fixed end and a revolute joint with a ground-fixed axis at the other end. How many joints do we need if we want to avoid redundancy?

A 2
B 3
C 4
D 5
E It is not possible

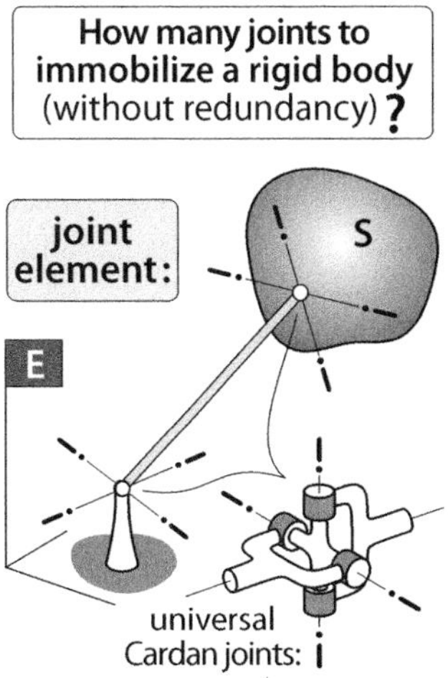

4.26 We want to immobilize a rigid body with respect to the ground by means of bars with universal Cardan joints at both ends. How many joints do we need if we want to avoid redundancy?

A 2
B 3
C 4
D 5
E 6

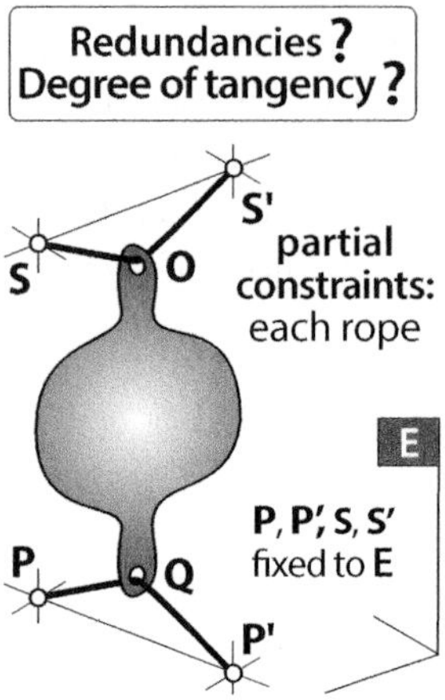

Fig. Q-4.27

4.27 The motion of the rigid body is constrained by two ropes **PP′** and **SS′**. If each rope is responsible for a partial constraint, what is the number of redundancies? If non-zero, what is the degree of tangent redundancy (DTR)?

	Redundancies	DTR
A	0	0
B	1	0
C	1	2
D	1	4
E	2	2

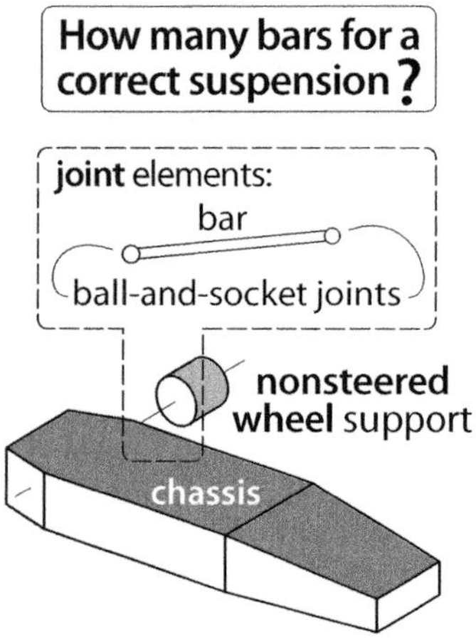

4.28 We want to link the support of a nonsteered wheel to the car chassis by means of bars with ball-and-socket joints at both ends. If the DoF associated with the wheel suspension has to be preserved, how many bars do we need if we want to avoid redundancy?

A 2
B 3
C 4
D 5
E The joint elements are not suitable

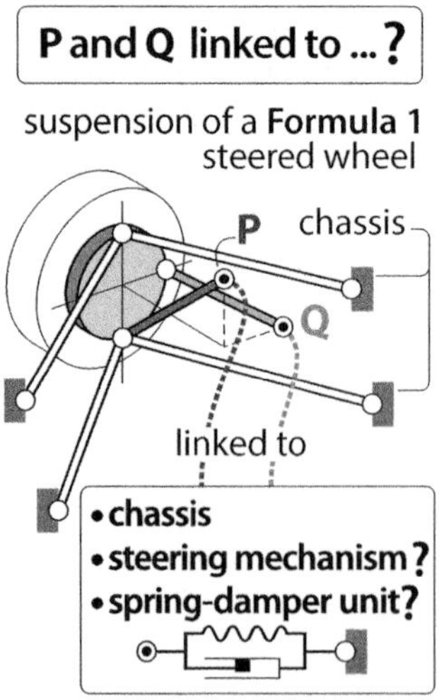

4.29 In a Formula 1 vehicle, there are six bars with ball-and-socket joints at the ends linked to the support of a steered wheel. Four bars are linked to the chassis. To what elements (chassis, steering mechanism, spring-and-damper unit) have to be linked **P** and **Q**?

A Both to the chassis
B **P** to the chassis and **Q** to the steering mechanism
C **P** to the steering mechanism and **Q** to the chassis
D **P** to the steering mechanism and **Q** to the spring-and-damper unit
E **P** to the spring-and-damper unit and **Q** to the steering mechanism

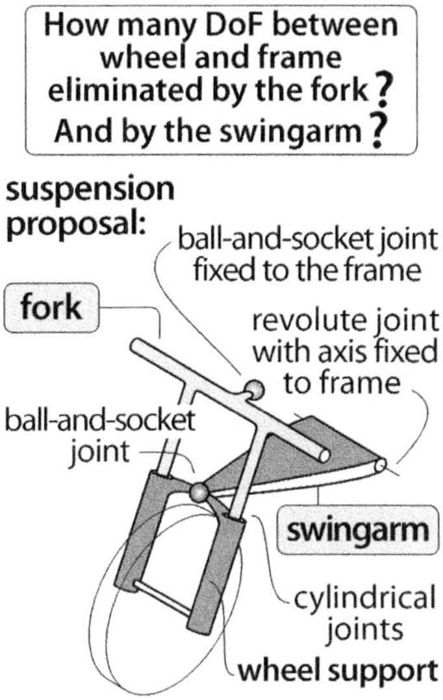

4.30 The design of the suspension of a motorcycle front wheel consists of a fork and a swingarm linking the wheel support to the frame. How many DoF between support and frame would eliminate the fork and the swingarm if just one of them were used?

	Fork	Swingarm
A	1	1
B	1	2
C	2	2
D	2	3
E	3	2

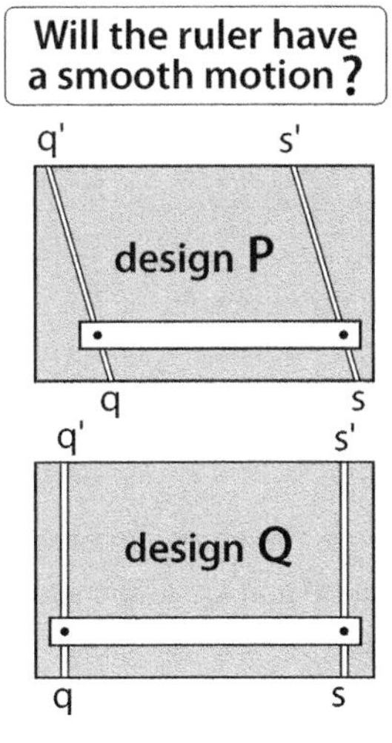

4.31 We propose two designs P and Q for a drawing board with a ruler to draw parallel lines. In both designs, the ruler has two pins, which slide in two parallel slots. Will they both work properly (with a smooth motion of the ruler)?

A Yes
B Only design P will work properly
C Only design Q will work properly
D Neither design P nor design Q will work properly
E Nothing can be said beforehand without trying them

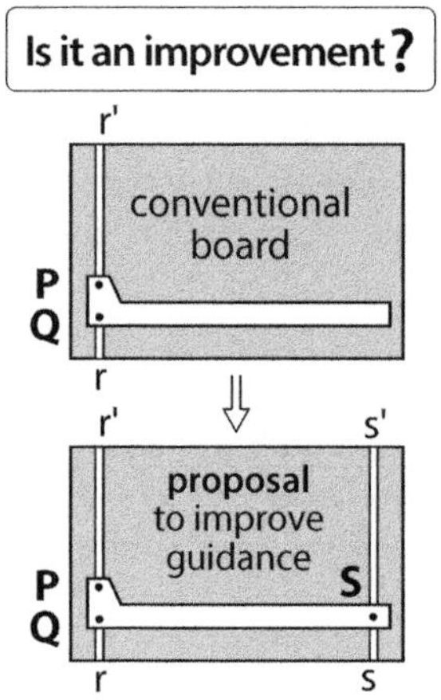

4.32 In a conventional drawing board, the two pins **P** and **Q** of the ruler slide in a same slot rr′. In order to improve the design, we propose adding a third pin **S** that would slide in a slot ss′ parallel to rr′. Is it really an improvement?

A It is, provided that there are no geometrical errors

B It is, provided that the line through **S** and perpendicular to ss′ goes through the midpoint between **P** and **Q**

C It is, because the ruler translation motion will be better controlled

D It is not, as there would be full redundancy

E It is not, as there would be tangent redundancy

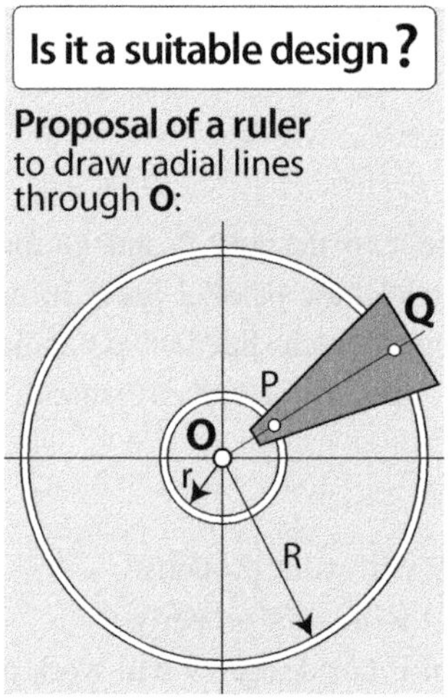

4.33 We propose a drawing board to draw radial lines through point **O**. The design consists on a plate with two pins (**P** and **Q**), which slide in two circular and concentric slots with radius R and r $\left(|\overline{\mathbf{PQ}}| = \mathrm{R} - \mathrm{r}\right)$. Is it a suitable design?

A It is not, because the system is non holonomic

B It is not, because there is tangent redundancy

C It is not, because one of the pins is redundant

D It is, provided that there is a high geometric accuracy

E It is a perfectly good design

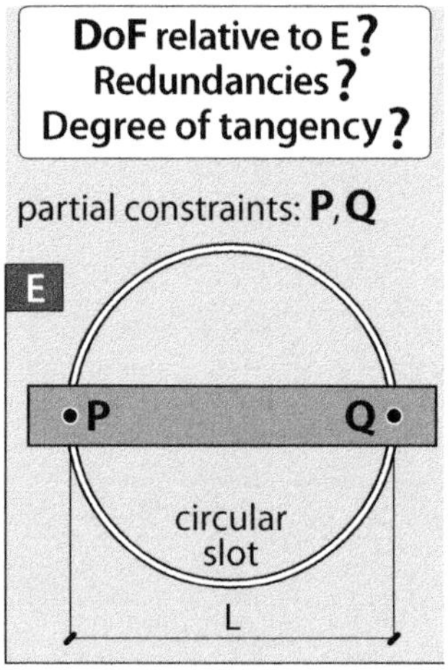

4.34 The motion of the ruler is constrained by the circular slot at points **P** and **Q**,. If the contacts at **P** and **Q** are taken as partial constraints, how many $\mathrm{DoF}_{\mathrm{geom}}$ are there between ruler and ground? What is the number of redundancies? If nonzero, what is the degree of tangent redundancy (DTR)?

	$\mathrm{DoF}_{\mathrm{geom}}$	Redundancies	DTR
A	1	0	—
B	1	1	0
C	1	1	1
D	0	2	0
E	0	2	2

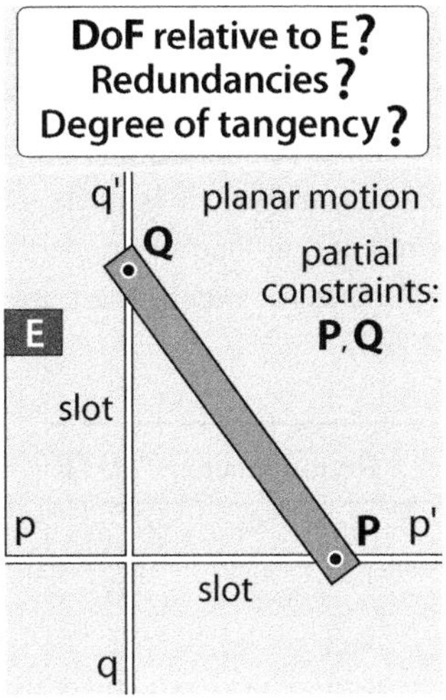

4.35 The motion of the rod is constrained by the straight slots at points **P** and **Q**. If the contacts at **P** and **Q** are taken as partial constraints, how many DoF_{geom} are there between rod and ground? What is the number of redundancies? If nonzero, what is the degree of tangent redundancy (DTR)?

	DoF_{geom}	Redundancies	DTR
A	1	0	—
B	1	1	0
C	1	1	1
D	0	2	0
E	1	2	2

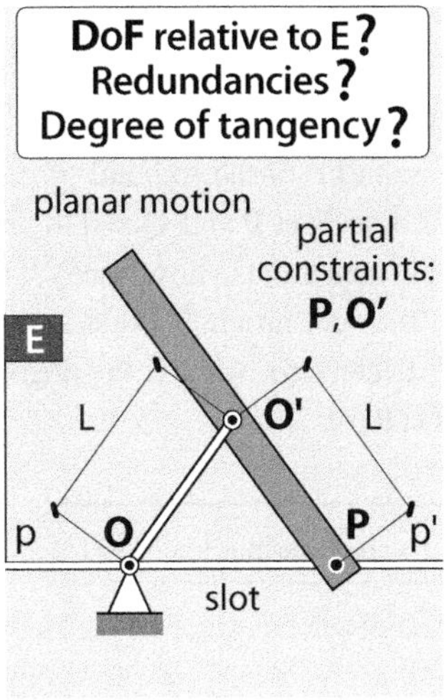

4.36 The motion of the rod is constrained by the straight slot at **P** and the bar with revolute joints linking point **O′** of the rod and of point **O** of the ground. If the contact at **P** and the link at **O′** are taken as partial constraints, how many DoF_{geom} are there between rod and ground? What is the number of redundancies? If nonzero, what is the degree of tangent redundancy (DTR)?

	DoF_{geom}	Redundancies	DTR
A	1	0	—
B	1	1	0
C	1	1	1
D	0	2	0
E	1	2	2

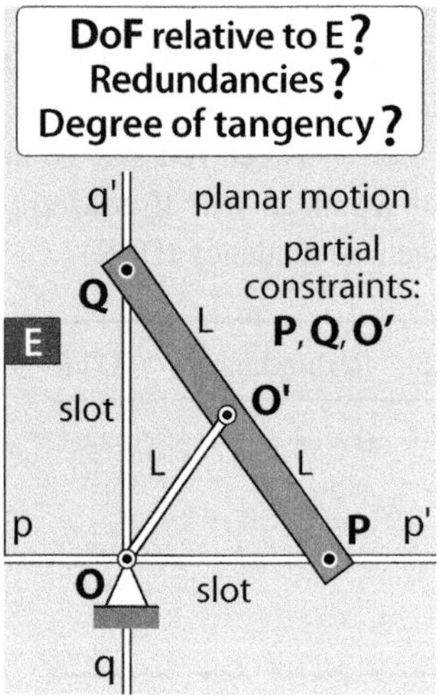

4.37 The motion of the rod is constrained by the straight slots at points **P** and **Q**, and the bar with revolute joints linking point **O′** of the rod and of point **O** of the ground. If the contacts at **P** and **Q** and the link at **O′** are taken as partial constraints, how many DoF_{geom} are there between rod and ground? What is the number of redundancies? If nonzero, what is the degree of tangent redundancy (DTR)?

	DoF_{geom}	Redundancies	DTR
A	1	0	—
B	1	1	0
C	1	1	1
D	1	2	0
E	0	2	2

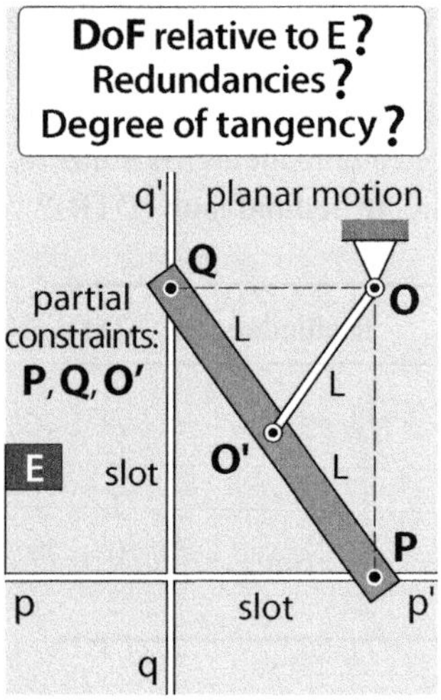

4.38 The motion of the rod is constrained by the straight slots at points **P** and **Q**, and the bar with revolute joints linking point **O′** of the rod and of point **O** of the ground. If the contacts at **P** and **Q** and the link at **O′** are taken as partial constraints, how many DoF_{geom} are there between rod and ground? What is the number of redundancies? If nonzero, what is the degree of tangent redundancy (DTR)?

	DoF_{geom}	Redundancies	DTR
A	1	0	—
B	1	1	0
C	0	0	—
D	0	1	1
E	0	2	0

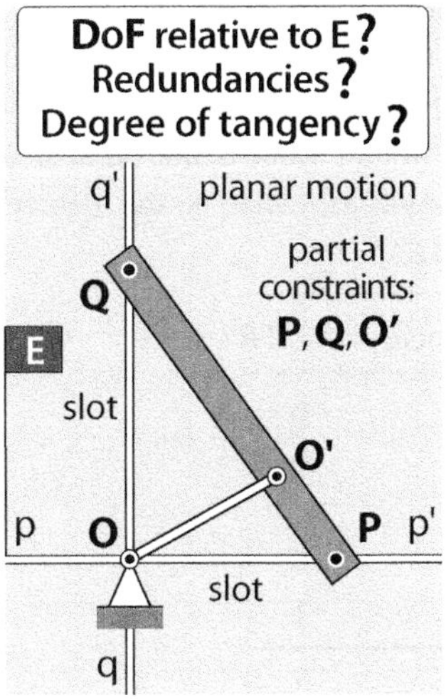

4.39 The motion of the rod is constrained by the straight slots at points **P** and **Q**, and the bar with revolute joints linking point **O′** of the rod and of point **O** of the ground. If the contacts at **P** and **Q** and the link at **O′** are taken as partial constraints, how many DoF-$_{geom}$ are there between rod and ground? What is the number of redundancies? If nonzero, what is the degree of tangent redundancy (DTR)?

	DoF_{geom}	Redundancies	DTR
A	1	0	—
B	1	1	0
C	0	0	—
D	0	1	1
E	0	2	0

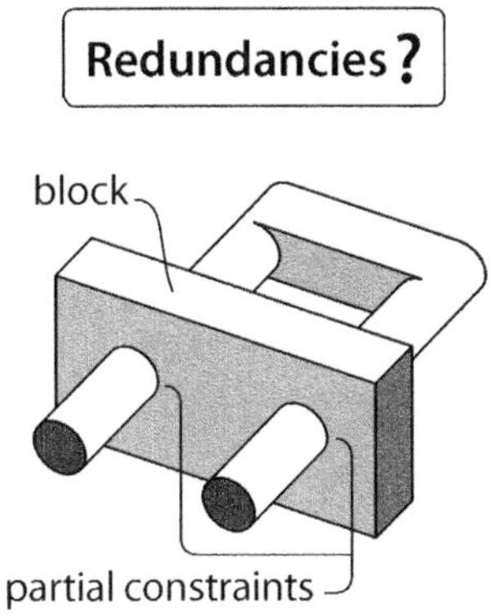

4.40 The U-shaped rigid body may have a translational motion relative to the block thanks to two cylindrical guides with parallel axes. If every guide is taken as a partial constraint, what is the number of redundancies?

A 0
B 1
C 2
D 3
E 4

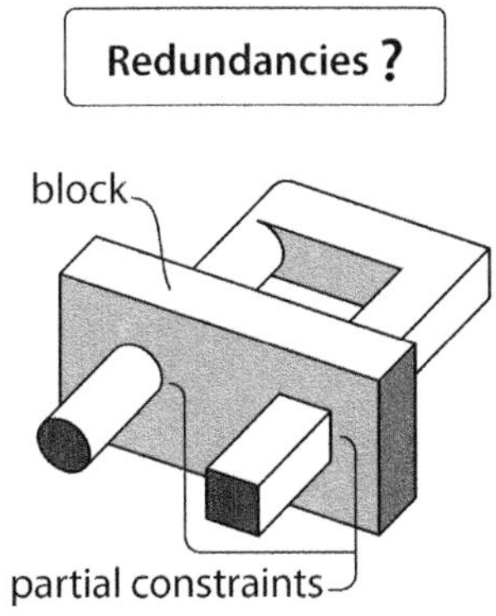

4.41 The U-shaped rigid body may have a translational motion relative to the block thanks to a cylindrical guide and a prismatic guide with parallel axes. If every guide is taken as a partial constraint, what is the number of redundancies?

A 0
B 1
C 2
D 3
E 4

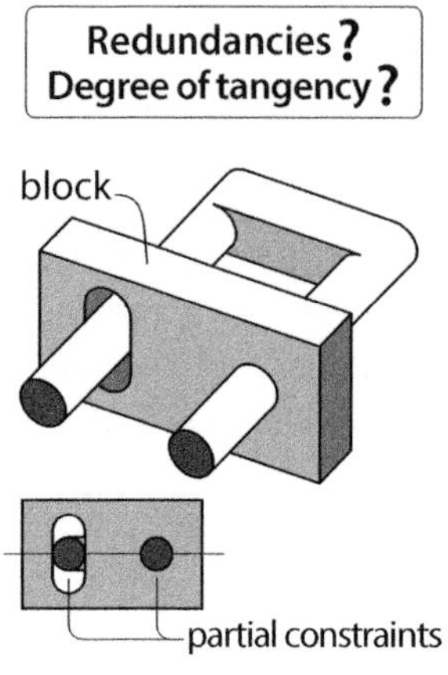

4.42 The U-shaped rigid body may have a translational motion relative to the block thanks to two guides (one of them cylindrical) with parallel axes. If every guide is taken as a partial constraint, what is the number of redundancies? If it is nonzero, what is the degree of tangent redundancy (DTR)?

	Redundancies	DTR
A	0	0
B	1	0
C	1	1
D	2	0
E	2	1

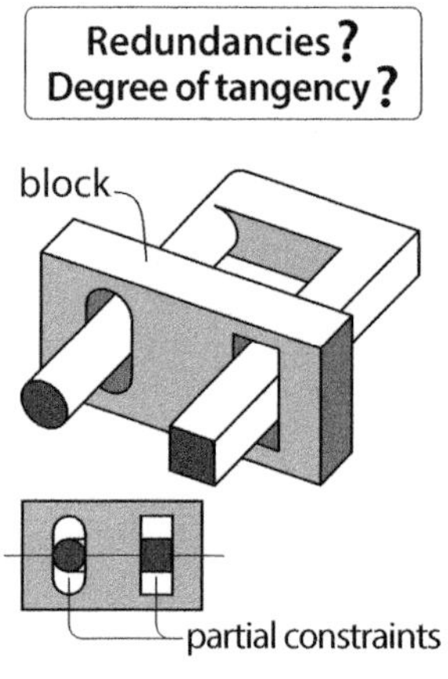

4.43 The U-shaped rigid body may have a translational motion relative to the block thanks to two guides with parallel axes. If every guide is taken as a partial constraint, what is the number of redundancies? If it is nonzero, what is the degree of tangent redundancy (DTR)?

	Redundancies	DTR
A	0	0
B	1	0
C	1	1
D	2	0
E	2	2

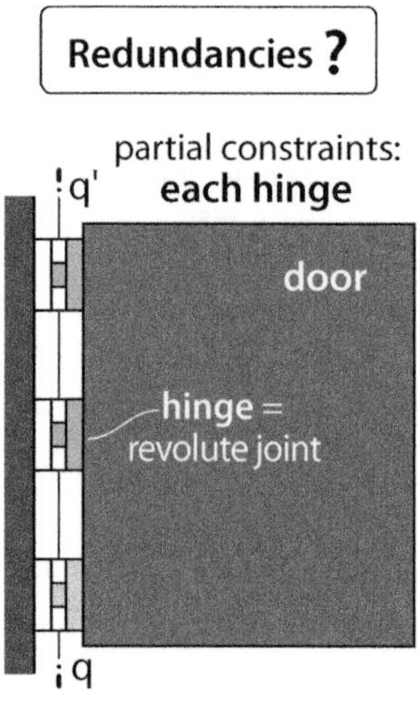

4.44 If every hinge (revolute joint) in a door is taken as a partial constraint, what is the number of redundancies?

A 4
B 6
C 9
D 10
E 14

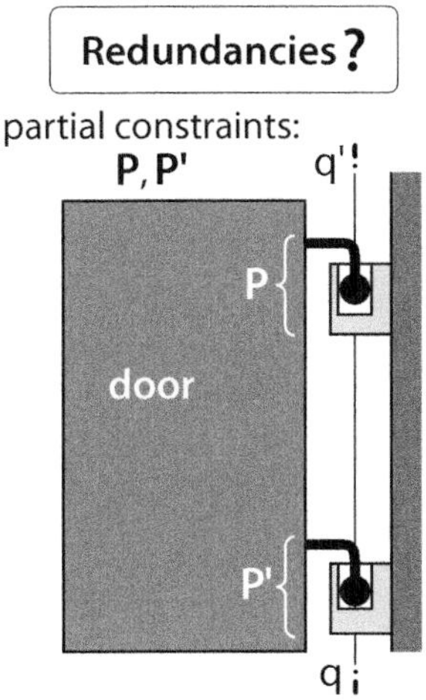

4.45 The door of a safety vault is linked to the wall by a combination of a cylindrical joint and a ball-and-socket joint at **P** and **P**′. If the links at **P** and **P**′ are taken as partial constraints, what is the number of redundancies?

A 0
B 1
C 2
D 3
E 4

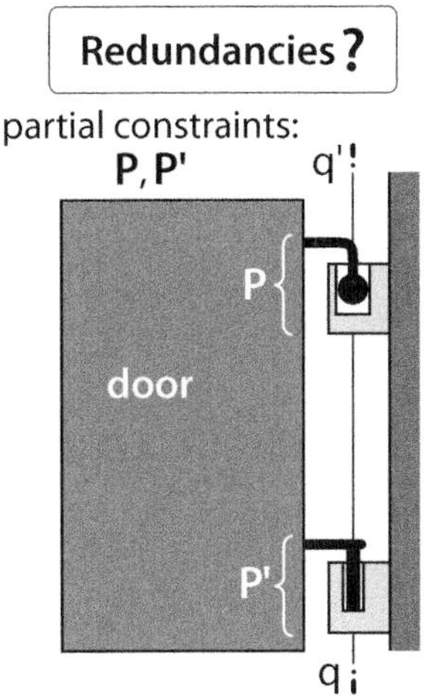

4.46 The door of a safety vault is linked to the wall by a combination of a cylindrical joint and a ball-and-socket joint at **P** and by a revolute joint at **P**′. If the links at **P** and **P**′ are taken as partial constraints, what is the number of redundancies?

A 0
B 1
C 2
D 3
E 4

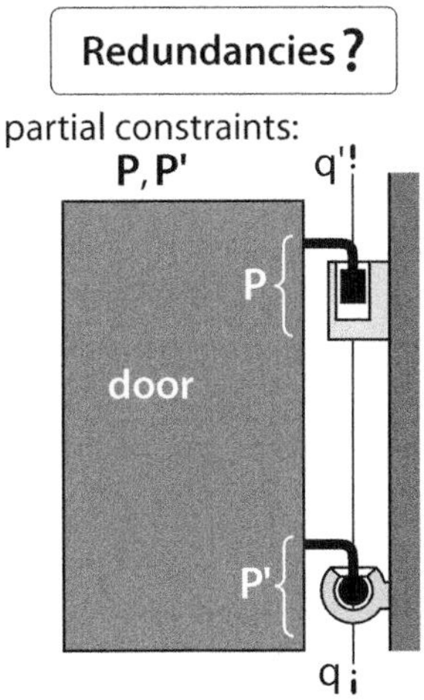

4.47 The door of a safety vault is linked to the wall by a cylindrical joint at **P** and by a ball-and-socket joint at **P**′. If the links at **P** and **P**′ are taken as partial constraints, what is the number of redundancies?

A 0
B 1
C 2
D 3
E 4

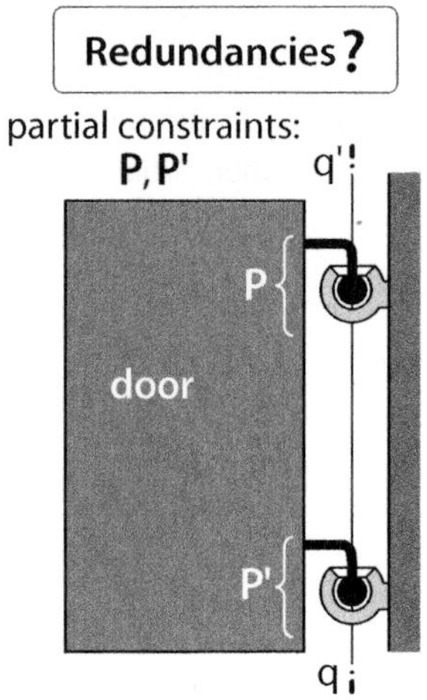

4.48 The door of a safety vault is linked to the wall by ball-and-socket joints at **P** and **P′**. If the links at **P** and **P′** are taken as partial constraints, what is the number of redundancies?

A 0
B 1
C 2
D 3
E 4

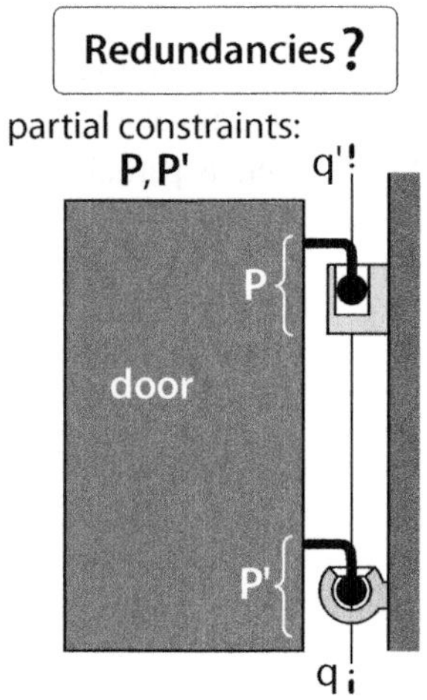

4.49 The door of a safety vault is linked to the wall by a combination of a cylindrical joint and a ball-and-socket joint at **P**, and by a ball-and-socket joint at **P′**. If the links at **P** and **P′** are taken as partial constraints, what is the number of redundancies?

A 0
B 1
C 2
D 3
E 4

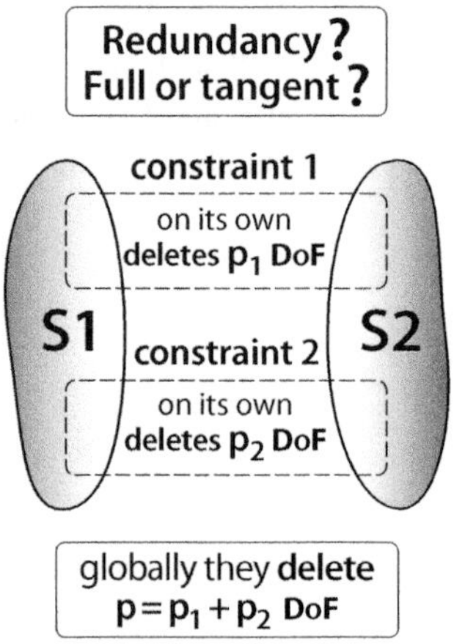

4.50 Two constraints delete p_1 and p_2 DoF on their own. Together, they delete $p = p_1 + p_2$ DoF_{geom}. Assess the possible redundancies in that design.

A There is no redundancy
B There is full redundancy
C There is tangent redundancy
D Either there is no redundancy or there is full redundancy
E Either there is no redundancy or there is tangent redundancy

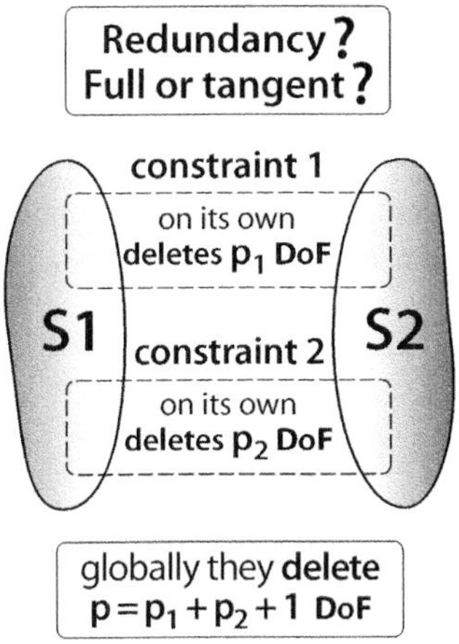

4.51 Two constraints delete p_1 and p_2 DoF on their own. Together, they delete $p = p_1 + p_2 + 1$ DoF_{geom}. Assess the possible redundancies in that design.

A There is one degree of tangent redundancy
B There is at least one degree of tangent redundancy
C There is one full redundancy
D There is at least one full redundancy
E It is impossible to assess the existence of redundancy

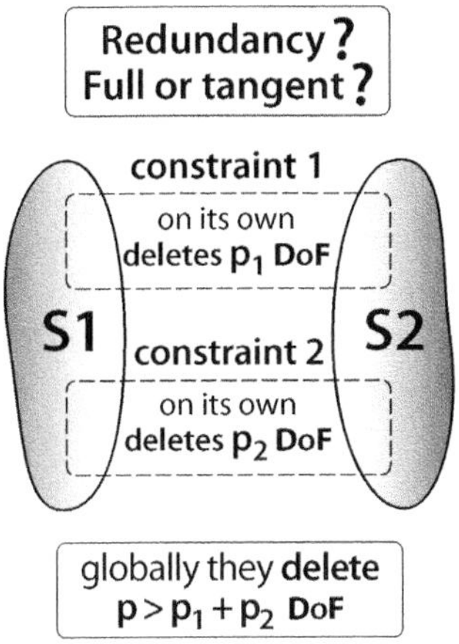

4.52 Two constraints delete p_1 and p_2 DoF on their own. Together, they delete $p > p_1 + p_2$ DoF_{geom}. Assess the possible redundancies in that design.

A There are $(p - (p_1 + p_2))$ redundancies
B The number of deleted DoF can never be greater than $(p_1 + p_2)$
C There is tangent redundancy
D The number of deleted DoF can be greater than $(p_1 + p_2)$ without implying any redundancy
E The system is nonholonomic

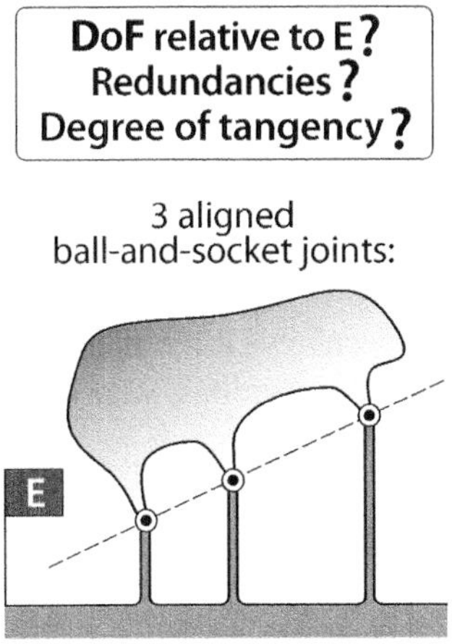

4.53 A rigid body is linked to the ground through three aligned ball-and-socket joints. If each joint is taken as a partial constraint, what is the number of redundancies? If it is nonzero, what is the degree of tangent redundancy (DTR)?

	Redundancies	DTR
A	1	0
B	3	0
C	3	3
D	4	4
E	4	0

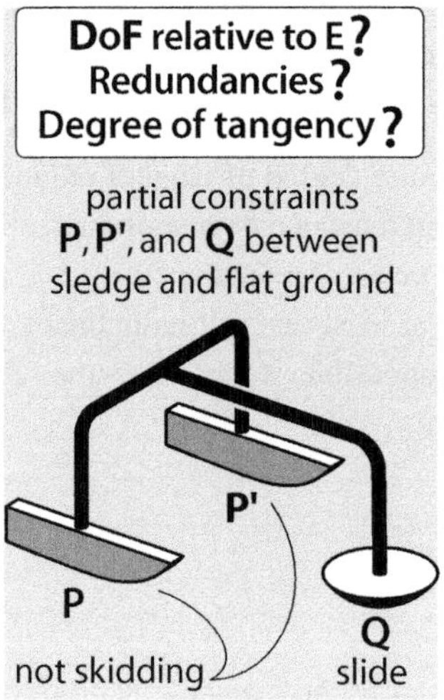

4.54 The sledge is provided with two skates P and P′, and with a spherical foot Q. When acting on their own, each skate prevents skidding while the foot allows sliding in all directions. If P, P′, and Q are taken as partial constraints, what is the the number of redundancies? If nonzero, what is the degree of tangent redundancy?

	Redundancies	DTR
A	0	0
B	1	0
C	1	1
D	1	0
E	2	0

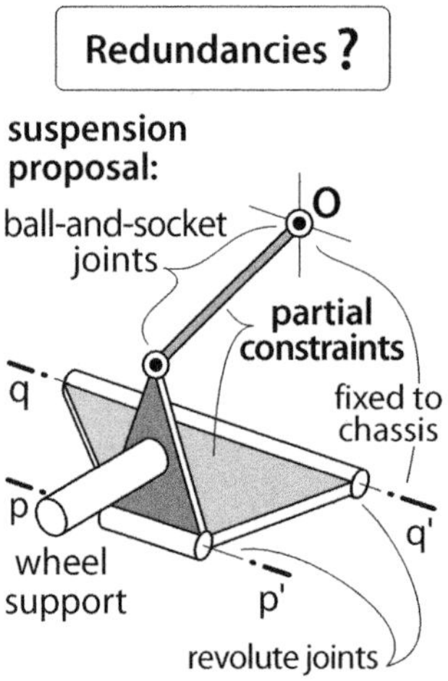

4.55 The design of a suspension consists of a trapezoid with a revolute joint with axis p–p′ fixed to the wheel support, another one with axis q–q′ fixed to the chassis, and a bar with ball-and-socket joints at the ends linking the wheel support and the chassis. If the trapezoid and the bar are taken as partial constraints, what is the number of redundancies?

A 0
B 1
C 2
D 3
E 4

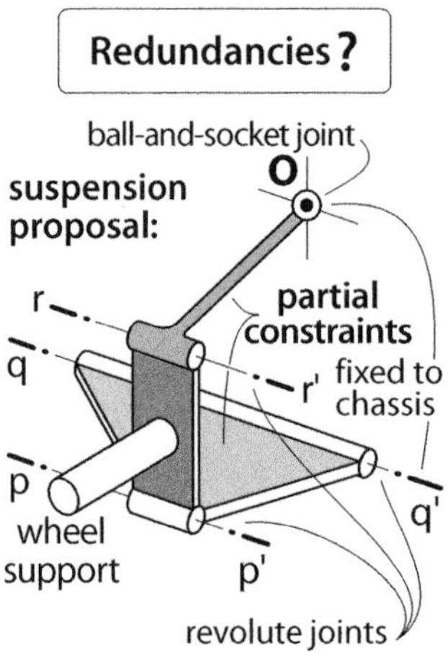

4.56 The design of a suspension consists of a trapezoid with a revolute joint with axis p–p′ fixed to the wheel support, another one with axis q–q′ fixed to the chassis, and a bar linked to the wheel support and the chassis through a revolute joint and a ball-and-socket joint, respectively. If the trapezoid and the bar are taken as partial constraints, what is the number of redundancies?

A 0
B 1
C 2
D 3
E 4

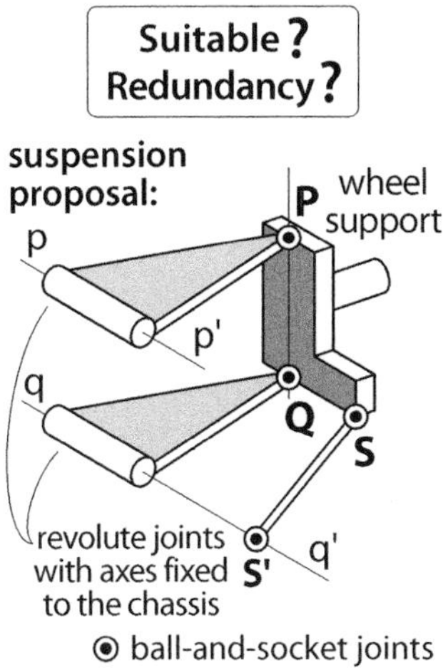

4.57 The design of a suspension consists of two trapezoids with revolute joints and ball-and-socket joints, plus a bar with ball-and-socket joints at the ends linking the wheel support to the chassis. Is it a suitable design?

A It is suitable and with no redundancy

B It is unsuitable because the wheel is blocked to the chassis, but there is no tangent redundancy

C It is unsuitable because the wheel is blocked to the chassis, and there is tangent redundancy

D It allows the required DoF between wheel and chassis, but there is tangent redundancy

E There is no redundancy, but the allowed DoF between wheel and chassis is not suitable for a suspension

Puzzles

4.1 Mobility of human hands

How many DoF does a human hand (including the fingers) have relative to the forearm?

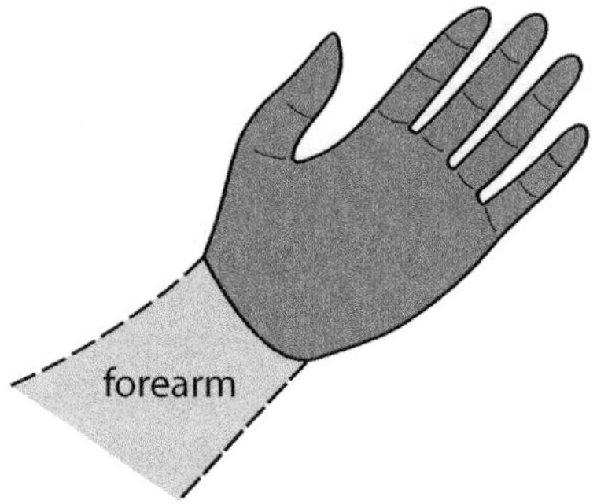

4.2 Secular displacements of rocking stones

Rocking stones oscillate when slightly pushed. Some of them have two nonsliding contact points with the ground, whereas some others have just one.

Is it possible that the contact points at the equilibrium configuration undergo a secular displacement (finite position change after a long time period)?

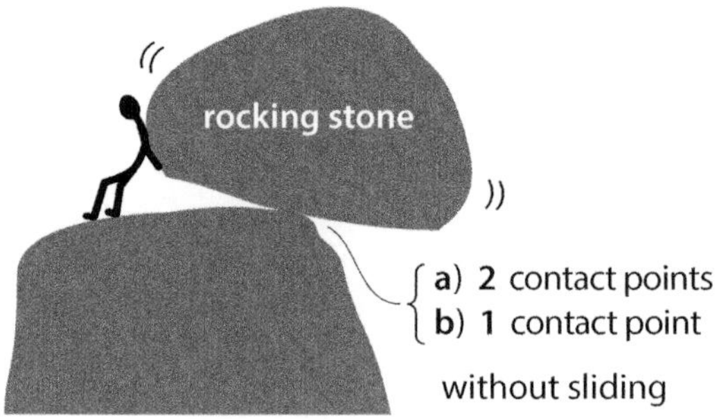

Quiz Questions: Answers

	1	2	3	4	5	6	7	8	9	10
	E	B	E	B	E	C	A	A	D	C
+10	B	D	A	B	C	B	E	A	C	C
+20	B	A	B	E	A	B	D	D	E	C
+30	B	D	B	C	A	A	B	D	C	D
+40	E	E	D	D	A	C	C	B	A	E
+50	B	C	E	B	A	B	A			

Puzzles: Solutions

4.1 Mobility of human hands

22 DoF.
Each finger has two revolute joints (with just one DoF) between every pair of phalanges plus a joint with two DoF linking the finger to the arm palm. Thus, the five fingers contribute $5 \times 4 = 20$ DoF.

The wrist joint adds two more DoF. So the total number is $20 + 2 = 22$ DoF.

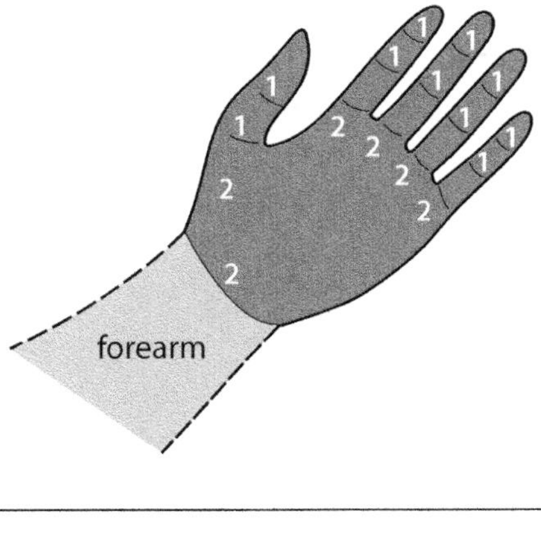

4.2 Secular displacements of rocking stones

Stones with Two Nonsliding Contact Points
Dynamical consideration: in an equilibrium configuration, the center of mass of the stone is located on a vertical line intersecting the line containing those two contact points.

The two contact points define the ISA, and the stone has one DoF; therefore, it is a holonomic system. The fixed and the moving axodes are univocally defined by the stone and the ground geometry.

If the stone starts its rocking motion from the equilibrium position, the center of mass will necessarily leave the vertical line going through the ISA, and thus the new configuration will not be an equilibrium configuration. The stone will roll back and forth, unable to maneuver to achieve a displacement of the equilibrium configuration.

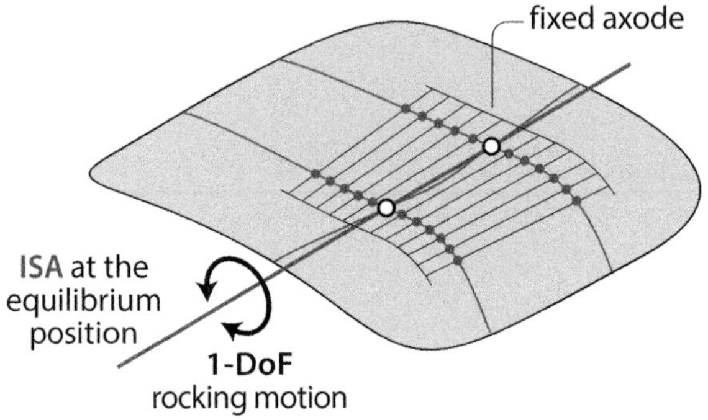

Stones with One Nonsliding Contact Point

Dynamical consideration: in the equilibrium configuration, the center of mass of the stone is located on the vertical line going through that contact point.

The stone has three DoF and is a nonholonomic system. Thus, it may maneuver to achieve a displacement of the equilibrium configuration.

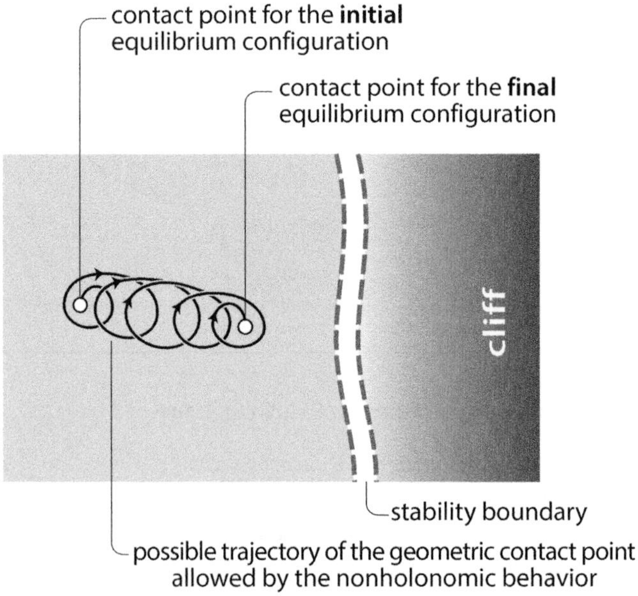

Index